Springer Series in Optical Sciences Volume 16

Edited by David L. MacAdam

Springer Series in Optical Sciences

Edited by David L. MacAdam

1 **Solid-State Laser Engineering**
By W. Koechner

2 **Table of Laser Lines in Gases and Vapors** 2nd Edition
By R. Beck, W. Englisch, and K. Gürs

3 **Tunable Lasers and Applications**
Editors: A. Mooradian, T. Jaeger, and P. Stokseth

4 **Nonlinear Laser Spectroscopy**
By V. S. Letokhov and V. P. Chebotayev

5 **Optics and Lasers** An Engineering Physics Approach
By M. Young

6 **Photoelectron Statistics** With Applications to Spectroscopy and Optical Communication
By B. Saleh

7 **Laser Spectroscopy III**
Editors: J. L. Hall and J. L. Carlsten

8 **Frontiers in Visual Science**
Editors: S. J. Cool and E. J. Smith III

9 **High-Power Lasers and Applications**
Editors: K.-L. Kompa and H. Walther

10 **Detection of Optical and Infrared Radiation**
By R. H. Kingston

11 **Matrix Theory of Photoelasticity**
By P. S. Theocaris and E. E. Gdoutos

12 **The Monte Carlo Method in Atmospheric Optics**
By G. I. Marchuk, G. A. Mikhailov, M. A. Nazaraliev, R. A. Darbinian, B. A. Kargin, and B. S. Elepov

13 **Physiological Optics**
By Y. Le Grand and S. G. El Hage

14 **Laser Crystals** Physics and Properties
By A. A. Kaminskii

15 **X-Ray Spectroscopy**
By B. K. Agarwal

16 **Holographic Interferometry** From the Scope of Deformation Analysis of Opaque Bodies
By W. Schumann and M. Dubas

17 **Nonlinear Optics of Free Atoms and Molecules**
By D. C. Hanna, M. A. Yuratich, D. Cotter

18 **Holography in Medicine and Biology**
By G. von Bally

19 **Color Theory and Its Application in Art and Design**
By G. A. Agoston

20 **Interferometry by Holography**
By J. Ostrowskij, M. Butussov, G. Ostrowskaja

21 **Laser Spectroscopy IV**
Editors: H. Walther, K. W. Rothe

W. Schumann M. Dubas

Holographic Interferometry

From the Scope of Deformation Analysis of Opaque Bodies

With 73 Figures

Springer-Verlag Berlin Heidelberg New York 1979

Professor Dr. WALTER SCHUMANN
Dr. MICHEL DUBAS

Laboratory of Photoelasticity, Swiss Federal Institute of Technology
Leonhardstrasse 33, CH-8092 Zurich, Switzerland

ISBN 3-540-09371-0 Springer-Verlag Berlin Heidelberg New York
ISBN 0-387-09371-0 Springer-Verlag New York Heidelberg Berlin

Library of Congress Cataloging in Publication Data. Schumann, Walter, 1927-. Holographic interferometry. (Springer series in optical sciences ; v. 16) Bibliography: p. Includes indexes. 1. Holographic interferometry. I. Dubas, M., joint author. II. Title. TA1555.S38 774 79-10555

Offset printing: Beltz Offsetdruck, Hemsbach/Bergstr. Bookbinding: J. Schäffer oHG, Grünstadt.
2153/3130-543210

To the memory of
Henry Favre

Preface

This small book intends to build a bridge between the aspects of Optics and of Mechanics that are involved in the application of holographic interferometry to deformation analysis of opaque bodies. As such, it follows in some way the footsteps of the late Prof. H. Favre, who already in 1927 proposed to use interferometry for deformation measurements (refer to his thesis "Sur une nouvelle méthode optique de détermination des tensions intérieures"). Many a concept also originates from the research and lectures of Prof. W. Prager in continuum mechanics, Profs. D. C. Drucker and C. Mylonas in experimental mechanics, Prof. C. R. Steele in shell theory and Prof. W. Lukosz in physical optics. Further stimulation arose in discussions about holography with Profs. R. Dändliker, J. Der Hovanesian and H. Tiziani as well as with Drs. B. Ineichen and F. M. Mottier. The contribution of Drs. W. Wüthrich, P. Bohler and G. Teichmann must also be acknowledged, the latter more particularly for rendering valuable assistance on the delicate points of tensor calculus as well as in the drawing of the figures. Full gratitude must also be expressed to those who made the publication of this book possible: Dr. D. MacAdam who openheartedly accepted it in his series, Dr. H. Lotsch and the collaborators of Springer-Verlag, Mr. P. Hagnauer who revised the original text and Mrs. L. Wehrli whose patience was tried in carefully typing the manuscript, which Mr. F. Dufour read over again.

Zurich, December 1978 *W. Schumann* *M. Dubas*

Contents

1. Introduction 1

2. Some Basic Concepts of Differential Geometry and Continuum Mechanics 3
2.1 Some Elements of Tensor Analysis and Differential Geometry of Surfaces 3
2.1.1 Vectors and Tensors, General Coordinates 3
2.1.2 Derivatives in Three-Dimensional Space, Normal Projections 10
2.1.3 Vectors and Tensors on a Curved Surface 15
2.1.4 Derivatives on a Surface, Curvature of the Surface 18
2.1.5 Connections Between Two Surfaces, Oblique Projections . . 21
2.2 Elements of Continuum Mechanics: The Kinematics of Deformation 27
2.2.1 Deformation, Strain and Rotation, and Their Relations to the Displacement 27
2.2.2 Strain and Rotation at the Surface of a Body 33

3. Principles of Image Formation in Holography 39
3.1 Image Formation in Standard Holography 39
3.1.1 Recording the Phase and the Amplitude of the Light Wave Emitted by a Point Source 39
3.1.2 Reconstruction of the Image of a Point Source 41
3.1.3 Holography of a Whole Object and Double Exposure 44
3.2 Image Formation in the Case of Modification of the Optical Arrangement at the Reconstruction 47
3.2.1 Large Modification of the Reference Source: Position of an Image Point, Aberrations 47
3.2.2 Small Modification of the Reference Source: Position of an Image Point 56
3.2.3 Modification of the Reference Source: Analysis in Terms of Transverse Ray Aberration 59
3.2.4 Movement of the Hologram: Position of an Image Point . . . 65
3.2.5 Movement of Two Sandwiched Holograms: Difference Between the Lateral Displacements of Two Image Points . . 69
3.3 Conclusion 74

4. Fringe Interpretation in Holographic Interferometry 75
4.1 Optical-Path Difference in Standard Holographic Interferometry . . 77
4.1.1 Basic Relation 77

4.1.2 Determination of the Displacement by Means of the Optical-Path Difference 80
4.1.3 Determination of the Displacement with Additional Equipment 85
4.1.4 Determination of Strain with Finite Differences 89
4.2 Derivatives of the Optical-Path Difference in Standard Holographic Interferometry 92
4.2.1 Fringe Spacing and Direction 92
4.2.2 Fringe Contrast or Visibility, Concepts of Localization 97
4.2.3 Fringe Visibility for Different Apertures, Partial Localization 110
4.2.4 Complete Localization, Normality Theorem, Shape of the Line of Complete Localization 115
4.2.5 Determination of Strain and Rotation by Means of the Derivatives of the Optical-Path Difference 127
4.3 Modifications of the Optical Arrangement at the Reconstruction . . 132
4.3.1 Basic Relations when the Reconstruction Sources are Shifted and when the Wavelength is Changed 132
4.3.2 Principal Relations that Result from the Movement of One Hologram or of Two Sandwiched Holograms 140
4.3.3 Generalization and Application of the Results Concerning the Modification of the Optical Arrangement 145
4.4 Conclusion 151

5. Second Derivatives of the Displacement and of the Optical-Path Difference 153
5.1 Some Additional Equations of Continuum Mechanics, Particularly Relations Containing the Second Derivative of the Displacement . . 153
5.1.1 Curvature Change of the Outer Surface of an Arbitrary Body or of the Middle Surface of a Shell 154
5.1.2 Integrability and Compatibility Conditions on a Curved Surface 157
5.1.3 Equations of Motion or of Equilibrium and Kinematic Relations in the Neighborhood of a Free Surface. Constitutive Equations for an Elastic Body 161
5.2 The Second Derivatives of the Optical-Path Difference and Some of Their Possible Applications 164
5.2.1 Derivative of the Fringe Spacing, Curvature of a Fringe, Singularities in the Fringe Pattern 165
5.2.2 Extrapolation of Mechanical Quantities into the Interior of a Nontransparent Body 169
5.3 Conclusion 170

References 171

Author Index 187

Subject Index 191

Chapter 1 Introduction

In the year 1948, D. Gabor described "a new microscopic principle" laying the foundation for the reconstruction of the image of an object from a diffraction pattern produced by that object. Because this technique made it possible to record not only the amplitude but also the phase of a light wave, its inventor named it "holography" (from $\dot{o}\lambda o\varsigma$ = entire and $\gamma\rho\alpha\phi\varepsilon\iota\nu$ = to write). At that time however, two obstacles had to be overcome: on the one hand, no sufficiently monochromatic and coherent light source was available and, on the other hand, observation of the image was hindered by overlapping of the wave fields at the reconstruction. In 1962, E. N. Leith and J. Upatnieks remedied this situation by using a laser beam (i.e. Light Amplification by Stimulated Emission of Radiation) as a light source and by providing the object wave and the reference background with two different directions. Since then, holography rapidly developed and found many an application in various domains of science. Three years later, the insight that a holographically produced wave field could be brought to interfere with another wave field, whether holographically produced or not, particularly fascinated several researchers.

Thus was born "holographic interferometry", which was applied to the subject in which we are interested, namely the analysis of the deformation undergone by the surface of a nontransparent, diffusely reflecting body. Indeed, because light waves can be recorded and reconstructed any time later, the wave fields reflected by the surface of a body in two slightly different and not simultaneously existing configurations can be superimposed, thus forming an interference pattern. So it becomes possible to measure the deformation that took place between the two configurations. Due particularly to the possibility of examining bodies of any shape, of investigating the object itself and not a model of it, of making really punctal measurements, and of obtaining a great amount of information, holographic interferometry soon became a well-established technique, along with others, such as photoelasticity, moiré and speckle methods, all of which, by the way, reciprocally influenced each other.

In the following pages, before we come to holographic interferometry itself, however, we shall lay down, in Chap. 2, some basic concepts of differential geometry and continuum mechanics, and, in Chap. 3, the principles of image formation in holography. The first chapter provides the elements we shall require all along, whereas the purpose of the second, on the one hand, is to make clear how the wave fields with which we have to deal are produced and, on the other hand, to examine more in detail the images as such, especially the aberrations they may suffer when at the reconstruction the optical arrangement is modified with respect to the recording disposition. Here already, the interaction between the concepts of Mechanics and those of Optics will appear. Then, in the central part of this book, i.e. Chap. 4, we shall investigate both the formation of the interference phenomenon that follows the superimposition of the wave fields pertaining to two given configurations of an object, and the opposite problem of measuring by means of a given fringe pattern the deformation undergone by that body. There, we shall first study the fringe order, i.e. the difference of the optical paths covered by two interfering rays, and the means of deducing from it the displacement vector. Second, we shall look at some properties of the fringes, such as their spacing, their direction, their visibility or their localization, which depend on first derivatives of the optical-path difference. We shall then show how the derivative of the displacement, i.e. the strain and the rotation, can be measured. Third, we will point out the effects of modification of the optical arrangement during the reconstruction on the interference phenomenon and on the methods of measurement. Chapter 5, finally, will sketch some further possible uses of holographic interferometry relating to higher-order derivatives of the optical-path difference and again to continuum mechanics.

Chapter 2 Some Basic Concepts of Differential Geometry and Continuum Mechanics

In this book, we shall frequently encounter vectors in the space of elementary geometry as well as linear transformations, also called tensors, applied onto them. Therefore, at the very beginning, we feel it advisable to sum up the reasonings we shall be using all along. They relate, on the one hand, to the geometry of three-dimensional space, particularly on curved surfaces, and, on the other hand, to the kinematics of deformation.

2.1 Some Elements of Tensor Analysis and Differential Geometry of Surfaces

Vector and tensor calculus can be established in an axiomatic way by means of the set theory (see e.g. [2.1–4]). Here, however, it will be sufficient to deduce briefly the basic concepts, beginning with the well-known elementary vector calculus in a three-dimensional Euclidean space $\mathbb{R}^3$ in a manner comparable to that of *Green* and *Zerna* [2.5] (see also [2.6–9]).

2.1.1 Vectors and Tensors, General Coordinates

Let $\boldsymbol{r}$ be the position vector of a point $\mathrm{P} \in \mathbb{R}^3$, its components in a cartesian coordinate system being x, y, z. Assume furthermore the three functions

$$x, y, z \to \theta^1(x, y, z), \theta^2(x, y, z), \theta^3(x, y, z),$$

so that they be reversible and derivable, and that one may then calculate

$$\theta^1, \theta^2, \theta^3 \to x(\theta^1, \theta^2, \theta^3), y(\theta^1, \theta^2, \theta^3), z(\theta^1, \theta^2, \theta^3).$$

Owing to these assumptions, there is a one-one correspondence between x, y, z and $\theta^1, \theta^2, \theta^3$ so that the position of point P may be defined by the three new variables $\theta^1, \theta^2, \theta^3$. When two of them are constant, e.g. θ^2 and θ^3, the three functions

$$\theta^1 \to x(\theta^1), y(\theta^1), z(\theta^1)$$

define a *curve* in space, on which the parameter θ^1 determines the position of any point on it. Therefore $\theta^1, \theta^2, \theta^3$ may be called *curvilinear coordinates*

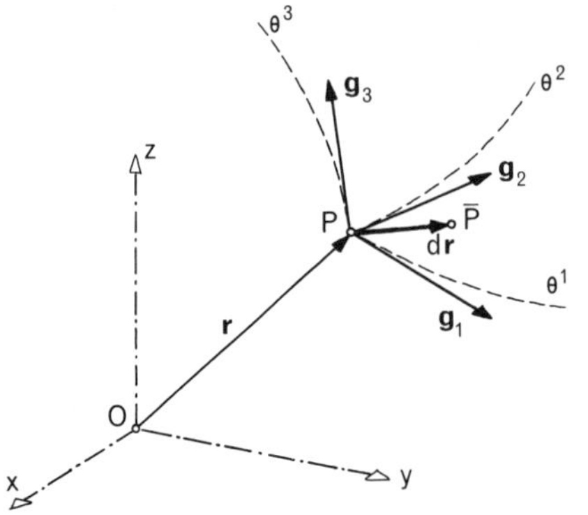

Fig. 2.1. Definition of the position of a point P by means of cartesian and curvilinear coordinates

(Fig. 2.1). Most frequently, we shall speak simply of the coordinates θ^i, the latin index i ranging from 1 to 3.

Let us now look at the *increment* $d\boldsymbol{r}$ that joins P to a neighboring point $\bar{\text{P}}$. We form the total differential

$$d\boldsymbol{r} = \frac{\partial \boldsymbol{r}}{\partial \theta^1} d\theta^1 + \frac{\partial \boldsymbol{r}}{\partial \theta^2} d\theta^2 + \frac{\partial \boldsymbol{r}}{\partial \theta^3} d\theta^3$$

and write it shortly in the alternative form

$$d\boldsymbol{r} = \boldsymbol{r}_{,i} d\theta^i = \boldsymbol{g}_i d\theta^i \tag{2.1}$$

where, as usual, the repetition of an index in each term means summation from 1 to 3, and $,i$ stands for the partial derivative. The vectors $\boldsymbol{g}_i = \boldsymbol{r}_{,i} = \partial \boldsymbol{r}/\partial \theta^i$ are called *covariant base* vectors; they are tangent to the curves of the coordinates θ^i (Fig. 2.1). This definition of $\boldsymbol{g}_i$ yields a continuous field of variable base vectors for a whole configuration {P}. When, however, attention is restricted to vector algebra, leaving out analysis, any system of three linearly independent vectors $\boldsymbol{g}_i$ can serve as a basis, and we need not know what derivative they are equal to. We may also encounter the particular case in which the base vectors are invariant in the whole space and represent any oblique system of three linearly independent vectors. For example, if $\theta^1 = x$, $\theta^2 = y$, $\theta^3 = z$, the $\boldsymbol{g}_i$ are constant unit vectors parallel to the axes of the cartesian system.

Furthermore, we shall use a *contravariant basis* $\boldsymbol{g}^k$ $(k = 1, 2, 3)$ defined by the *orthogonality* relations (Fig. 2.2)

$$\boldsymbol{g}_i \cdot \boldsymbol{g}^k = \delta_i^k = \begin{cases} 1, & \text{if} \quad i = k \\ 0, & \text{if} \quad i \neq k, \end{cases} \tag{2.2}$$

where the dot signifies the scalar product, and δ_i^k denotes the so-called *Kronecker* symbol.

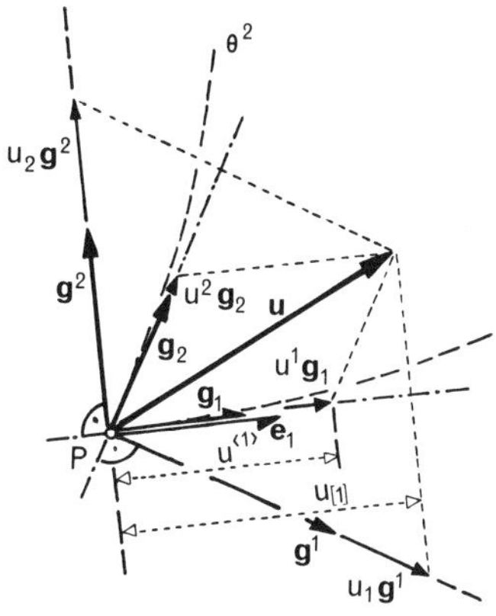

Fig. 2.2. Covariant and contravariant basis vectors, related components of a vector $\boldsymbol{u}$ in two dimensions

Any vector $\boldsymbol{u}$ referred to P may then be expressed by either of these two bases

$$\boldsymbol{u} = u^i \boldsymbol{g}_i = u_k \boldsymbol{g}^k, \tag{2.3}$$

with contravariant components u^i and covariant components u_k. The prefixes co- and contra- distinguish the two types of transformation, when the coordinates are changed; this is hinted by the position of the indices; the vector $\boldsymbol{u}$ as such is invariant. By scalar multiplication with the base vectors, we obtain the value of the components

$$u^j = u^i \boldsymbol{g}_i \cdot \boldsymbol{g}^j = \boldsymbol{u} \cdot \boldsymbol{g}^j, \qquad u_l = \boldsymbol{u} \cdot \boldsymbol{g}_l. \tag{2.4}$$

It is easy to pass from the one type of components to the other; with the definition $\boldsymbol{g}_i \cdot \boldsymbol{g}_k = g_{ik}$ we have

$$u^i g_{ik} = u^i \boldsymbol{g}_i \cdot \boldsymbol{g}_k = \boldsymbol{u} \cdot \boldsymbol{g}_k = u_k, \tag{2.5}$$

which means that indices can be lowered or raised.

As the base vectors are generally not unit vectors, so-called *physical* quantities are also used in addition to the above components (Fig. 2.2). Let $\boldsymbol{e}_i$ be a unit vector parallel to $\boldsymbol{g}_i$, then we may define *physical* (*oblique*) *components* (i not summed, see, e.g., [Ref. 2.2, p. 85])

$$u^{\langle i \rangle} = \sqrt{g_{ii}}\, u^i, \tag{2.6}$$

so that (i summed)

$$\boldsymbol{u} = u^i \boldsymbol{g}_i = u^i \sqrt{g_{ii}}\, \boldsymbol{e}_i = u^{\langle i \rangle} \boldsymbol{e}_i.$$

Further, we may consider *normal projections* (k not summed)

$$u_{[k]} = \frac{1}{\sqrt{g_{kk}}} u_k, \tag{2.7}$$

where $u_{[k]} = \boldsymbol{u} \cdot \boldsymbol{e}_k$ is generally not a component of $\boldsymbol{u}$.

In addition to scalars and vectors, we shall encounter *linear transformations* (mappings) or *tensors*, transforming vectors into other vectors. For instance, let us apply such a tensor $\boldsymbol{T}$ to a vector $\boldsymbol{u}$, the image of $\boldsymbol{u}$ under $\boldsymbol{T}$ is then a new vector $\boldsymbol{w}$. The transformation

$$\boldsymbol{u} \rightarrow \boldsymbol{w} = \boldsymbol{T}\boldsymbol{u},$$

is called linear if, for two vectors $\boldsymbol{u}$, $\boldsymbol{v}$,

$$\boldsymbol{T}(\lambda \boldsymbol{u} + \mu \boldsymbol{v}) = \lambda \boldsymbol{T}\boldsymbol{u} + \mu \boldsymbol{T}\boldsymbol{v},$$

where λ and μ are any scalars. In particular, there exists a linear transformation obtained by two given vectors $\boldsymbol{p}$, $\boldsymbol{q}$ (Fig. 2.3)

$$\boldsymbol{u} \rightarrow \boldsymbol{w} = (\boldsymbol{q} \cdot \boldsymbol{u})\boldsymbol{p},$$

mapping $\boldsymbol{u}$ into a vector $\boldsymbol{w}$ parallel to $\boldsymbol{p}$. The transformation created by these vectors may be written as a *tensor product* or *dyadic product* $\boldsymbol{p} \otimes \boldsymbol{q}$ and constitutes an element of the tensor-product space $\mathbb{R}^3 \otimes \mathbb{R}^3$. The transpose of $\boldsymbol{p} \otimes \boldsymbol{q}$ is $(\boldsymbol{p} \otimes \boldsymbol{q})^{\mathrm{T}} = \boldsymbol{q} \otimes \boldsymbol{p}$, so that in summary

$$(\boldsymbol{p} \otimes \boldsymbol{q})\boldsymbol{u} = \boldsymbol{p}(\boldsymbol{q} \cdot \boldsymbol{u}), \qquad (\boldsymbol{p} \otimes \boldsymbol{q})^{\mathrm{T}}\boldsymbol{u} = (\boldsymbol{q} \otimes \boldsymbol{p})\boldsymbol{u} = \boldsymbol{q}(\boldsymbol{p} \cdot \boldsymbol{u}). \tag{2.8}$$

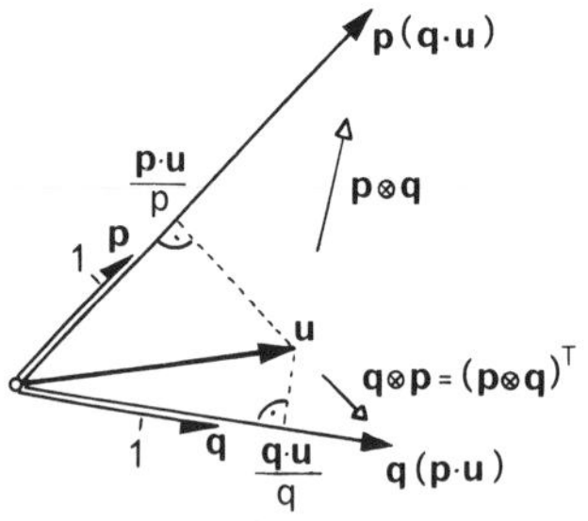

Fig. 2.3. Application of a dyadic product onto a vector $\boldsymbol{u}$

Dyadics thus "assimilate" the vector on which they are applied. It is to be remarked here that general linear transformations are mostly *not symmetrical* (even the tensor product is not commutative). For reasons that will become evident later, it is sometimes convenient to let transformation symbols follow the vector to which they are applied, as in

$$t = uT = T^{\mathrm{T}}u,$$

or in

$$t \cdot v = uT \cdot v = u \cdot Tv,$$

the result of the two latter combinations being the same, regardless of the position of the dot.

It remains to be explained how two tensors may be combined. First, they can succeed one another in the sense of two following transformations; so the dyadics $T = p \otimes q$ and $S = r \otimes s$ give

$$TS = (p \otimes q)(r \otimes s) = p \otimes s(q \cdot r).$$

Second, they may be contracted to form a scalar, namely the *trace* (i.e. the sum of diagonal terms in the matrix of mixed components)

$$\mathrm{tr}\,\{TS\} = T \cdot S = (p \otimes q) \cdot (r \otimes s) = (p \cdot s)(q \cdot r).$$

Similarly to vectors, general tensors may be expressed by means of base vectors. Several linear combinations of dyadics are possible, namely

$$T = T^{ij} g_i \otimes g_j = T_{kl} g^k \otimes g^l = T^i{}_l g_i \otimes g^l = T_k{}^j g^k \otimes g_j. \tag{2.9}$$

T^{ij} are the contravariant, T_{kl}, the covariant and $T^i{}_l$, $T_k{}^j$, the mixed components of the tensor T. In terms of components, a linear transformation then reads

$$Tu = (T^i{}_l g_i \otimes g^l)(u^h g_h) = T^i{}_l u^h g_i \delta^l_h = T^i{}_l u^l g_i.$$

Applying T in particular to the vector g_n, and thereafter contracting the resulting vector Tg_n with g_m (scalar product) we obtain the covariant components

$$T_{mn} = g_m \cdot Tg_n. \tag{2.10}$$

Again, the indices may be lowered or raised by the method used for vector components. Furthermore, we can define *physical quantities* that differ by scale factors $\sqrt{g_{ii}}$ from the above components. For instance, we have (no summation)

$$T_{[mn]} = e_m \cdot Te_n = \frac{1}{\sqrt{g_{mm} g_{nn}}} T_{mn}, \tag{2.11}$$

or

$$T^{\langle i}{}_{l]} = T^i{}_l \sqrt{\frac{g_{ii}}{g_{ll}}}. \tag{2.12}$$

The set of these latter quantities forms the physical *components* of $\boldsymbol{T}$, because the linear transformation on $\boldsymbol{u}$ may be written [Ref. 2.2, p. 85]

$$T^i{}_l u^l \boldsymbol{g}_i = T^i{}_l \frac{\sqrt{g_{ii}}}{\sqrt{g_{ll}}}(u^l \sqrt{g_{ll}}) \frac{\boldsymbol{g}_i}{\sqrt{g_{ii}}} = T^{\langle i}{}_{l]} u^{\langle l \rangle} \boldsymbol{e}_i.$$

Sometimes, [Ref. 2.5, p. 61] quantities are also encountered that decompose the vector $\boldsymbol{T}\boldsymbol{e}^n$ with regard to the unit vectors $\boldsymbol{e}_m$

$$\boldsymbol{T}\boldsymbol{e}^n = T^{mj}\sqrt{g_{mm}}(\boldsymbol{e}_m \otimes \boldsymbol{g}_j)\frac{\boldsymbol{g}^n}{\sqrt{g^{nn}}} = T^{\langle mn]}\boldsymbol{e}_m,$$

so that

$$T^{\langle m\,n]} = \sqrt{\frac{g_{mm}}{g^{nn}}}\,T^{mn}. \tag{2.13}$$

In a cartesian (i.e. orthonormal) system, all of these quantities are equal and are the components of the tensor $\boldsymbol{T}$. The different physical quantities reflect only particular aspects of the meaning of a vector or a tensor; therefore, one does not use them until this specific meaning must be pointed out.

Apart from the second-order tensors of which we have just spoken, there also exist tensors of higher order. Here we shall encounter third-order tensors, particularly those formed by three vectors in the sense of a *triadic product*, as

$$\mathscr{T} = \boldsymbol{p} \otimes \boldsymbol{q} \otimes \boldsymbol{s},$$

or more generally those as a product of a second-order tensor and a vector

$$\mathscr{S} = \boldsymbol{T} \otimes \boldsymbol{s}, \text{ etc.}$$

Such tensors may act on vectors from the left, from the right or from the "middle" to form a second-order tensor; for this reason, they may also be *partially transposed*; so

$$[\boldsymbol{p} \otimes \boldsymbol{q} \otimes \boldsymbol{s})^{\mathrm{T}}]\boldsymbol{u} = (\boldsymbol{p} \otimes \boldsymbol{s} \otimes \boldsymbol{q})\boldsymbol{u} = \boldsymbol{p} \otimes \boldsymbol{s}(\boldsymbol{q}\cdot\boldsymbol{u}),$$

or

$$[(\boldsymbol{p} \otimes \boldsymbol{q} \otimes \boldsymbol{s})^{\mathrm{T}}]\boldsymbol{u} = (\boldsymbol{s} \otimes \boldsymbol{q} \otimes \boldsymbol{p})\boldsymbol{u} = \boldsymbol{s} \otimes \boldsymbol{q}(\boldsymbol{p}\cdot\boldsymbol{u}). \tag{2.14}$$

As we see from the above, in general calculus it seems to become a matter of personal taste whether one uses the usual notation with the indices-summation convention and co- or contravariant components as in $T^{ij}u_j$ or the corresponding symbolic notation $\boldsymbol{T}\boldsymbol{u}$. However, because in holography we have to change the system of coordinates often, especially when passing from the space to the curved (!) object surface or to the plane hologram plate, the abstract symbolic notation is preferable, the more so as a great number of indices in a succession of linear transformations overshadows the physical reality which, in fact, does not depend upon specific coordinates (see also [Ref. 2.2, p. 31]). The rules for the calculus, including the derivatives coming later are in fact very simple and bring out the *geometrical meaning*. For greater ease, but still in accordance with part of the literature of mechanics, we shall also systematically use ordinary letters for *scalars*, small bold-face letters for *vectors*, capital bold-face latin or bold-face greek letters for *second-order tensors* and round-hand capital bold-face letters for *third-order tensors*.

Let us finish this paragraph by having a look at some special tensors. First, the *identity* is the *symmetric* tensor

$$\boldsymbol{I} = \boldsymbol{g}_i \otimes \boldsymbol{g}^i, \tag{2.15}$$

because

$$\boldsymbol{I}\boldsymbol{u} = \boldsymbol{g}_i(\boldsymbol{g}^i \cdot \boldsymbol{u}) = u^i \boldsymbol{g}_i = \boldsymbol{u}.$$

Its cartesian components would be

$$\boldsymbol{I} \mathrel{\hat{=}} \begin{bmatrix} 1 & 0 & 0 \\ 0 & 1 & 0 \\ 0 & 0 & 1 \end{bmatrix}.$$

Second, the *permutation* is the third-order tensor (see e.g. [Ref. 2.5, p. 11]),

$$\boldsymbol{\mathscr{E}} = E_{ijk} \boldsymbol{g}^i \otimes \boldsymbol{g}^j \otimes \boldsymbol{g}^k, \tag{2.16}$$

with $E_{ijk} = +\sqrt{g}, -\sqrt{g}, 0$, accordingly as the indices i, j, k form an even permutation or an odd permutation of 1, 2, 3 or contain the same index twice, and where g is the determinant of g_{ij}. The cartesian components of $\boldsymbol{\mathscr{E}}$ would be

$$\boldsymbol{\mathscr{E}} \mathrel{\hat{=}} \left[\begin{array}{ccc:ccc:ccc} 0 & 0 & 0 & 0 & 0 & -1 & 0 & 1 & 0 \\ 0 & 0 & 1 & 0 & 0 & 0 & -1 & 0 & 0 \\ 0 & -1 & 0 & 1 & 0 & 0 & 0 & 0 & 0 \end{array}\right].$$

If $\boldsymbol{\mathscr{E}}$ is applied onto a vector $\boldsymbol{\psi}$, we obtain a *skew-symmetric* second-order tensor

$$\boldsymbol{\Psi} = \boldsymbol{\mathscr{E}}\boldsymbol{\psi} = \boldsymbol{\psi}\boldsymbol{\mathscr{E}} = -\boldsymbol{\Psi}^{\mathrm{T}}, \tag{2.17}$$

with the cartesian components

$$\boldsymbol{\Psi} \mathrel{\hat{=}} \begin{bmatrix} 0 & \psi_z & -\psi_y \\ -\psi_z & 0 & \psi_x \\ \psi_y & -\psi_x & 0 \end{bmatrix}.$$

If we apply this linear transformation to any vector $\boldsymbol{s}$, the result is

$$\boldsymbol{s}\boldsymbol{\Psi} = \boldsymbol{\Psi}^{\mathrm{T}}\boldsymbol{s} = \boldsymbol{s}\mathscr{E}\boldsymbol{\psi} = \boldsymbol{\psi} \times \boldsymbol{s}, \tag{2.18}$$

the *vector product* of $\boldsymbol{\psi}$ and $\boldsymbol{s}$.

2.1.2 Derivatives in Three-Dimensional Space, Normal Projections

Let us first look at the derivative of a *scalar function* $\Phi(\theta^1, \theta^2, \theta^3)$. The total differential may be written as

$$d\Phi = \sum_{i=1}^{3} \frac{\partial \Phi}{\partial \theta^i} d\theta^i = \Phi_{,i} d\theta^i,$$

or, because of (2.2),

$$d\Phi = d\theta^i \boldsymbol{g}_i \cdot \boldsymbol{g}^k \Phi_{,k} = d\boldsymbol{r} \cdot \boldsymbol{\nabla}\Phi. \tag{2.19}$$

We have thus interpreted the differential by means of the increment $d\boldsymbol{r} = \boldsymbol{g}_i d\theta^i$ and of the *gradient* $\boldsymbol{\nabla}\Phi$ formed with the *derivative operator* (sometimes called "nabla")

$$\boldsymbol{\nabla} = \boldsymbol{g}^k \frac{\partial}{\partial \theta^k}. \tag{2.20}$$

$\boldsymbol{\nabla}$ is thus formally a "vector": it can be combined with scalars, vectors or tensors as well as an ordinary vector; furthermore it differentiates each part of the quantities with which it is associated from the left.

The differential of a *vector-valued function* $\boldsymbol{u}(\theta^1, \theta^2, \theta^3)$ is similarly

$$d\boldsymbol{u} = d\theta^i(\boldsymbol{g}_i \cdot \boldsymbol{g}^k) \frac{\partial \boldsymbol{u}}{\partial \theta^k} = d\boldsymbol{r}(\boldsymbol{\nabla} \otimes \boldsymbol{u}). \tag{2.21}$$

The derivative $\boldsymbol{\nabla} \otimes \boldsymbol{u}$ is thus the sum of the dyadic products $\boldsymbol{g}^k \otimes \partial \boldsymbol{u}/\partial \theta^k$. This linear transformation is here applied to the left onto $d\boldsymbol{r}$, the latter being combined with $\boldsymbol{\nabla}$ (acting as a vector) in order to form the increment $d\boldsymbol{u}$ related to $d\boldsymbol{r}$. On the other hand, the derivative action of $\boldsymbol{\nabla}$, i.e. $\partial/\partial \theta^k$, always involves the elements placed on its right, as here the vector $\boldsymbol{u}$.

Before carrying on with the calculus, let us write out the components of the derivative of a vector. With $\boldsymbol{u} = u_l \boldsymbol{g}^l$, we have

$$\boldsymbol{\nabla} \otimes \boldsymbol{u} = u_{l,k} \boldsymbol{g}^k \otimes \boldsymbol{g}^l + u_l \boldsymbol{g}^k \otimes \boldsymbol{g}^l_{,k}.$$

In the second term, we may exchange the dummy index l with j and then decompose the derivative of the base (see e.g. [Ref. 2.5, p. 25])

$$\boldsymbol{g}^j_{,k} = -\Gamma^j_{kl}\boldsymbol{g}^l. \tag{2.22}$$

The coefficients Γ^j_{kl} are the so-called *Christoffel symbols of the second kind.* By contraction with $\boldsymbol{g}_i$ we obtain

$$\Gamma^j_{ki} = -\boldsymbol{g}^j_{,k}\cdot\boldsymbol{g}_i = \boldsymbol{g}^j\cdot\boldsymbol{g}_{i,k} = \boldsymbol{g}^j\cdot\boldsymbol{g}_{k,i} = \Gamma^j_{ik}. \tag{2.23}$$

The derivative thus becomes with respect to the contravariant basis

$$\boldsymbol{\nabla}\otimes\boldsymbol{u} = (u_{l,k} - \Gamma^j_{kl}u_j)\boldsymbol{g}^k\otimes\boldsymbol{g}^l. \tag{2.24}$$

The components of this tensor,

$$u_{l,k} - \Gamma^j_{kl}u_j = u_{l|k}, \tag{2.25}$$

are called the *covariant derivatives* of the components of the vector $\boldsymbol{u}$. In the special case of a cartesian system, where $\boldsymbol{g}^j_{,k} = 0$, $\Gamma^j_{kl} = 0$, the covariant derivative is equal to the ordinary partial derivative $u_{l,k}$. Similarly, we may also calculate the covariant derivatives of the contravariant components u^j

$$u^j_{|k} = u^j_{,k} + \Gamma^j_{kl}u^l. \tag{2.26}$$

As a special case thereof, the *divergence* of a vector $\boldsymbol{u}$ is the trace

$$\mathrm{tr}\,\{\boldsymbol{\nabla}\otimes\boldsymbol{u}\} = \boldsymbol{\nabla}\cdot\boldsymbol{u} = u^j_{|k}\boldsymbol{g}^k\cdot\boldsymbol{g}_j = u^j_{|j}. \tag{2.27}$$

Finally, the *curl* (rotation) of a vector would be, according to (2.18) and (2.16)

$$\boldsymbol{\nabla}\times\boldsymbol{u} = -\boldsymbol{\nabla}\mathscr{E}\boldsymbol{u} = E^{ikl}u_{l|k}\boldsymbol{g}_i. \tag{2.28}$$

The derivatives of *tensor-valued functions* may be formed in the same manner, e.g.

$$\boldsymbol{\nabla}\otimes\boldsymbol{T} = T^{ij}_{|k}\boldsymbol{g}^k\otimes\boldsymbol{g}_i\otimes\boldsymbol{g}_j, \tag{2.29}$$

where

$$T^{ij}_{|k} = T^{ij}_{,k} + \Gamma^i_{kl}T^{lj} + \Gamma^j_{km}T^{im};$$

the number of Christoffel symbols corresponds to the number of base vectors to be differentiated.

In addition to the foregoing derivatives, we shall often be confronted with the problem of differentiating *combinations* of vectors and of tensors. The general rule is the well-known rule for differentiating scalar functions of one variable

$$\frac{d(uv)}{d\theta} = \frac{du}{d\theta}v + \frac{dv}{d\theta}u.$$

Note also that

a)—the operator ∇ always stands on the left of the function that it differentiates, and acts successively upon each element of that function;

b)— ∇ may be applied to vectors and tensors by means of a dyadic product, in which case the order is increased, or by means of a contraction, in which case the order is decreased;

c)—the links (products or contractions) between the elements, as well as their sequence, have to be maintained, sometimes by tensor transpositions.

Thus, the derivatives of products of vectors $\boldsymbol{u}, \boldsymbol{v}, \ldots$, or of tensors $\boldsymbol{T}, \boldsymbol{S}, \ldots$ are

$$\begin{aligned}
&\nabla(\boldsymbol{u}\cdot\boldsymbol{v}) = (\nabla \otimes \boldsymbol{u})\boldsymbol{v} + (\nabla \otimes \boldsymbol{v})\boldsymbol{u},\\
&\nabla(\boldsymbol{T}\cdot\boldsymbol{S}) = (\nabla \otimes \boldsymbol{T})\cdot\boldsymbol{S} + (\nabla \otimes \boldsymbol{S})\cdot\boldsymbol{T},\\
&\nabla\cdot(\boldsymbol{T}\boldsymbol{u}) = (\nabla\boldsymbol{T})\cdot\boldsymbol{u} + (\nabla \otimes \boldsymbol{u})\cdot\boldsymbol{T}^{\mathrm{T}},\\
&\nabla \otimes (\boldsymbol{T}\boldsymbol{u}) = (\nabla \otimes \boldsymbol{T})\boldsymbol{u} + (\nabla \otimes \boldsymbol{u})\boldsymbol{T}^{\mathrm{T}},\\
&\nabla(\boldsymbol{u} \otimes \boldsymbol{v}) = (\nabla\cdot\boldsymbol{u})\boldsymbol{v} + (\nabla \otimes \boldsymbol{v})^{\mathrm{T}}\boldsymbol{u},\\
&\nabla(\boldsymbol{T}\boldsymbol{S}) = (\nabla\boldsymbol{T})\boldsymbol{S} + (\nabla \otimes \boldsymbol{S})^{\mathrm{T}}\cdot\boldsymbol{T},\\
&\nabla \otimes (\boldsymbol{u} \otimes \boldsymbol{v}) = (\nabla \otimes \boldsymbol{u}) \otimes \boldsymbol{v} + (\nabla \otimes \boldsymbol{v}) \otimes \boldsymbol{u}]^{\mathrm{T}}.
\end{aligned} \tag{2.30}$$

In terms of components, the last of these relations would be written

$$\begin{aligned}
\boldsymbol{g}^k\frac{\partial}{\partial\theta^k} \otimes (u^i v^j \boldsymbol{g}_i \otimes \boldsymbol{g}_j) &= (u^i{}_{|k} v^j + u^i v^j{}_{|k})(\boldsymbol{g}^k \otimes \boldsymbol{g}_i \otimes \boldsymbol{g}_j)\\
&= (u^i v^j)_{|k}(\boldsymbol{g}^k \otimes \boldsymbol{g}_i \otimes \boldsymbol{g}_j).
\end{aligned}$$

Higher derivatives are formed in exactly the same way as first derivatives; thus, in the Taylor-series development of a scalar function $\Phi(\theta^1, \theta^2, \theta^3)$,

$$\begin{aligned}
\bar{\Phi} &= \Phi + d\boldsymbol{r}\cdot\nabla\Phi + \frac{1}{2!}d\boldsymbol{r}\cdot\nabla(d\boldsymbol{r}\cdot\nabla\Phi) + \cdots\\
&= \Phi + d\boldsymbol{r}\cdot\nabla\Phi + \frac{1}{2}[d\boldsymbol{r}\cdot(\nabla \otimes \nabla\Phi)d\boldsymbol{r} + d^2\boldsymbol{r}\cdot\nabla\Phi] + \cdots
\end{aligned} \tag{2.31}$$

since $d\boldsymbol{r}(\boldsymbol{\nabla} \otimes d\boldsymbol{r}) = d^2\boldsymbol{r}$. Because $\boldsymbol{r}$ is considered here as an independent variable, $d^2\boldsymbol{r} = 0$. The second-order term contains the tensor

$$\boldsymbol{\nabla} \otimes \boldsymbol{\nabla}\Phi = \Phi_{|lk}\boldsymbol{g}^k \otimes \boldsymbol{g}^l, \tag{2.32}$$

which is first applied to one $d\boldsymbol{r}$ as a linear transformation, the resulting vector being then contracted with the other $d\boldsymbol{r}$. The *trace* is obtained by contraction and represents the *Laplace operator* on Φ

$$\Delta\Phi = \boldsymbol{\nabla}\cdot\boldsymbol{\nabla}\Phi = \Phi_{|lk}\boldsymbol{g}^k\cdot\boldsymbol{g}^l = \Phi|_k^k. \tag{2.33}$$

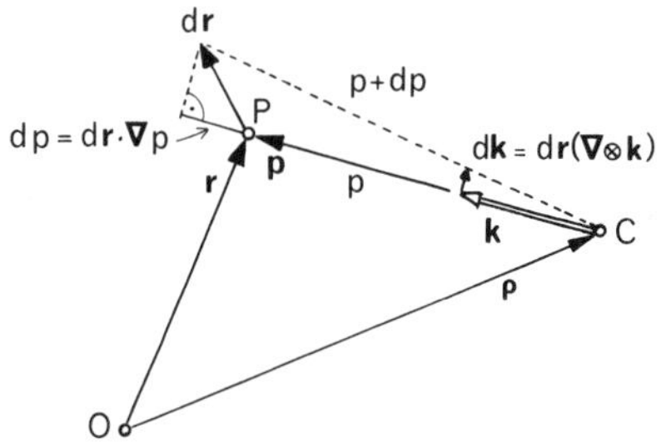

Fig. 2.4. Derivation of the length p and of the unit vector $\boldsymbol{k}$ between a fixed point C and a variable point P

Concluding this section, let us investigate some particular derivatives that we shall often encounter in holography. It concerns the vector between a *fixed center* C with position vector $\boldsymbol{\rho}$ and a variable point P (Fig. 2.4). If $\boldsymbol{p}$ denotes the vector from C to P, we have first

$$\boldsymbol{\nabla} \otimes \boldsymbol{p} = \boldsymbol{\nabla} \otimes (\boldsymbol{r} - \boldsymbol{\rho}) = \boldsymbol{\nabla} \otimes \boldsymbol{r} = \boldsymbol{I}. \tag{2.34}$$

Let us now put $\boldsymbol{p} = p\boldsymbol{k}$ where $p = |\boldsymbol{p}|$ is the length of $\boldsymbol{p}$ and $\boldsymbol{k}$, a *unit vector* parallel to $\boldsymbol{p}$. The gradient of p becomes

$$\boldsymbol{\nabla}p = \boldsymbol{\nabla}\sqrt{\boldsymbol{p}\cdot\boldsymbol{p}} = \frac{1}{2p}\boldsymbol{\nabla}(\boldsymbol{p}\cdot\boldsymbol{p}) = \frac{1}{p}(\boldsymbol{\nabla} \otimes \boldsymbol{p})\boldsymbol{p},$$

that is to say

$$\boldsymbol{\nabla}p = \boldsymbol{k}. \tag{2.35}$$

Thus, the *derivative of the length is a direction unit vector*. The derivative of this unit vector is, next,

$$\boldsymbol{\nabla} \otimes \boldsymbol{k} = \boldsymbol{\nabla} \otimes \left(\frac{\boldsymbol{p}}{p}\right) = \frac{1}{p}\boldsymbol{\nabla} \otimes \boldsymbol{p} - \frac{1}{p^2}\boldsymbol{\nabla}p \otimes \boldsymbol{p} = \frac{1}{p}(\boldsymbol{I} - \boldsymbol{k} \otimes \boldsymbol{k}),$$

where we should take into account that the tensor

$$\boldsymbol{K} = \boldsymbol{I} - \boldsymbol{k} \otimes \boldsymbol{k},$$

is the *normal projection* onto the plane perpendicular to $\boldsymbol{k}$. Indeed, if we apply this transformation onto a vector, the result is $\boldsymbol{Ku} = \boldsymbol{u} - \boldsymbol{k}(\boldsymbol{k} \cdot \boldsymbol{u})$ (Fig. 2.5). Let us indicate also the cartesian components of this symmetrical tensor

$$\boldsymbol{K} \triangleq \begin{bmatrix} 1 - k_x^2 & -k_x k_y & -k_x k_z \\ -k_y k_x & 1 - k_y^2 & -k_y k_z \\ -k_z k_x & -k_z k_y & 1 - k_z^2 \end{bmatrix}.$$

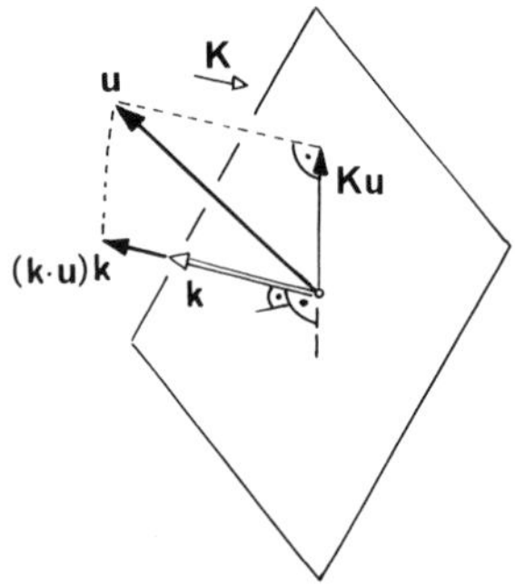

Fig. 2.5. Normal projection of a vector $\boldsymbol{u}$ onto the plane perpendicular to $\boldsymbol{k}$

As a result, we have then

$$\nabla \otimes \boldsymbol{k} = \frac{1}{p}\boldsymbol{K}. \tag{2.36}$$

In other words, the *derivative of the direction unit vector is a projection combined with a shortening*. Further, the derivative of the latter quantity is

$$\begin{aligned} \nabla \otimes \left(\frac{1}{p}\boldsymbol{K}\right) &= -\frac{1}{p^2}\nabla p \otimes \boldsymbol{K} - \frac{1}{p}(\nabla \otimes \boldsymbol{k}) \otimes \boldsymbol{k} - \frac{1}{p}(\nabla \otimes \boldsymbol{k}) \otimes \boldsymbol{k}]^{\mathrm{T}} \\ &= -\frac{1}{p^2}[\boldsymbol{k} \otimes \boldsymbol{K} + \boldsymbol{K} \otimes \boldsymbol{k} + \boldsymbol{K} \otimes \boldsymbol{k})^{\mathrm{T}}]. \end{aligned}$$

In analogy with the preceding, we shall name $\boldsymbol{k} \otimes \boldsymbol{K} + \boldsymbol{K} \otimes \boldsymbol{k} + \boldsymbol{K} \otimes \boldsymbol{k})^{\mathrm{T}} = \mathcal{K}$ a *superprojection*, so that

$$\nabla \otimes \left(\frac{1}{p}\boldsymbol{K}\right) = -\frac{1}{p^2}\mathcal{K}. \tag{2.37}$$

Thus, the *derivative of a reduced projection is a square reduced superprojection.* For higher-order derivatives, we could, of course, continue in the same manner.

2.1.3 Vectors and Tensors on a Curved Surface

As mentioned earlier, holography involves phenomena on several surfaces. These are, for instance, the surfaces of a deformed and the same underformed object, or the planes of the holograms during recording and during reconstruction. Decomposition of vectors and tensors with respect to a curved surface will thus be useful; we shall deal with this in this section.

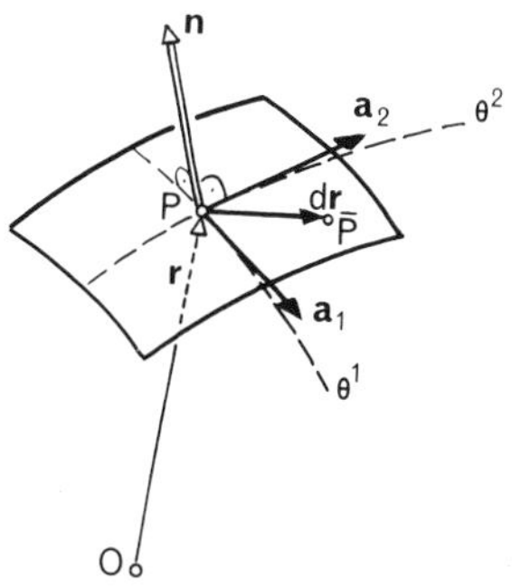

Fig. 2.6. Coordinates and base vectors related to a curved surface

A surface is given by a vector-valued function

$$\theta^1, \theta^2 \rightarrow \boldsymbol{r} = \boldsymbol{r}(\theta^1, \theta^2),$$

where θ^1, θ^2 are curvilinear coordinates on this surface (Fig. 2.6). The covariant base vectors $\boldsymbol{a}_\alpha = \boldsymbol{g}_\alpha = \partial \boldsymbol{r}/\partial \theta^\alpha$ $(\alpha = 1, 2)$ are of course situated in the *tangent plane T*. If we take θ^3 as the distance from the surface, $\boldsymbol{g}_3 = \boldsymbol{g}^3 = \boldsymbol{n}$ becomes the *unit normal*. This vector is expressed by the formula

$$\boldsymbol{n} = \frac{\boldsymbol{a}_1 \times \boldsymbol{a}_2}{|\boldsymbol{a}_1 \times \boldsymbol{a}_2|}. \tag{2.38}$$

Any other vector on the surface $\boldsymbol{u}(\theta^1, \theta^2)$ may be decomposed into (the repetition of greek indices in any term implies summation from 1 to 2)

$$\boldsymbol{u} = u^\alpha \boldsymbol{g}_\alpha + u^3 \boldsymbol{g}_3 = v^\alpha \boldsymbol{a}_\alpha + w\boldsymbol{n} = \boldsymbol{v} + w\boldsymbol{n}. \tag{2.39}$$

$\boldsymbol{v}$ can be called the tangential or *interior* part and $w\boldsymbol{n}$, the normal or *exterior* part (Fig. 2.7). $\boldsymbol{v}(\theta^1, \theta^2)$ is thus a two-dimensional vector, whereas $w(\theta^1, \theta^2)$ is to be considered a scalar function on the surface.

Similarly, a second-order tensor $\boldsymbol{T}$ may be written

$$\boldsymbol{T} = T^{\alpha\beta} \boldsymbol{a}_\alpha \otimes \boldsymbol{a}_\beta + T^{\alpha 3} \boldsymbol{a}_\alpha \otimes \boldsymbol{n} + T^{3\beta} \boldsymbol{n} \otimes \boldsymbol{a}_\beta + T^{33} \boldsymbol{n} \otimes \boldsymbol{n},$$

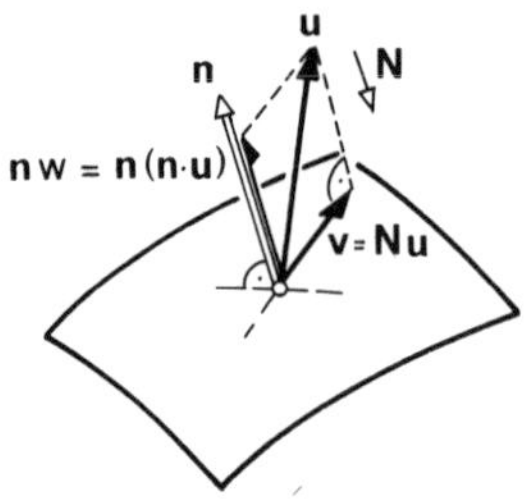

Fig. 2.7. Decomposition of a vector $\boldsymbol{u}$ into its interior part $\boldsymbol{v}$ (projection $\boldsymbol{N}\boldsymbol{u}$) and its exterior part $w\boldsymbol{n}$

or

$$\boldsymbol{T} = \boldsymbol{D} + \boldsymbol{t} \otimes \boldsymbol{n} + \boldsymbol{n} \otimes \boldsymbol{s} + \sigma\boldsymbol{n} \otimes \boldsymbol{n}. \tag{2.40}$$

In matrix notation, this corresponds to the separation of the third row and the third column from the remainder

$$\boldsymbol{T} \mathrel{\hat{=}} \left[\begin{array}{c:c} \boldsymbol{D} & \boldsymbol{t} \\ \hdashline \boldsymbol{s} & \sigma \end{array}\right].$$

Again, we may call the two-dimensional tensor $\boldsymbol{D}$ the *interior* part, $\boldsymbol{t} \otimes \boldsymbol{n} + \boldsymbol{n} \otimes \boldsymbol{s}$ the *semiexterior* part and $\sigma\boldsymbol{n} \otimes \boldsymbol{n}$ the *full exterior* part. $\boldsymbol{t}$ and $\boldsymbol{s}$ themselves are interior vectors and σ is a scalar.

As a particular case thereof, the decomposition of the identity becomes

$$\boldsymbol{I} = \boldsymbol{a}_\alpha \otimes \boldsymbol{a}^\alpha + \boldsymbol{n} \otimes \boldsymbol{n},$$

whose interior part is ($\boldsymbol{a}^1$, $\boldsymbol{a}^2$ are interior, like $\boldsymbol{a}_1$, $\boldsymbol{a}_2$)

$$\boldsymbol{N} = \boldsymbol{I} - \boldsymbol{n} \otimes \boldsymbol{n} = \boldsymbol{a}_\alpha \otimes \boldsymbol{a}^\alpha = a_{\alpha\beta}\boldsymbol{a}^\alpha \otimes \boldsymbol{a}^\beta, \tag{2.41}$$

where $a_{\alpha\beta} = \boldsymbol{a}_\alpha \cdot \boldsymbol{a}_\beta$. On the one hand, this tensor constitutes the *metric tensor* of the surface, the word "metric" relating to the role of the components $a_{\alpha\beta}$. As a matter of fact, the square of a line element $ds = |d\boldsymbol{r}|$ in the tangent plane is given by the so-called *first fundamental form* I

$$(ds)^2 = d\boldsymbol{r} \cdot d\boldsymbol{r} = a_{\alpha\beta} d\theta^\alpha d\theta^\beta. \tag{2.42}$$

On the other hand, (2.41) shows that $\boldsymbol{N}$ acts as a *normal projection* onto the tangent plane, which, of course, varies from point to point. By use of the decomposition (2.39) we obtain

$$\begin{aligned} \boldsymbol{N}\boldsymbol{u} &= \boldsymbol{v} && \text{(projection of a vector, see Fig. 2.7)} \\ \boldsymbol{N}\boldsymbol{T} &= \boldsymbol{D} + \boldsymbol{t} \otimes \boldsymbol{n} && \text{(left semiprojection of a tensor)} \\ \boldsymbol{N}\boldsymbol{T}\boldsymbol{N} &= \boldsymbol{D} && \text{(full projection of a tensor).} \end{aligned} \tag{2.43}$$

Projections onto the unit normal $\boldsymbol{n}$ yield

$$\begin{aligned} \boldsymbol{N T n} &= \boldsymbol{t}, \\ \boldsymbol{n T N} &= \boldsymbol{s}, \\ \boldsymbol{n} \cdot \boldsymbol{T n} &= \sigma. \end{aligned} \tag{2.44}$$

In a similar way, the parts of higher-order tensors may be obtained. Such operations are especially

$$\mathscr{E}\boldsymbol{n} = E_{\alpha\beta 3}\boldsymbol{a}^{\alpha} \otimes \boldsymbol{a}^{\beta} = \boldsymbol{E} = \boldsymbol{n}\mathscr{E},$$

and

$$\mathscr{E} = \boldsymbol{E} \otimes \boldsymbol{n} + \boldsymbol{n} \otimes \boldsymbol{E} - \boldsymbol{E} \otimes \boldsymbol{n})^{\mathrm{T}}. \tag{2.45}$$

Application of the three-dimensional permutation $\mathscr{E}$ onto the normal $\boldsymbol{n}$ results thus in the *two-dimensional permutation tensor* $\boldsymbol{E}$, with the components

$$E_{\alpha\beta} = \sqrt{a}\begin{bmatrix} 0 & 1 \\ -1 & 0 \end{bmatrix}, \quad \text{or} \quad E^{\alpha\beta} = \frac{1}{\sqrt{a}}\begin{bmatrix} 0 & 1 \\ -1 & 0 \end{bmatrix},$$

where $a = a_{11}a_{22} - a_{12}^2$ denotes the determinant of $a_{\alpha\beta}$. This skew-symmetric tensor is often used to *rotate* a vector $\boldsymbol{v} \in T$ through an angle of 90° in the tangent plane (pivot motion, Fig. 2.8). In fact, we have

$$\boldsymbol{v} \cdot \boldsymbol{E v} = 0 \quad \text{and} \quad \boldsymbol{E E} = E_{\alpha\lambda}E^{\lambda\beta}\boldsymbol{a}^{\alpha} \otimes \boldsymbol{a}_{\beta} = -\boldsymbol{N}. \tag{2.46}$$

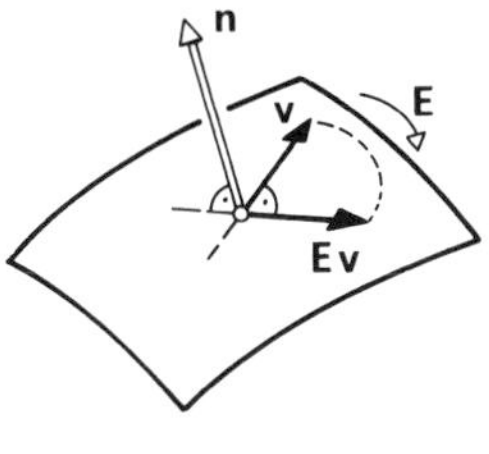

Fig. 2.8. Application of the tensor $\boldsymbol{E}$ onto an interior vector $\boldsymbol{v}$ (rotation of $\boldsymbol{v}$)

The tensor $\boldsymbol{E}$ may also serve to decompose any two-dimensional tensor $\boldsymbol{D}$ additively into a symmetric part $\boldsymbol{S}$ and a skew-symmetric part $\Lambda\boldsymbol{E}$, Λ being a scalar,

$$\boldsymbol{D} = \boldsymbol{S} + \Lambda\boldsymbol{E}. \tag{2.47}$$

So, the left semiprojection of any three-dimensional tensor $\boldsymbol{T}$ may be expressed in the following way, which will be of importance in holography

$$\boldsymbol{N}\boldsymbol{T} = \boldsymbol{S} + \Lambda \boldsymbol{E} + \boldsymbol{t} \otimes \boldsymbol{n}. \tag{2.48}$$

2.1.4 Derivatives on a Surface, Curvature of the Surface

The decomposition into interior and exterior parts that we have described in the preceding paragraph can also be applied to the derivative operator

$$\nabla = \boldsymbol{g}^k \frac{\partial}{\partial \theta^k} = \boldsymbol{a}^\alpha \frac{\partial}{\partial \theta^\alpha} + \boldsymbol{n} \frac{\partial}{\partial \theta^3}.$$

Operating with the projection $\boldsymbol{N}$ onto the formal vector ∇, we obtain a *two-dimensional derivative operator on the surface*

$$\nabla_{\mathrm{n}} = \boldsymbol{N}\nabla = \boldsymbol{a}^\alpha \frac{\partial}{\partial \theta^\alpha}. \tag{2.49}$$

The symbol ∇ is often used in the literature, not so, however, ∇_{n}. The index n specifies which of several (!) surfaces is considered, by indicating its normal, i.e. *the direction along which ∇ is projected.* With the help of ∇_{n}, the total differential of any scalar function $\Phi(\theta^1, \theta^2)$ defined on the surface becomes

$$d\Phi = d\theta^\lambda \boldsymbol{a}_\lambda \cdot \boldsymbol{a}^\alpha \Phi_{,\alpha} = d\boldsymbol{r} \cdot \nabla_{\mathrm{n}} \Phi, \tag{2.50}$$

where $d\boldsymbol{r} \in T$ represents an increment in the tangent plane.

∇_{n} is also used to form the derivatives of vectors on the surface, as for example that of the unit normal $\boldsymbol{n}$. As a result, the tensor

$$\boldsymbol{B} = -\nabla_{\mathrm{n}} \otimes \boldsymbol{n} = -\boldsymbol{a}^\alpha \otimes \frac{\partial \boldsymbol{n}}{\partial \theta^\alpha} = \boldsymbol{a}^\alpha \otimes \Gamma^3_{\alpha\beta} \boldsymbol{a}^\beta \tag{2.51}$$

is obtained by use of the definition (2.22), with $\boldsymbol{g}^3 = \boldsymbol{g}_3 = \boldsymbol{n}$, and $\Gamma^3_{33} = \Gamma^3_{\alpha 3} = 0$. Usually, the Christoffel symbols are not the components of a tensor; but in the present case, they show up as the covariant components $\Gamma^3_{\alpha\beta} \equiv B_{\alpha\beta}$ of the tensor $\boldsymbol{B}$ that describes the *curvature of the surface.* (In advanced differential geometry, the term *tensor of curvature* is used for another fourth-order tensor, e.g.: [Ref. 2.6, p. 129].) The geometrical significance of $\boldsymbol{B}$ is shown by the so-called *second fundamental form* II of the surface, which expresses the normal curvature B in any direction $d\boldsymbol{r}$ (Fig. 2.9, see e.g. [Ref. 2.5, p. 35; Ref. 2.8, p. 209])

$$B(ds)^2 = d^2\boldsymbol{r} \cdot \boldsymbol{n} = -d\boldsymbol{r} \cdot d\boldsymbol{n} = -d\boldsymbol{r} \cdot (\nabla_{\mathrm{n}} \otimes \boldsymbol{n}) d\boldsymbol{r} = d\boldsymbol{r} \cdot \boldsymbol{B} d\boldsymbol{r}. \tag{2.52}$$

As $\Gamma^3_{\alpha\beta} = \Gamma^3_{\beta\alpha}$, $\boldsymbol{B}$ is *symmetric* and thus may be diagonalized. Its physical com-

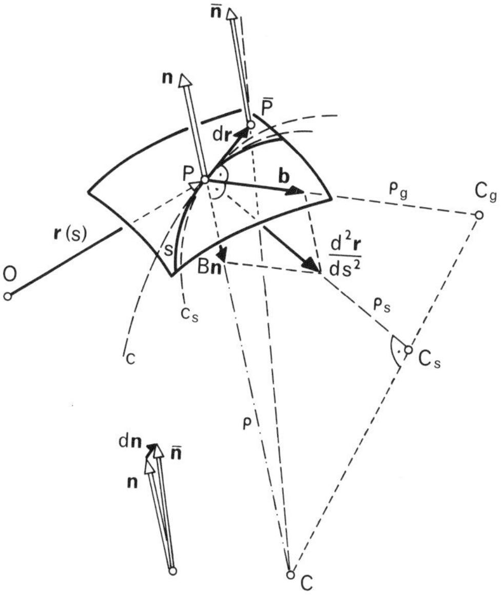

Fig. 2.9. Curvatures of an arbitrary curve $\boldsymbol{r}(s)$ on a surface. c_s: circle of curvature of $\boldsymbol{r}(s)$ with center C_s and radius $\rho_s = 1/|d^2\boldsymbol{r}/ds^2|$. c: circle of normal curvature with center C and radius $\rho = 1/B$. C_g: center of the geodesic curvature of $\boldsymbol{r}(s)$, with the radius $\rho_g = 1/|\boldsymbol{b}|$

ponents are then the reciprocals of the principal radii of curvature R_1, R_2. For certain considerations, the two *invariants* are also of importance, namely the *mean curvature* H and the *Gaussian curvature* K

$$H = \frac{1}{2} \operatorname{tr} \boldsymbol{B} = \frac{1}{2}\left(\frac{1}{R_1} + \frac{1}{R_2}\right), \qquad K = \det \boldsymbol{B} = \frac{1}{R_1 R_2}. \tag{2.53}$$

The "curvature tensor" is a central element with regard to the surface; for example it occurs in any differentiation of vectors and tensors on the surface. Thus, $\boldsymbol{B}$ is required to form [by the rule of (2.30)]

$$\boldsymbol{\nabla} \otimes \boldsymbol{N} = -(\boldsymbol{\nabla} \otimes \boldsymbol{n}) \otimes \boldsymbol{n} - (\boldsymbol{\nabla} \otimes \boldsymbol{n}) \otimes \boldsymbol{n}]^{\mathrm{T}},$$

which, when projected from the left, becomes

$$\boldsymbol{\nabla}_{\mathrm{n}} \otimes \boldsymbol{N} = \boldsymbol{B} \otimes \boldsymbol{n} + \boldsymbol{B} \otimes \boldsymbol{n})^{\mathrm{T}}. \tag{2.54}$$

The derivative of any vector-valued function $\boldsymbol{u} = \boldsymbol{v} + w\boldsymbol{n}$ may be developed as

$$\boldsymbol{\nabla}_{\mathrm{n}} \otimes \boldsymbol{u} = \boldsymbol{\nabla}_{\mathrm{n}} \otimes \boldsymbol{v} + \boldsymbol{\nabla}_{\mathrm{n}} \otimes (w\boldsymbol{n}),$$

or, utilizing $\boldsymbol{v} \equiv \boldsymbol{N}\boldsymbol{v}$ in order to let interior parts appear, and again with (2.30),

$$\begin{aligned}\nabla_{\mathrm{n}} \otimes \boldsymbol{u} &= (\nabla_{\mathrm{n}} \otimes \boldsymbol{N})\boldsymbol{v} + (\nabla_{\mathrm{n}} \otimes \boldsymbol{v})\boldsymbol{N}^{\mathrm{T}} + \nabla_{\mathrm{n}} w \otimes \boldsymbol{n} + w\nabla_{\mathrm{n}} \otimes \boldsymbol{n} \\ &= [\boldsymbol{B} \otimes \boldsymbol{n} + \boldsymbol{B} \otimes \boldsymbol{n})^{\mathrm{T}}]\boldsymbol{v} + (\nabla_{\mathrm{n}} \otimes \boldsymbol{v})\boldsymbol{N} + \nabla_{\mathrm{n}} w \otimes \boldsymbol{n} - w\boldsymbol{B}, \\ \nabla_{\mathrm{n}} \otimes \boldsymbol{u} &= (\nabla_{\mathrm{n}} \otimes \boldsymbol{v})\boldsymbol{N} - \boldsymbol{B}w + (\boldsymbol{B}\boldsymbol{v} + \nabla_{\mathrm{n}} w) \otimes \boldsymbol{n}. \end{aligned} \tag{2.55}$$

The *derivative of a vector on the surface* is the left-semiprojection of the tensor $\nabla \otimes \boldsymbol{u}$. The first two terms in (2.55) form its interior part, whereas the last two are semiexterior. Let us indicate, furthermore, the component notation of this important result (see e.g. [Ref. 2.5, p. 38])

$$\nabla_{\mathrm{n}} \otimes \boldsymbol{u} = (v_{\beta|\alpha} - B_{\alpha\beta} w)\boldsymbol{a}^{\alpha} \otimes \boldsymbol{a}^{\beta} + (B_{\alpha\lambda} v^{\lambda} + w_{,\alpha})\boldsymbol{a}^{\alpha} \otimes \boldsymbol{n}. \tag{2.56}$$

Similarly, the *derivative on the surface of any tensor* $\boldsymbol{T} = \boldsymbol{D} + \boldsymbol{t} \otimes \boldsymbol{n} + \boldsymbol{n} \otimes \boldsymbol{s} + \sigma \boldsymbol{n} \otimes \boldsymbol{n}$ is written [always according to (2.30)]

$$\begin{aligned}\nabla_{\mathrm{n}} \boldsymbol{T} = {} & \nabla_{\mathrm{n}} \boldsymbol{D} + (\nabla_{\mathrm{n}} \cdot \boldsymbol{t})\boldsymbol{n} + (\nabla_{\mathrm{n}} \otimes \boldsymbol{n})^{\mathrm{T}} \boldsymbol{t} + (\nabla_{\mathrm{n}} \cdot \boldsymbol{n})\boldsymbol{s} + (\nabla_{\mathrm{n}} \otimes \boldsymbol{s})^{\mathrm{T}} \boldsymbol{n} \\ & + (\nabla_{\mathrm{n}} \sigma \cdot \boldsymbol{n})\boldsymbol{n} + \sigma(\nabla_{\mathrm{n}} \cdot \boldsymbol{n})\boldsymbol{n} + \sigma(\nabla_{\mathrm{n}} \otimes \boldsymbol{n})^{\mathrm{T}} \boldsymbol{n}.\end{aligned}$$

Observing that

$$\begin{aligned}\nabla_{\mathrm{n}} \boldsymbol{D} &= \nabla_{\mathrm{n}}(\boldsymbol{D}\boldsymbol{N}) = (\nabla_{\mathrm{n}} \boldsymbol{D})\boldsymbol{N} + (\nabla_{\mathrm{n}} \otimes \boldsymbol{N})^{\mathrm{T}} \cdot \boldsymbol{D} \\ &= (\nabla_{\mathrm{n}} \boldsymbol{D})\boldsymbol{N} + [\boldsymbol{n} \otimes \boldsymbol{B} + \boldsymbol{B} \otimes \boldsymbol{n})^{\mathrm{T}}] \cdot \boldsymbol{D} = (\nabla_{\mathrm{n}} \boldsymbol{D})\boldsymbol{N} + (\boldsymbol{B} \cdot \boldsymbol{D})\boldsymbol{n},\end{aligned}$$

and that $\nabla_{\mathrm{n}} \cdot \boldsymbol{n} = \operatorname{tr}\{\nabla_{\mathrm{n}} \otimes \boldsymbol{n}\} = -2H$ and paying attention to the vanishing scalar products of perpendicular vectors, we obtain

$$\nabla_{\mathrm{n}} \boldsymbol{T} = (\nabla_{\mathrm{n}} \boldsymbol{D})\boldsymbol{N} - \boldsymbol{B}\boldsymbol{t} - 2H\boldsymbol{s} + (\boldsymbol{B} \cdot \boldsymbol{D} + \nabla_{\mathrm{n}} \cdot \boldsymbol{t} - 2H\sigma)\boldsymbol{n}. \tag{2.57}$$

This vector possesses thus again an interior and an exterior part. Results similar to those of (2.55) and (2.57) would appear in other derivatives, of any order and with any type of operation.

Particularly, the series development of a scalar function on the surface is [similar to (2.31)]

$$\bar{\Phi} = \Phi + d\boldsymbol{r} \cdot \nabla_{\mathrm{n}} \Phi + \frac{1}{2}[d\boldsymbol{r} \cdot (\nabla_{\mathrm{n}} \otimes \nabla_{\mathrm{n}} \Phi) d\boldsymbol{r} + d^2\boldsymbol{r} \cdot \nabla_{\mathrm{n}} \Phi] + \cdots. \tag{2.58}$$

If Φ is taken along a curve $\boldsymbol{r}(s)$ on the surface, $d^2\boldsymbol{r}$ is expressed by (Fig. 2.9)

$$\frac{d^2\boldsymbol{r}}{ds^2} = \boldsymbol{b} + B\boldsymbol{n} \tag{2.59}$$

in terms of the normal curvature B [see (2.52)] and the geodesic (interior) curvature $\boldsymbol{b}$. Therefore, it may be written

$$\bar{\Phi} = \Phi + d\boldsymbol{r}\cdot\boldsymbol{\nabla}_{\mathrm{n}}\Phi + \frac{1}{2}[d\boldsymbol{r}\cdot(\boldsymbol{\nabla}_{\mathrm{n}} \otimes \boldsymbol{\nabla}_{\mathrm{n}}\Phi)d\boldsymbol{r} + (ds)^2\boldsymbol{b}\cdot\boldsymbol{\nabla}_{\mathrm{n}}\Phi] + \cdots. \tag{2.60}$$

2.1.5 Connections Between Two Surfaces, Oblique Projections

Let us now consider two surfaces, given by the vector-valued functions

$$\boldsymbol{r}(\theta^1, \theta^2) \quad \text{and} \quad \hat{\boldsymbol{r}}(\theta^1, \theta^2).$$

We have used here the same parameters θ^1, θ^2 on both surfaces, thus assuming a correspondence between them, or in other words a one-one mapping between the sets $\{\mathrm{P}\}$ and $\{\hat{\mathrm{P}}\}$; $\hat{\boldsymbol{r}}$ can then be expressed as a function of $\boldsymbol{r}$:

$$\boldsymbol{r} \rightarrow \hat{\boldsymbol{r}} = \hat{\boldsymbol{r}}(\boldsymbol{r}).$$

One may also say that the curvilinear coordinates are transferred when one surface is changed so that it becomes the other. Such coordinates are called *convected coordinates* (e.g. [Ref. 2.3, p. 423]). This interpretation is obvious when thinking for example of the surfaces of a body before and after a deformation; but in the following we shall relate surfaces of different bodies, as for instance those of the hologram and those of the recorded body.

The increments situated in each of the tangent planes may also be related. As $\hat{\boldsymbol{r}}$ is assumed to be a function of $\boldsymbol{r}$, we may write

$$d\hat{\boldsymbol{r}} = d\boldsymbol{r}(\boldsymbol{\nabla}_{\mathrm{n}} \otimes \hat{\boldsymbol{r}}). \tag{2.61}$$

The tensor $\boldsymbol{\nabla}_{\mathrm{n}} \otimes \hat{\boldsymbol{r}}$ acts here from the right on the vector $d\boldsymbol{r}$, as a linear transformation. It must be pointed out that this transformation is valid only for increments and that it changes from point to point.

Of course, there is a large variety of mappings $\{\mathrm{P}\} \rightarrow \{\hat{\mathrm{P}}\}$; in holography however, we encounter mostly transformations between tangent planes with respect to a *collineation center* (Fig. 2.10). For this case, let us determine the tensor $\boldsymbol{\nabla}_{\mathrm{n}} \otimes \hat{\boldsymbol{r}}$. Looking at the figure, we guess that it contains an oblique projection along the straight line $\mathrm{CP\hat{P}}$.

Let us again call $\boldsymbol{p}$ the vector from C to P, $\hat{\boldsymbol{p}}$ that from C to $\hat{\mathrm{P}}$, and $\boldsymbol{k}$ a unit vector parallel to $\boldsymbol{p}$. The collineation consists in the fact that $\boldsymbol{p}$ and $\hat{\boldsymbol{p}}$ are aligned, not only in P, but also in the vicinity of this point. Thus we have with

$$\hat{\boldsymbol{r}} = \boldsymbol{\rho} + \hat{p}\boldsymbol{k}$$

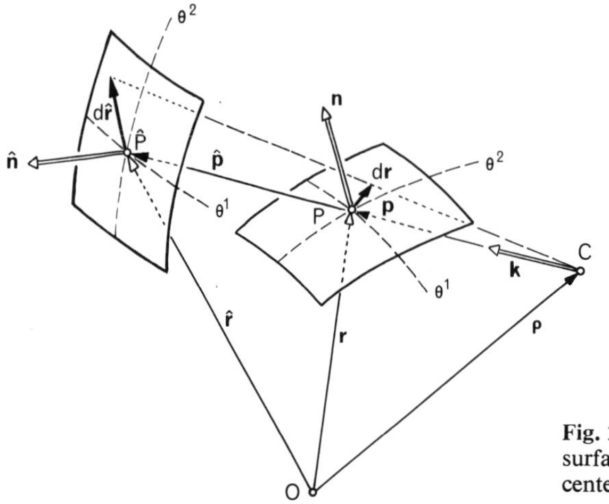

Fig. 2.10. Connection between two surfaces established by a collineation center C

the relation of *stationary behavior*

$$\boldsymbol{\nabla}_{\mathrm{n}} \otimes \hat{\boldsymbol{r}} = \boldsymbol{\nabla}_{\mathrm{n}} \otimes (\hat{p}\boldsymbol{k}) = \boldsymbol{\nabla}_{\mathrm{n}}\hat{p} \otimes \boldsymbol{k} + \frac{\hat{p}}{p}\boldsymbol{N}\boldsymbol{K}, \tag{2.62}$$

where we used the result (2.36) concerning the derivative of $\boldsymbol{k}$ and (2.49) for the projection $\boldsymbol{N}\boldsymbol{\nabla}$ of $\boldsymbol{\nabla}$. On the other hand,

$$0 = d\hat{\boldsymbol{r}} \cdot \hat{\boldsymbol{n}} = d\boldsymbol{r} \cdot (\boldsymbol{\nabla}_{\mathrm{n}} \otimes \hat{\boldsymbol{r}})\hat{\boldsymbol{n}},$$

that is, because $d\boldsymbol{r}$ is arbitrary,

$$0 = (\boldsymbol{\nabla}_{\mathrm{n}} \otimes \hat{\boldsymbol{r}})\hat{\boldsymbol{n}} = \boldsymbol{\nabla}_{\mathrm{n}}\hat{p}(\boldsymbol{k} \cdot \hat{\boldsymbol{n}}) + \frac{\hat{p}}{p}\boldsymbol{N}\boldsymbol{K}\hat{\boldsymbol{n}},$$

therefore

$$\boldsymbol{\nabla}_{\mathrm{n}}\hat{p} = -\frac{\hat{p}}{p(\boldsymbol{k} \cdot \boldsymbol{n})}\boldsymbol{N}\boldsymbol{K}\hat{\boldsymbol{n}}. \tag{2.63}$$

Substituting this result for $\boldsymbol{\nabla}_{\mathrm{n}}\hat{p}$ into (2.62), we find

$$\boldsymbol{\nabla}_{\mathrm{n}} \otimes \hat{\boldsymbol{r}} = \frac{\hat{p}}{p}\boldsymbol{N}\boldsymbol{K}\left(\boldsymbol{I} - \frac{1}{\hat{\boldsymbol{n}} \cdot \boldsymbol{k}}\hat{\boldsymbol{n}} \otimes \boldsymbol{k}\right).$$

Taking into account $\boldsymbol{K} = \boldsymbol{I} - \boldsymbol{k} \otimes \boldsymbol{k}$, we obtain finally

$$\boldsymbol{\nabla}_{\mathrm{n}} \otimes \hat{\boldsymbol{r}} = \frac{\hat{p}}{p}\boldsymbol{N}\left(\boldsymbol{I} - \frac{1}{\hat{\boldsymbol{n}} \cdot \boldsymbol{k}}\hat{\boldsymbol{n}} \otimes \boldsymbol{k}\right). \tag{2.64}$$

Let us then introduce the tensor

$$\hat{\boldsymbol{M}} = \boldsymbol{I} - \frac{1}{\hat{\boldsymbol{n}}\cdot\boldsymbol{k}}\hat{\boldsymbol{n}} \otimes \boldsymbol{k}. \tag{2.65}$$

Its transpose,

$$\hat{\boldsymbol{M}}^{\mathrm{T}} = \boldsymbol{I} - \frac{1}{\boldsymbol{k}\cdot\hat{\boldsymbol{n}}}\boldsymbol{k} \otimes \hat{\boldsymbol{n}}, \tag{2.66}$$

allows us to write (2.61), in the case of a collineation between the tangent planes, as

$$d\hat{\boldsymbol{r}} = \frac{\hat{p}}{p} d\boldsymbol{r}\boldsymbol{N}\hat{\boldsymbol{M}} = \frac{\hat{p}}{p}\hat{\boldsymbol{M}}^{\mathrm{T}}\boldsymbol{N}d\boldsymbol{r} = \frac{\hat{p}}{p}\hat{\boldsymbol{M}}^{\mathrm{T}}d\boldsymbol{r}. \tag{2.67}$$

More generally, it is worth while to note that the tensor $\hat{\boldsymbol{M}}^{\mathrm{T}}$ acts as an *oblique projection along* **k** *onto the plane perpendicular to* $\hat{\boldsymbol{n}}$ when it is applied to the right onto any vector (not necessarily normal to $\boldsymbol{n}$, see Figs. 2.11, 12). Conversely, $\hat{\boldsymbol{M}}$ acts as an *oblique projection along* $\hat{\boldsymbol{n}}$ *onto the plane perpendicular to* **k**. Let us remark here that, with a Lagrangean identity [Ref. 2.11, Eq. (I.6.12)], one can also write $\hat{\boldsymbol{M}} \ldots = -[\boldsymbol{k} \times (\hat{\boldsymbol{n}} \times \ldots)]/\hat{\boldsymbol{n}}\cdot\boldsymbol{k}$.

Similarly, if the roles of the two surfaces are exchanged, i.e. if $\boldsymbol{r}$ is expressed as a function of $\hat{\boldsymbol{r}}$, there appear the *oblique projection along* **n** *onto the plane perpendicular to* **k**

$$\boldsymbol{M} = \boldsymbol{I} - \frac{1}{\boldsymbol{n}\cdot\boldsymbol{k}}\boldsymbol{n} \otimes \boldsymbol{k} \tag{2.68}$$

and its transpose

$$\boldsymbol{M}^{\mathrm{T}} = \boldsymbol{I} - \frac{1}{\boldsymbol{k}\cdot\boldsymbol{n}}\boldsymbol{k} \otimes \boldsymbol{n}, \tag{2.69}$$

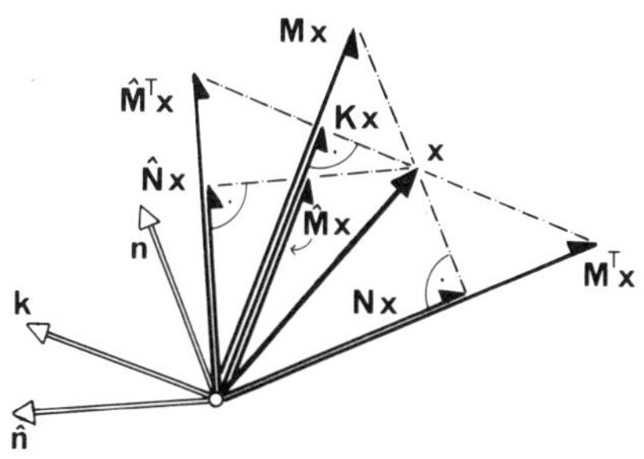

Fig. 2.11. Normal and oblique projections applied onto an arbitrary vector $\boldsymbol{x}$. This sketch represents the particular case where all vectors are in the same plane

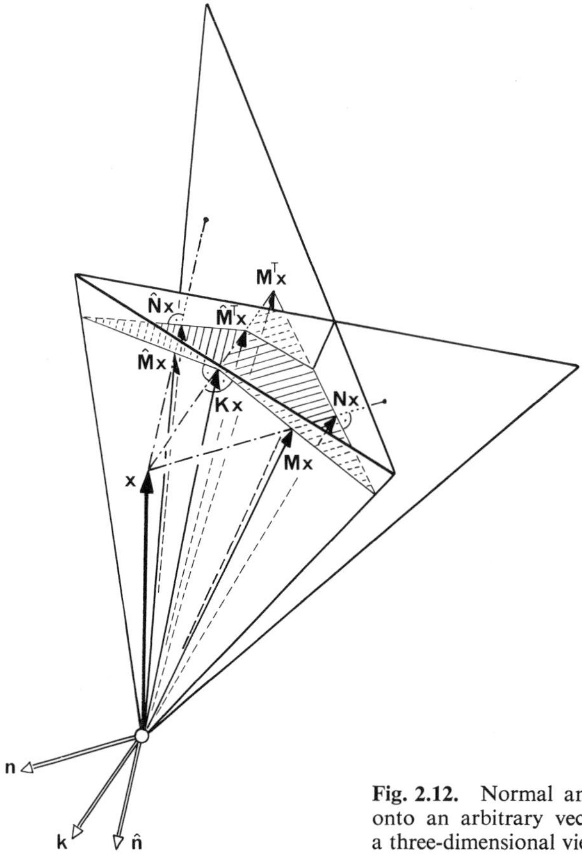

Fig. 2.12. Normal and oblique projections applied onto an arbitrary vector $\boldsymbol{x}$. This sketch represents a three-dimensional view

which *projects along* **k** *onto the plane perpendicular to* $\boldsymbol{n}$. Instead of (2.67), we obtain

$$dr = \frac{p}{\hat{p}} \boldsymbol{M}^{\mathrm{T}} \hat{\boldsymbol{N}} d\hat{\boldsymbol{r}} = \frac{p}{\hat{p}} \boldsymbol{M}^{\mathrm{T}} d\hat{\boldsymbol{r}}. \tag{2.70}$$

Furthermore, the reader may easily verify the important formulae

$$\begin{aligned} \boldsymbol{M}\hat{\boldsymbol{M}} &= \hat{\boldsymbol{M}}, & \hat{\boldsymbol{M}}\boldsymbol{M} &= \boldsymbol{M}, \\ \boldsymbol{M}\boldsymbol{N} &= \boldsymbol{M}, & \boldsymbol{N}\boldsymbol{M} &= \boldsymbol{N}, \\ \boldsymbol{M}\boldsymbol{K} &= \boldsymbol{K}, & \boldsymbol{K}\boldsymbol{M} &= \boldsymbol{M}, \ldots, \end{aligned} \tag{2.71}$$

either by use of the geometrical meaning of the projections (see Fig. 2.11, where $\hat{\boldsymbol{n}}$, $\boldsymbol{n}$ and $\boldsymbol{k}$ are drawn in the same plane) or by using the rule for combining successive dyadics. Finally, Fig. 2.12 shows the whole system of the projections $\boldsymbol{N}$, $\boldsymbol{M}$, $\boldsymbol{M}^{\mathrm{T}}$, $\hat{\boldsymbol{N}}$, $\hat{\boldsymbol{M}}$, $\hat{\boldsymbol{M}}^{\mathrm{T}}$ and $\boldsymbol{K}$ in the general case where $\boldsymbol{n}$, $\hat{\boldsymbol{n}}$ and $\boldsymbol{k}$ are not coplanar.

Let us now consider a scalar-valued function $\Phi(\theta^1, \theta^2)$ defined on both surfaces $\{\mathrm{P}\}$ and $\{\hat{\mathrm{P}}\}$. Its differential is

$$d\Phi = d\boldsymbol{r}\cdot\boldsymbol{\nabla}_{\mathrm{n}}\Phi = d\hat{\boldsymbol{r}}\cdot\boldsymbol{\nabla}_{\hat{\mathrm{n}}}\Phi.$$

The relations (2.67, 70) can be used to transform $d\boldsymbol{r}$ and $d\hat{\boldsymbol{r}}$ into each other and also to write

$$d\Phi = \frac{p}{\hat{p}}d\hat{\boldsymbol{r}}\cdot\boldsymbol{M}\boldsymbol{\nabla}_{\mathrm{n}}\Phi = \frac{\hat{p}}{p}d\boldsymbol{r}\cdot\hat{\boldsymbol{M}}\boldsymbol{\nabla}_{\hat{\mathrm{n}}}\Phi.$$

Because $d\boldsymbol{r}$ and $d\hat{\boldsymbol{r}}$ are arbitrary, but of course only in their respective tangent planes, we find by comparison of the above relations for $d\Phi$ the following transformations between the operators $\boldsymbol{\nabla}_{\mathrm{n}}$ and $\boldsymbol{\nabla}_{\hat{\mathrm{n}}}$ (expressing in fact the chain rule)

$$\begin{aligned} \boldsymbol{\nabla}_{\mathrm{n}} &= \frac{\hat{p}}{p}\boldsymbol{N}\hat{\boldsymbol{M}}\boldsymbol{\nabla}_{\hat{\mathrm{n}}}, \\ \boldsymbol{\nabla}_{\hat{\mathrm{n}}} &= \frac{p}{\hat{p}}\hat{\boldsymbol{N}}\boldsymbol{M}\boldsymbol{\nabla}_{\mathrm{n}}. \end{aligned} \tag{2.72}$$

Contrary to the relations with $d\boldsymbol{r}$ and $d\hat{\boldsymbol{r}}$, the projections $\boldsymbol{N}$ and $\hat{\boldsymbol{N}}$ cannot be omitted here. As the main feature of (2.72) the following may be asserted: *passing from the derivative on one surface to the derivative on another surface involves double projections*, here $\boldsymbol{N}\hat{\boldsymbol{M}}$ and $\hat{\boldsymbol{N}}\boldsymbol{M}$.

Double projections also appear when a unit direction vector on a surface is derived. Therefore, we have used in (2.62) the result

$$\boldsymbol{\nabla}_{\mathrm{n}} \otimes \boldsymbol{k} = \frac{1}{p}\boldsymbol{N}\boldsymbol{K}, \tag{2.73}$$

which follows from (2.36, 49). In fact, double projections will repeatedly appear in holography and in holographic interferometry.

One particular connection between two tangent planes will also be of importance; if one of the two involved surfaces, for instance $\{\hat{\mathrm{P}}\}$, is a *unit sphere* on which the vector $\boldsymbol{k}$ varies, $\hat{p}$ is equal to 1, $\hat{\boldsymbol{n}}$ is identical to $\boldsymbol{k}$ and consequently also $\hat{\boldsymbol{M}}$, to $\boldsymbol{K}$. The preceding relations (2.67, 70, 72) then become

$$d\boldsymbol{k} = \frac{1}{p}\boldsymbol{K}d\boldsymbol{r}, \qquad d\boldsymbol{r} = p\boldsymbol{M}^{\mathrm{T}}d\boldsymbol{k}, \tag{2.74}$$

$$\boldsymbol{\nabla}_{\mathrm{n}} = \frac{1}{p}\boldsymbol{N}\boldsymbol{\nabla}_{\mathrm{k}}, \qquad \boldsymbol{\nabla}_{\mathrm{k}} = p\boldsymbol{M}\boldsymbol{\nabla}_{\mathrm{n}}, \tag{2.75}$$

and contain only *single projections*.

At this point, let us add a remark about the derivatives of functions with several variables. Suppose for example a scalar function $\Phi(\hat{\boldsymbol{r}}, \boldsymbol{k})$ with the two, at the moment independent, vector variables $\hat{\boldsymbol{r}}$ and $\boldsymbol{k}$; the total differential may be written in the form

$$d\Phi = d\hat{\boldsymbol{r}} \cdot \partial_{\hat{r}}\Phi + d\boldsymbol{k} \cdot \partial_k \Phi, \tag{2.76}$$

where $\partial_{\hat{r}}$ and ∂_k are vector operators of partial derivative with respect to $\hat{\boldsymbol{r}}$ or $\boldsymbol{k}$. In our considerations, $\hat{\boldsymbol{r}}$ and $\boldsymbol{k}$ are in fact varying position vectors on surfaces $\in \mathbb{R}^3$, as in Fig. 2.10. The corresponding operators are thus two-dimensional, i.e. $\hat{\boldsymbol{N}}\partial_{\hat{r}} = \partial_{\hat{r}}$ and $\boldsymbol{K}\partial_k = \partial_k$. Moreover, because in most cases there exists a connection to a third surface given by a position vector $\boldsymbol{r}$, we have $\hat{\boldsymbol{r}} = \hat{\boldsymbol{r}}(\boldsymbol{r})$ and $\boldsymbol{k} = \boldsymbol{k}(\boldsymbol{r})$; on the other hand, the increments are related similarly to (2.61) so that

$$d\Phi = d\boldsymbol{r} \cdot \boldsymbol{\nabla}_{\mathrm{n}}\Phi = d\boldsymbol{r} \cdot (\boldsymbol{\nabla}_{\mathrm{n}} \otimes \hat{\boldsymbol{r}})\partial_{\hat{r}}\Phi + d\boldsymbol{r} \cdot (\boldsymbol{\nabla}_{\mathrm{n}} \otimes \boldsymbol{k})\partial_k\Phi. \tag{2.77}$$

The derivative of Φ is then expressed by

$$\boldsymbol{\nabla}_{\mathrm{n}}\Phi = (\boldsymbol{\nabla}_{\mathrm{n}} \otimes \hat{\boldsymbol{r}})\partial_{\hat{r}}\Phi + (\boldsymbol{\nabla}_{\mathrm{n}} \otimes \boldsymbol{k})\partial_k\Phi. \tag{2.78}$$

When the connection between the three surfaces is established by means of a collineation center (Fig. 2.10), the derivatives of $\hat{\boldsymbol{r}}$ and $\boldsymbol{k}$ are given by (2.64, 73).

Another connection between two surfaces is the "*jump*" *between parallel near surfaces*. Let us then put

$$\bar{\boldsymbol{r}} = \boldsymbol{r} + z\boldsymbol{n},$$

where $z = \theta^3$ denotes the small distance between the given surface $\boldsymbol{r}(\theta^1, \theta^2)$ and the parallel surface $\bar{\boldsymbol{r}}(\theta^1, \theta^2)$(we write $\bar{\boldsymbol{r}}$ instead of $\hat{\boldsymbol{r}}$ to avoid confusion with the preceding). Now, referring to (2.51, 61) we have

$$\begin{aligned} &\boldsymbol{\nabla}_{\mathrm{n}} \otimes \bar{\boldsymbol{r}} = \boldsymbol{N} - \boldsymbol{B}z, \\ &d\bar{\boldsymbol{r}} = d\boldsymbol{r}(\boldsymbol{N} - \boldsymbol{B}z), \\ &\boldsymbol{\nabla}_{\mathrm{n}} = (\boldsymbol{N} - \boldsymbol{B}z)\bar{\boldsymbol{\nabla}}_{\mathrm{n}}, \end{aligned} \tag{2.79}$$

because the tangent planes have the same normal and $\boldsymbol{B}$ is symmetric. For the same reason, in the present case we can invert the transformation by a multiplication

$$\bar{\boldsymbol{\nabla}}_{\mathrm{n}} = (\boldsymbol{N} + \boldsymbol{B}z + \boldsymbol{B}^2z^2 + \cdots)\boldsymbol{\nabla}_{\mathrm{n}}, \tag{2.80}$$

where the terms containing z^3 and higher are negligible. This series converges if z is sufficiently small. Equations (2.79, 80) will be useful in the investigation of the deformation beneath the surface of a body in Chap. 5.

2.2 Elements of Continuum Mechanics: The Kinematics of Deformation

As was already mentioned in the Introduction, the purpose of this book is to show on the one hand how holographic interferometry may be used to determine the deformation of a body, and on the other hand how the principles of kinematics help to understand certain concepts in holography. It will prove worth while to recall briefly what the "deformation" consists in. Again we will look first at the general case of a three-dimensional body and then adapt the results to the surface of a body. The means that we are going to use are similar to those of modern continuum mechanics (see particularly [Ref. 2.2, Chap. 6; Ref. 2.3, Chap. 3], [2.4, 5, and also 10–15]).

2.2.1 Deformation, Strain and Rotation, and Their Relations to the Displacement

Let us consider two states of a body, for instance the "undeformed" state, which serves as the *reference* state, and a "deformed" state, i.e. the form of the body at a given time t (Fig. 2.13). P and P′ denote the respective positions of one *material point* or *particle*, the two sets {P} and {P′} are then called *configurations* of the object. For the sake of simplicity we assume that there are *no dislocations.*

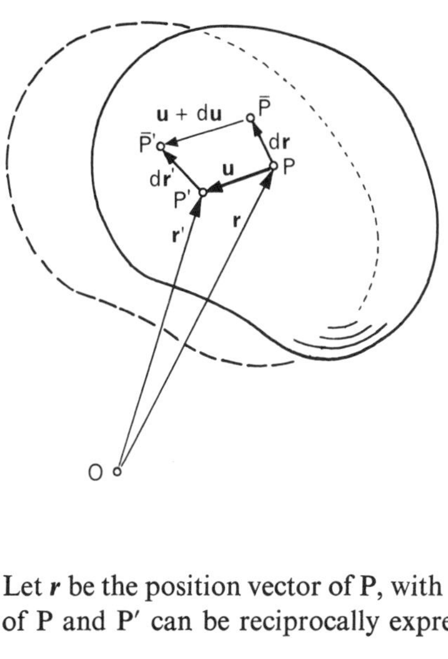

Fig. 2.13. Deformation of a body. P: one material point of the body in the reference configuration. $\bar{P}$: point in the vicinity of P. P′, $\bar{P}'$: positions of the material points P and $\bar{P}$ in the deformed configuration

Let $\boldsymbol{r}$ be the position vector of P, with cartesian coordinates x, y, z. The positions of P and P′ can be reciprocally expressed as a function of the other, that is to say by a one-one transformation $\boldsymbol{r} \rightleftarrows \boldsymbol{r}'$. The mapping

$$\boldsymbol{r} \to \boldsymbol{r}'(\boldsymbol{r}, t),$$

is called the *deformation* of the body. The position vector $\boldsymbol{r}$ is the independent variable, $\boldsymbol{r}'$ is the dependent variable, and t, a parameter. This is often called a *Lagrangean representation.* For the vicinity of a particle, we write, similarly to Sect. 2.1.5, the linear transformation

$$d\boldsymbol{r}' = d\boldsymbol{r}(\boldsymbol{\nabla} \otimes \boldsymbol{r}') = \boldsymbol{F}d\boldsymbol{r}, \tag{2.81}$$

which maps the elemental vectors $\{d\boldsymbol{r}\}$ around P into the corresponding vectors $\{d\boldsymbol{r}'\}$ around P′. The invertible tensor $\boldsymbol{F}$ is called the *deformation gradient.* Let us also define the *displacement vector* by

$$\boldsymbol{u} = \boldsymbol{r}' - \boldsymbol{r}, \tag{2.82}$$

with cartesian components u, v, w. $\boldsymbol{F}$ may then be calculated by deriving the displacement. In fact, (2.81, 82) imply

$$\boldsymbol{F} = \boldsymbol{I} + (\boldsymbol{\nabla} \otimes \boldsymbol{u})^{\mathrm{T}}. \tag{2.83}$$

In a cartesian coordinate system, the matrix of the components of $\boldsymbol{\nabla} \otimes \boldsymbol{u}$ would be

$$\boldsymbol{\nabla} \otimes \boldsymbol{u} \mathrel{\hat{=}} \begin{pmatrix} \dfrac{\partial u}{\partial x} & \dfrac{\partial v}{\partial x} & \dfrac{\partial w}{\partial x} \\ \dfrac{\partial u}{\partial y} & \dfrac{\partial v}{\partial y} & \dfrac{\partial w}{\partial y} \\ \dfrac{\partial u}{\partial z} & \dfrac{\partial v}{\partial z} & \dfrac{\partial w}{\partial z} \end{pmatrix}. \tag{2.84}$$

Let us introduce further a unit vector $\boldsymbol{e}$ in the direction of $d\boldsymbol{r}$ so that $d\boldsymbol{r} = \boldsymbol{e}ds$, $ds = |d\boldsymbol{r}|$ meaning the length of the increment $d\boldsymbol{r}$. The square of the corresponding length ds' in the deformed configuration is then

$$(ds')^2 = d\boldsymbol{r}' \cdot d\boldsymbol{r}' = \boldsymbol{e} \cdot \boldsymbol{F}^{\mathrm{T}}\boldsymbol{F}\boldsymbol{e}(ds)^2. \tag{2.85}$$

In the particular case of the motion of a rigid body, $\boldsymbol{F}^{\mathrm{T}}\boldsymbol{F} = \boldsymbol{I}$ since $ds' = ds$ for any $\boldsymbol{e}$, but $\boldsymbol{F}$ itself still differs from the identity as shown by (2.81). Hence, the *linear dilatation* of the elemental distances around P must be described by the so-called *Cauchy–Green* tensor $\boldsymbol{F}^{\mathrm{T}}\boldsymbol{F}$. Moreover, (2.85) shows that $\boldsymbol{F}^{\mathrm{T}}\boldsymbol{F}$ must be positive definite, so that there exists a *symmetric* tensor $\boldsymbol{U}$ defined by the equation

$$U^2 = F^T F. \tag{2.86}$$

Owing to this, there is another tensor Q given by

$$Q = FU^{-1},$$

which must be *orthogonal*, because we may write

$$Q^T Q = U^{-1} F^T F U^{-1} = I. \tag{2.87}$$

Thus, we arrive at what is called the polar decomposition

$$F = QU = U'Q, \tag{2.88}$$

where U' is symmetric, similar to U. Therefore, the deformation gradient consists of a *stretch tensor* U (or U') which is related to the linear dilatation and which here is followed (or preceded) by a *rotation tensor* Q. Let us note also that these two parts give a *multiplicative* decomposition of the tensor F.

Concerning the dilatation, it is worth while adding further explanations. We define the *strain tensor*

$$\tilde{\mathcal{E}} = \frac{1}{2}(F^T F - I) = \frac{1}{2}(U^2 - I), \tag{2.89}$$

for which, with (2.83), we may also write the *kinematic relation*

$$\tilde{\mathcal{E}} = \frac{1}{2}[\nabla \otimes u + (\nabla \otimes u)^T + (\nabla \otimes u)(\nabla \otimes u)^T]. \tag{2.90}$$

The meaning of this tensor becomes clear in the case of a "small" deformation. Indeed, the linear dilatation or strain of an elemental distance ds in the direction e is

$$\varepsilon = \frac{ds' - ds}{ds}$$

or, because $|\varepsilon| \ll 1$,

$$\varepsilon \simeq \frac{(ds')^2 - (ds)^2}{2(ds)^2} = \frac{1}{2} e \cdot (F^T F - I) e = e \cdot \tilde{\mathcal{E}} e.$$

In the same manner, the angular dilatation or shearing strain, i.e. the change of angle between two perpendicular directions $\boldsymbol{e}_1$ and $\boldsymbol{e}_2$, is

$$\gamma_{12} \simeq \frac{d\boldsymbol{r}'_1 \cdot d\boldsymbol{r}'_2}{ds'_1 ds'_2} \simeq \boldsymbol{e}_1 \cdot \boldsymbol{F}^{\mathrm{T}} \boldsymbol{F} \boldsymbol{e}_2 = 2\boldsymbol{e}_1 \cdot \tilde{\boldsymbol{\mathcal{E}}} \boldsymbol{e}_2,$$

because $\boldsymbol{e}_1 \cdot \boldsymbol{e}_2 = 0$. In the case of a cartesian coordinate system, the unit vectors $\boldsymbol{e}$ may be replaced by the unit base vectors, so that the components of the tensor $\tilde{\boldsymbol{\mathcal{E}}}$ are identical with the strains and shearing strains relative to the axes

$$\tilde{\boldsymbol{\mathcal{E}}} \triangleq \begin{pmatrix} \tilde{\varepsilon}_x & \frac{1}{2}\tilde{\gamma}_{xy} & \frac{1}{2}\tilde{\gamma}_{xz} \\ \frac{1}{2}\tilde{\gamma}_{yx} & \tilde{\varepsilon}_y & \frac{1}{2}\tilde{\gamma}_{yz} \\ \frac{1}{2}\tilde{\gamma}_{zx} & \frac{1}{2}\tilde{\gamma}_{zy} & \tilde{\varepsilon}_z \end{pmatrix}.$$

In cases where certain quantities are *small*, it seems convenient to introduce provisionally some parameters that characterize this smallness to be able to discuss the deformation properly. Therefore, let us introduce the expansion

$$\tilde{\boldsymbol{\mathcal{E}}} = \eta \boldsymbol{\mathcal{E}}_1 + \eta^2 \ldots .$$

In other words, the strain tensor is assumed to be of some order of magnitude $O(\eta)$. From (2.89) we see that $\boldsymbol{U}$ has the form

$$\boldsymbol{U} = \boldsymbol{I} + \eta \boldsymbol{\mathcal{E}}_1 + \eta^2 \boldsymbol{\mathcal{E}}_2 + \cdots,$$

which shows that $\boldsymbol{U}$ differs little from the identity for $\eta \ll 1$. If, in addition to small strains, the rotation is small, $\boldsymbol{Q}$ may be developed in the same way

$$\boldsymbol{Q} = \boldsymbol{I} + \chi \boldsymbol{\Omega}_1 + \chi^2 \boldsymbol{\Omega}_2 + \cdots,$$

with another small parameter χ. The product of this series with its transpose gives

$$\boldsymbol{Q}^{\mathrm{T}} \boldsymbol{Q} = \boldsymbol{I} + \chi(\boldsymbol{\Omega}_1 + \boldsymbol{\Omega}_1^{\mathrm{T}}) + \chi^2 \ldots,$$

which, with (2.87), shows that the first-order tensor is skew symmetric

$$\boldsymbol{\Omega}_1 = -\boldsymbol{\Omega}_1^{\mathrm{T}}.$$

The polar decomposition of the deformation gradient becomes now

$$\begin{aligned} \boldsymbol{F} = \boldsymbol{Q}\boldsymbol{U} &= \boldsymbol{I} + (\eta \boldsymbol{\mathcal{E}}_1 + \chi \boldsymbol{\Omega}_1) + (\eta^2 \boldsymbol{\mathcal{E}}_2 + \eta\chi \boldsymbol{\mathcal{E}}_1 \boldsymbol{\Omega}_1 + \chi^2 \boldsymbol{\Omega}_2) + \cdots \\ &= \boldsymbol{I} + (\boldsymbol{\nabla} \otimes \boldsymbol{u})^{\mathrm{T}}. \end{aligned}$$

If the parameters η and χ are not independent but are, for instance, both of the same order of magnitude, i.e. $\eta = O(\chi)$, and if terms smaller than this order are neglected, we may decompose $(\boldsymbol{\nabla} \otimes \boldsymbol{u})^{\mathrm{T}}$ into a symmetric and a skew-symmetric part and write the *linear kinematic relations*

$$\begin{aligned} \eta \boldsymbol{\mathcal{E}}_1 &\simeq \frac{1}{2}[\boldsymbol{\nabla} \otimes \boldsymbol{u} + (\boldsymbol{\nabla} \otimes \boldsymbol{u})^{\mathrm{T}}] = \boldsymbol{\mathcal{E}}, \\ -\chi \boldsymbol{\Omega}_1 &\simeq \frac{1}{2}[\boldsymbol{\nabla} \otimes \boldsymbol{u} - (\boldsymbol{\nabla} \otimes \boldsymbol{u})^{\mathrm{T}}] = \boldsymbol{\Omega}, \end{aligned} \tag{2.91}$$

which, in cartesian components in the usual notation, is

$$\begin{aligned} &\varepsilon_x = \frac{\partial u}{\partial x}, \quad \varepsilon_y = \frac{\partial v}{\partial y}, \ldots && \Omega_{xx} = \Omega_{yy} = \Omega_{zz} = 0 \\ &\gamma_{xy} = \frac{\partial v}{\partial x} + \frac{\partial u}{\partial y}, \ldots, \ldots && \Omega_{xy} = \frac{1}{2}\left(\frac{\partial v}{\partial x} - \frac{\partial u}{\partial y}\right), \ldots, \ldots . \end{aligned}$$

Consequently, we also have

$$\begin{aligned} &\boldsymbol{U} \simeq \boldsymbol{I} + \boldsymbol{\mathcal{E}}, && \boldsymbol{Q} \simeq \boldsymbol{I} - \boldsymbol{\Omega}, \\ &\boldsymbol{\nabla} \otimes \boldsymbol{u} = \boldsymbol{\mathcal{E}} + \boldsymbol{\Omega}, && \boldsymbol{F} = \boldsymbol{I} + \boldsymbol{\mathcal{E}} - \boldsymbol{\Omega}. \end{aligned} \tag{2.92}$$

Thus, the deformation gradient is now *additively decomposed*: $\boldsymbol{\mathcal{E}}$ and $\boldsymbol{\Omega}$ approximate the strain and the rotation; for this reason, they are called, in the case of small deformation, the *symmetric strain tensor* $\boldsymbol{\mathcal{E}}$ and the *skew-symmetric rotation tensor* $\boldsymbol{\Omega}$; this should be compared with the exact multiplicative decomposition of (2.88).

It should be added that the tensor $\boldsymbol{\Omega}$ can also be expressed by means of a *rotation vector* $\boldsymbol{\omega}_{\mathrm{r}}$ and the three-dimensional permutation $\mathscr{E}$ (see Sect. 2.1.1)

$$\boldsymbol{\Omega} = \mathscr{E}\boldsymbol{\omega}_{\mathrm{r}} = \boldsymbol{\omega}_{\mathrm{r}}\mathscr{E}. \tag{2.93}$$

In the case of a *small motion of a rigid body*, the linear transformation (2.81) becomes therefore ($\Delta \boldsymbol{r}$ may be finite)

$$\begin{aligned} \Delta \boldsymbol{r}' &= \boldsymbol{Q}\Delta \boldsymbol{r} \simeq \Delta \boldsymbol{r} - \boldsymbol{\Omega}\Delta \boldsymbol{r} = \Delta \boldsymbol{r} - \boldsymbol{\omega}_{\mathrm{r}}\mathscr{E}\Delta \boldsymbol{r} \\ \Delta \boldsymbol{r}' &\simeq \Delta \boldsymbol{r} + \boldsymbol{\omega}_{\mathrm{r}} \times \Delta \boldsymbol{r}, \end{aligned} \tag{2.94}$$

or, because $\Delta \boldsymbol{r}' = \Delta \boldsymbol{r} + \bar{\boldsymbol{u}} - \boldsymbol{u}$ (Fig. 2.14),

$$\bar{\boldsymbol{u}} \simeq \boldsymbol{u} + \boldsymbol{\omega}_{\mathrm{r}} \times \Delta \boldsymbol{r},$$

which is the well-known formula of elementary kinematics.

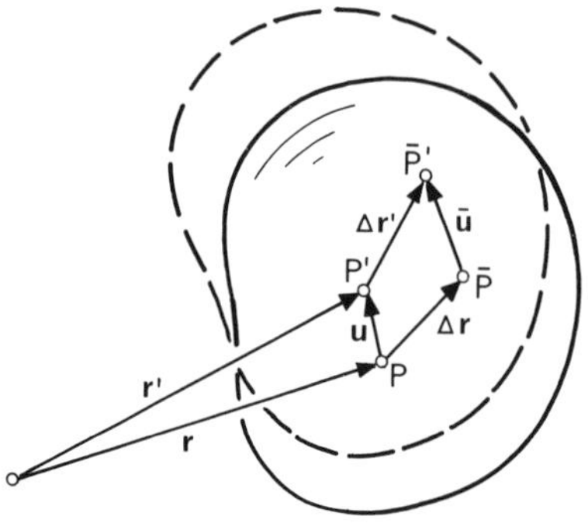

Fig. 2.14. Motion of a rigid body. P, P̄: arbitrary points of the body in the reference configuration. P′, P̄′: corresponding points in the deformed configuration. $|\Delta \boldsymbol{r}'| = |\Delta \boldsymbol{r}|$

Generally however, the parameters η and χ need not be of the same order. For example in thin bodies, such as plates and shells, sometimes very *small strains as well as moderate rotations* occur, when large displacements take place. In such cases, we have, e.g., $\eta = O(\chi^2)$ and thus

$$\boldsymbol{\nabla} \otimes \boldsymbol{u} = O(\chi^2)\boldsymbol{\varepsilon}_1 - \chi\boldsymbol{\Omega}_1 + \chi^2\boldsymbol{\Omega}_2^{\mathrm{T}} + O(\chi^3)\ldots$$

so that with the kinematic equations (2.90)

$$\tilde{\boldsymbol{\varepsilon}} \simeq \boldsymbol{\varepsilon} + \frac{1}{2}[-\chi\boldsymbol{\Omega}_1 + O(\chi^2)\ldots][\chi\boldsymbol{\Omega}_1 + O(\chi^2)\ldots],$$

$$\tilde{\boldsymbol{\varepsilon}} \simeq \boldsymbol{\varepsilon} - \frac{1}{2}\boldsymbol{\Omega}^2. \tag{2.95}$$

This does not mean that $\tilde{\boldsymbol{\varepsilon}}$ depends now on $\boldsymbol{\Omega}$ because its definition (2.89) does not contain $\boldsymbol{Q}$. On the contrary, it makes appear that $\boldsymbol{\varepsilon}$ is only the linear part of the strain (including in fact the rotation), and that, on the other hand, the nonlinear part depends on the rotation alone.

Let us make here a remark about the reference configuration. Instead of the position vector of the undeformed body, we can take that of the deformed body as an independent variable (*Euler representation*) and then write the inverse transformation of (2.81)

$$d\boldsymbol{r} = \boldsymbol{F}^{-1}d\boldsymbol{r}' = d\boldsymbol{r}' - d\boldsymbol{u} = d\boldsymbol{r}'(\boldsymbol{I} - \boldsymbol{\nabla}' \otimes \boldsymbol{u}),$$

so that now

$$\boldsymbol{F}^{-1} = \boldsymbol{I} - (\boldsymbol{\nabla}' \otimes \boldsymbol{u})^{\mathrm{T}}. \tag{2.96}$$

The derivative operator $\boldsymbol{\nabla}'$ is related to the deformed state. Passing from (2.83) to (2.96) is basically a question of transforming $\boldsymbol{\nabla}$. Starting from alternative forms of the differential of any scalar function Φ

$$d\Phi = d\boldsymbol{r}\cdot\nabla\Phi = d\boldsymbol{r}'\cdot\nabla'\Phi,$$

we get the linear transformations

$$\begin{aligned}\nabla' &= (\boldsymbol{F}^{\mathrm{T}})^{-1}\nabla,\\ \nabla &= \boldsymbol{F}^{\mathrm{T}}\nabla'.\end{aligned} \tag{2.97}$$

These formulae correspond to (2.72) which we established for the collinear transformations between planes that are tangent to two surfaces.

2.2.2 Strain and Rotation at the Surface of a Body

The strain tensor and the rotation tensor that we have discussed in the preceding section describe the modifications undergone by a (three-dimensional) volume element. Now holography is sensitive to what happens at the surface of an opaque body. Therefore, although in some circumstances extrapolation into the interior of a body may be possible (see Chap. 5), it is important to direct attention to the two-dimensional strain and the rotation of a surface element.

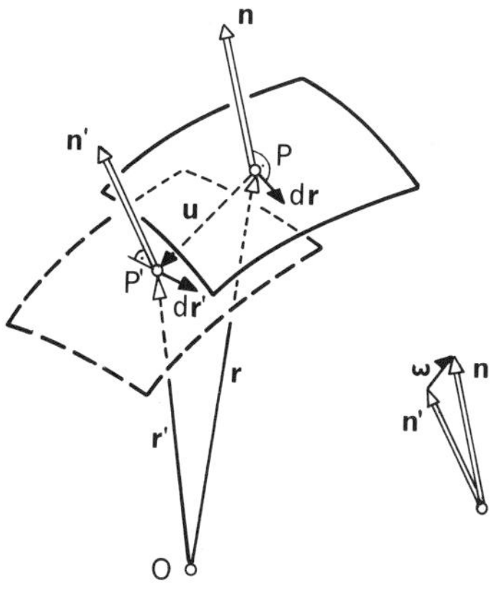

Fig. 2.15. Deformation of a surface. P: one point of the surface in the reference configuration. P′: position of this material point in the deformed configuration

We consider here increments $d\boldsymbol{r}$ and $d\boldsymbol{r}'$ situated respectively in the plane T tangential to the undeformed body (with unit normal $\boldsymbol{n}$) and in the plane T' tangential to the deformed body (with unit normal $\boldsymbol{n}'$, see Fig. 2.15). Therefore, (2.81) may be written with additional projections to indicate what part of the tensor is of importance

$$d\boldsymbol{r}' = \boldsymbol{N}'d\boldsymbol{r}' = \boldsymbol{F}d\boldsymbol{r} = \boldsymbol{F}\boldsymbol{N}d\boldsymbol{r} = \boldsymbol{N}'\boldsymbol{F}\boldsymbol{N}d\boldsymbol{r}. \tag{2.98}$$

The relevant tensor is here, contrary to (2.83),

$$NF^{T}N' = N(I + \nabla \otimes u)N' = NN' + (\nabla_{n} \otimes u)N'. \tag{2.99}$$

The *surface-deformation gradient* is thus a mixed projection of the three-dimensional deformation gradient. Similar to (2.85), the squares of the elemental length relative to the unit vector $e \in T$ are

$$(ds')^2 = dr' \cdot dr' = e \cdot NF^{T}FNe(ds)^2. \tag{2.100}$$

The *strain tensor of a surface element* consists then in the two-dimensional tensor

$$\tilde{\gamma} = \frac{1}{2}(NF^{T}FN - N) = N\tilde{\varepsilon}N, \tag{2.101}$$

which is the full projection of $\tilde{\varepsilon}$.

The right semiprojection may also be applied to the polar decomposition (2.88)

$$FN = QUN = Q(NUN + n \otimes nUN). \tag{2.102}$$

$NUN \equiv V$ is a two-dimensional symmetric tensor. If n is a principal direction of U, as for instance at the *free surface* of an isotropic elastic body, then $nU = \lambda n$, and so $n \otimes nUN = 0$. If the rotation is resolved into an in-plane or *pivot rotation* Q_p around n and an out-of-plane rotation or *inclination* Q_i, the polar decomposition becomes, in this special case,

$$FN = Q_i Q_p V, \tag{2.103}$$

where V describes here the linear dilatation in the surface.

In the case of *small strains* and *small rotations*, we have, according to (2.92),

$$\begin{aligned} &V \simeq N + N\varepsilon N, && Q \simeq I - \Omega, \\ &\nabla_{n} \otimes u = N\varepsilon + N\Omega, && FN = N + \varepsilon N - \Omega N. \end{aligned} \tag{2.104}$$

As is Sects. 2.1.3, 4, we shall distinguish between interior and exterior parts. First, the *symmetric tensor of small surface strain* is found to be, with (2.91),

$$\gamma = N\varepsilon N = \frac{1}{2}N[\nabla \otimes u + (\nabla \otimes u)^{T}]N. \tag{2.105}$$

In a coordinate system with the third axis parallel to n, the components of this tensor are obtained from those of ε by tracing out the third column and the third row in the 3 × 3 matrix. Using the decomposition (2.39)

$$u = v + wn$$

of the displacement vector, as well as the decomposition (2.55) of its derivative, we may also write the kinematic relations at the surface

$$\boldsymbol{\gamma} = \frac{1}{2}\boldsymbol{N}[\boldsymbol{\nabla} \otimes \boldsymbol{v} + (\boldsymbol{\nabla} \otimes \boldsymbol{v})^{\mathrm{T}}]\boldsymbol{N} - \boldsymbol{B}w. \tag{2.106}$$

Second, let us look at the skew-symmetric rotation tensor $\boldsymbol{\Omega}$. As we know from (2.93), it may be expressed by a three-dimensional rotation vector $\boldsymbol{\omega}_{\mathrm{r}}$, which we decompose now as (Fig. 2.16)

$$\boldsymbol{\omega}_{\mathrm{r}} = \boldsymbol{E}\boldsymbol{\omega}_{\mathrm{i}} + \Omega\boldsymbol{n}. \tag{2.107}$$

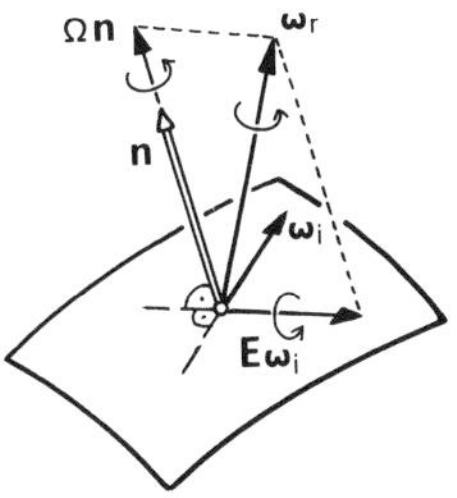

Fig. 2.16. Decomposition of the three-dimensional rotation vector $\boldsymbol{\omega}_{\mathrm{r}}$ into the inclination $\boldsymbol{\omega}_{\mathrm{i}}$ and the pivot motion $\Omega\boldsymbol{n}$

The interior part $\boldsymbol{E}\boldsymbol{\omega}_{\mathrm{i}}$ [$\boldsymbol{E}$ is the permutation tensor defined in (2.45)] takes into account the out-of-plane rotation or *inclination* of the normal $\boldsymbol{n}$ around $\boldsymbol{E}\boldsymbol{\omega}_{\mathrm{i}}$, that is to say in the direction of $-\boldsymbol{\omega}_{\mathrm{i}}$, whereas the exterior part $\Omega\boldsymbol{n}$ describes the in-plane rotation or *pivot motion* around $\boldsymbol{n}$. For the tensor $\boldsymbol{\Omega}$, we have with (2.45) the decomposition

$$\boldsymbol{\Omega} = \boldsymbol{\mathcal{E}}\boldsymbol{\omega}_{\mathrm{r}} = \boldsymbol{\mathcal{E}}\boldsymbol{E}\boldsymbol{\omega}_{\mathrm{i}} + \Omega\boldsymbol{\mathcal{E}}\boldsymbol{n} = \boldsymbol{\omega}_{\mathrm{i}} \otimes \boldsymbol{n} - \boldsymbol{n} \otimes \boldsymbol{\omega}_{\mathrm{i}} + \Omega\boldsymbol{E}. \tag{2.108}$$

In local cartesian coordinates with the z axis along $\boldsymbol{n}$, this would be written

$$\boldsymbol{\Omega} \triangleq \begin{bmatrix} 0 & \omega_z^{\mathrm{r}} & -\omega_y^{\mathrm{r}} \\ -\omega_z^{\mathrm{r}} & 0 & \omega_x^{\mathrm{r}} \\ \omega_y^{\mathrm{r}} & -\omega_x^{\mathrm{r}} & 0 \end{bmatrix} = \begin{bmatrix} 0 & 0 & \omega_x^{\mathrm{i}} \\ 0 & 0 & \omega_y^{\mathrm{i}} \\ -\omega_x^{\mathrm{i}} & -\omega_y^{\mathrm{i}} & 0 \end{bmatrix} + \begin{bmatrix} 0 & \Omega & 0 \\ -\Omega & 0 & 0 \\ 0 & 0 & 0 \end{bmatrix}.$$

Thus, the *additive decomposition* of the surface deformation gradient [see the third relation of (2.104)] leads to

$$\begin{aligned} \boldsymbol{\nabla}_{\mathrm{n}} \otimes \boldsymbol{u} &= \boldsymbol{N}\boldsymbol{\varepsilon}\boldsymbol{N} + \boldsymbol{N}\boldsymbol{\Omega}\boldsymbol{N} + (\boldsymbol{N}\boldsymbol{\varepsilon}\boldsymbol{n} + \boldsymbol{N}\boldsymbol{\Omega}\boldsymbol{n}) \otimes \boldsymbol{n} \\ &= \boldsymbol{\gamma} + \Omega\boldsymbol{E} + (\boldsymbol{N}\boldsymbol{\varepsilon}\boldsymbol{n} + \boldsymbol{\omega}_{\mathrm{i}}) \otimes \boldsymbol{n}, \end{aligned}$$

$$\boldsymbol{\nabla}_{\mathrm{n}} \otimes \boldsymbol{u} = \boldsymbol{\gamma} + \Omega\boldsymbol{E} + \boldsymbol{\omega} \otimes \boldsymbol{n}, \tag{2.109}$$

with a new vector $\omega = \omega_i + N\mathcal{E}n$. According to (2.48) one has here an additive decomposition into three parts instead of the multiplicative decomposition of (2.103). Let us further illustrate the effect of the rotation components. Suppose that $\gamma = 0$, $\omega = 0$, $\Omega \neq 0$; then the linear transformation (2.98) is (see also Fig. 2.17)

$$dr' = dr - \Omega E dr.$$

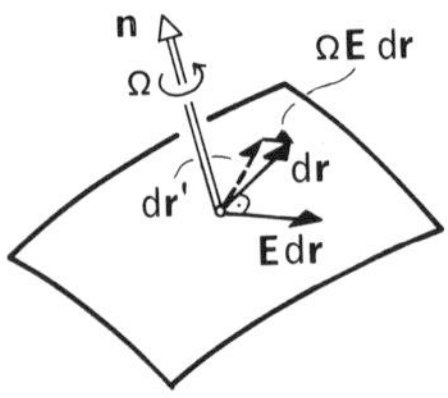

Fig. 2.17. Pivot motion of a surface element

Thus, dr is rotated in the tangent plane; the scalar Ω quantifies the pivot motion. If $\gamma = 0$, $\Omega = 0$, $\omega \neq 0$, then

$$dr' = dr + n(\omega \cdot dr),$$

which, because of $|dr|' \simeq |dr|$, means an inclination or out-of-plane rotation round $E\omega$ (Fig. 2.18); ω itself is an interior vector,

$$\omega = n - n'. \tag{2.110}$$

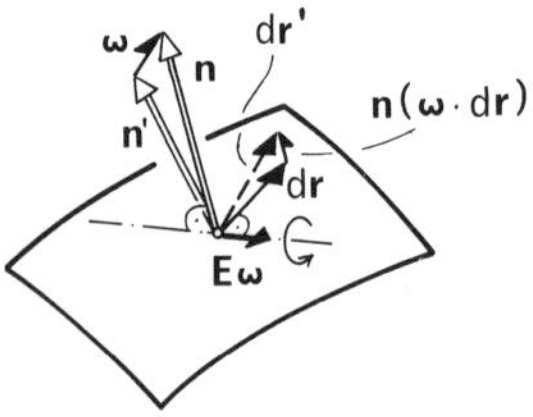

Fig. 2.18. Inclination of a surface element

Due attention must be paid to the fact that $\omega = \omega_i$ if and only if n is a principal direction of $\mathcal{E}$, i.e. $\mathcal{E}n = \lambda n$ (or $\gamma_{zx} = \gamma_{zy} = 0$; this is similar to what we have said about $n \otimes nUN$). As a matter of fact, the inclination of the surface element is caused by the inclination of the volume element under it, i.e. ω_i, and also by the eventual shearing strain between the normal and the tangent plane, i.e. $N\mathcal{E}n$.

The linear kinematic relations for the rotation, from (2.91, 55) and the relation $\boldsymbol{E}\cdot\boldsymbol{E} = -2$, are

$$\Omega = \frac{1}{4}\{\boldsymbol{N}[\nabla \otimes \boldsymbol{v} - (\nabla \otimes \boldsymbol{v})^{\mathrm{T}}]\boldsymbol{N}\}\cdot\boldsymbol{E}, \qquad \boldsymbol{\omega} = \boldsymbol{B}\boldsymbol{v} + \nabla_{\mathrm{n}} w. \tag{2.111}$$

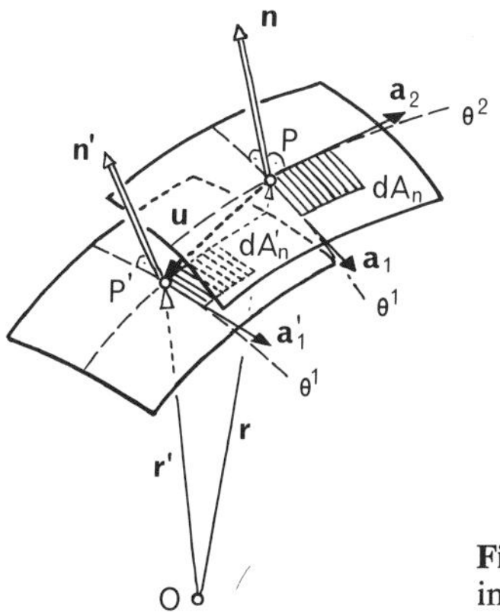

Fig. 2.19. Transformation of an elemental surface dA_{n} into an elemental surface dA'_{n}

In addition to determining the connection between elemental vectors on the undeformed and the deformed surface, we also have to transform surface elements. In the reference configuration, such an element is (Fig. 2.19)

$$dA_{\mathrm{n}} = |\boldsymbol{a}_1 \times \boldsymbol{a}_2| d\theta^1 d\theta^2, \tag{2.112}$$

where $\boldsymbol{a}_1 d\theta^1$ and $\boldsymbol{a}_2 d\theta^2$ are elemental vectors along the coordinate curves and where the index n reminds us of the importance of the normal. Instead of the vector product and the norm, we may write

$$|\boldsymbol{a}_1 \times \boldsymbol{a}_2| = (\boldsymbol{a}_1 \times \boldsymbol{a}_2)\cdot\boldsymbol{n} = \boldsymbol{a}_1\cdot(\boldsymbol{a}_2 \times \boldsymbol{n}) = \boldsymbol{a}_1\cdot\boldsymbol{E}\boldsymbol{a}_2,$$

and

$$dA_{\mathrm{n}} = \boldsymbol{a}_1\cdot\boldsymbol{E}\boldsymbol{a}_2 d\theta^1 d\theta^2.$$

Similarly, the deformed surface element is expressed by

$$dA'_{\mathrm{n}} = \boldsymbol{a}'_1\cdot\boldsymbol{E}'\boldsymbol{a}'_2 d\theta^1 d\theta^2 = \boldsymbol{a}_1\cdot\boldsymbol{N}\boldsymbol{F}^{\mathrm{T}}\boldsymbol{E}'\boldsymbol{F}\boldsymbol{N}\boldsymbol{a}_2 d\theta^1 d\theta^2$$
$$= \boldsymbol{a}_1\cdot(-\boldsymbol{N}\boldsymbol{F}^{\mathrm{T}}\boldsymbol{E}'\boldsymbol{F}\boldsymbol{E})\boldsymbol{E}\boldsymbol{a}_2 d\theta^1 d\theta^2,$$

since $\boldsymbol{E}^2 = -\boldsymbol{N}$. Because dA_{n} and dA'_{n} are (pseudo-) scalars, there must exist a relation

$$dA'_{\mathrm{n}} = \mu dA_{\mathrm{n}} \tag{2.113}$$

that is independent of the choice of coordinates. Hence, because $\boldsymbol{a}_1 d\theta^1$ and $\boldsymbol{a}_2 d\theta^2$ are arbitrary, we conclude that

$$-\boldsymbol{N}\boldsymbol{F}^{\mathrm{T}}\boldsymbol{E}'\boldsymbol{F}\boldsymbol{E}\boldsymbol{N} = \mu \boldsymbol{N},$$

$$\operatorname{tr}\{-\boldsymbol{N}\boldsymbol{F}^{\mathrm{T}}\boldsymbol{E}'\boldsymbol{F}\boldsymbol{E}\boldsymbol{N}\} = \mu \operatorname{tr} \boldsymbol{N} = 2\mu,$$

$$\mu = -\frac{1}{2}\boldsymbol{N}\boldsymbol{F}^{\mathrm{T}}\cdot\boldsymbol{E}'\boldsymbol{F}\boldsymbol{E}\boldsymbol{N}. \tag{2.114}$$

In the case of a small deformation, with $\boldsymbol{E}' = \mathscr{E}\boldsymbol{n}' \simeq \boldsymbol{E} - \mathscr{E}\boldsymbol{\omega} = \boldsymbol{E} + \boldsymbol{E}\boldsymbol{\omega} \otimes \boldsymbol{n} - \boldsymbol{n} \otimes \boldsymbol{E}\boldsymbol{\omega}$, we obtain

$$\begin{aligned}\boldsymbol{N}\boldsymbol{F}^{\mathrm{T}}\cdot\boldsymbol{E}'\boldsymbol{F}\boldsymbol{E}\boldsymbol{N} &= \boldsymbol{N}(\boldsymbol{I} + \boldsymbol{\nabla} \otimes \boldsymbol{u})\cdot\boldsymbol{E}'[\boldsymbol{I} + (\boldsymbol{\nabla} \otimes \boldsymbol{u})^{\mathrm{T}}]\boldsymbol{E}\boldsymbol{N}\\ &\simeq -\boldsymbol{N}\cdot\boldsymbol{N} - \boldsymbol{N}(\boldsymbol{\nabla} \otimes \boldsymbol{u})\cdot\boldsymbol{N} + \boldsymbol{N}\cdot\boldsymbol{E}(\boldsymbol{\nabla} \otimes \boldsymbol{u})^{\mathrm{T}}\boldsymbol{E}\boldsymbol{N} + \cdots,\end{aligned}$$

which means

$$\mu \simeq 1 + \operatorname{tr} \gamma.$$

Here, the transformation of a surface element is therefore

$$dA'_{\mathrm{n}} = (1 + \operatorname{tr} \gamma) dA_{\mathrm{n}}. \tag{2.115}$$

Chapter 3 Principles of Image Formation in Holography

Lately holography has become a broad subject, which has found many applications, as shown in the review [3.1] recently published by *Gabor*, the inventor of this technique [3.2–4]. There already exists a wide range of literature dealing with this topic so that, for the optical background as well as for a complete theoretical treatment, the reader may refer to books such as [3.5–20]. Further information concerning equipment, the photographic materials or the lasers may be found in [3.21–25].

For our purposes, it will suffice to make clear how holographic images are formed and particularly where they are formed. In a first part, we shall look at the case of standard holography, i.e. the case where the image is reconstructed with the same optical arrangement as was used for recording. We shall then study in a second part the changes undergone by an image point when the arrangement is modified during the reconstruction. We shall then have laid the foundations for holography in general, as they are required particularly for investigating the fringes and their modifications in Chap. 4.

3.1 Image Formation in Standard Holography

In this section, we first deal with the problem of recording and reconstructing the image of a point source, because that is what we shall need later on. Afterwards, we see how images of a whole object and of several objects are obtained.

3.1.1 Recording the Phase and the Amplitude of the Light Wave Emitted by a Point Source

The *spherical wave* emitted by a point source P, which we intend to record, is commonly written as

$$\mathrm{Re}\left\{\frac{A_\mathrm{P}}{p}\exp\left[2\pi \mathrm{i}\left(\frac{p}{\lambda}-\nu t\right)\right]\right\}.$$

$A_\mathrm{P} = |A_\mathrm{P}| \exp(\mathrm{i}\phi)$ takes into account the amplitude $|A_\mathrm{P}|$ and the phase ϕ of the source. p is the distance from P to the point where the vibration is considered, λ denotes the wavelength and ν, the frequency. As usual in optics, Re, meaning the real part, and the time harmonic factor $\exp(-2\pi \mathrm{i}\nu t)$, which is the same for all waves because we are dealing with monochromatic light, will be omitted.

We shall speak simply of the *complex amplitude* of the vibration caused by the source P at the distance p of it

$$U = \frac{A_P}{p} \exp\left(\frac{2\pi i}{\lambda} p\right). \tag{3.1}$$

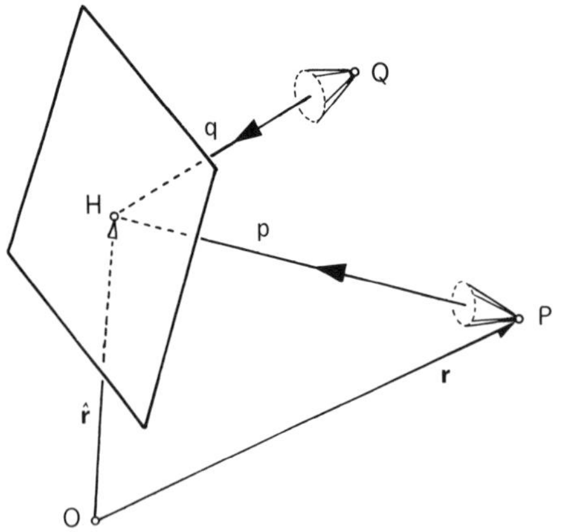

Fig. 3.1. Recording the object wave field produced by a point source P by means of a reference wave field produced by a point source Q. O: origin of the position vectors

In order to record not only the amplitude but also the phase of this *object wave*, let us add a *reference wave*, emitted by a point source Q and which has the same wavelength and is coherent with the first wave (roughly expressed, coherent means that the waves are able to interfere). A point H (with vector coordinate $\hat{\mathbf{r}}$) of a photographic plate located in front of both sources (Fig. 3.1) receives thus the total amplitude

$$U_{\text{tot}}(\hat{\mathbf{r}}) = \frac{A_Q}{q} \exp\left(\frac{2\pi i}{\lambda} q\right) + \frac{A_P}{p} \exp\left(\frac{2\pi i}{\lambda} p\right)$$

where $q = q(\hat{\mathbf{r}})$ and $p = p(\hat{\mathbf{r}})$ are the distances between the fixed points Q and P and a variable point H, and $A_Q = |A_Q| \exp(i\psi)$ is the complex amplitude of the reference source. The intensity produced by this field is

$$I(\hat{\mathbf{r}}) = \frac{1}{2}\left[\left(\frac{|A_Q|}{q}\right)^2 + \left(\frac{|A_P|}{p}\right)^2 + \frac{A_Q^*}{q} \exp\left(\frac{-2\pi i}{\lambda} q\right)\frac{A_P}{p} \exp\left(\frac{2\pi i}{\lambda} p\right)\right.$$
$$\left. + \frac{A_Q}{q} \exp\left(\frac{2\pi i}{\lambda} q\right)\frac{A_P^*}{p} \exp\left(\frac{-2\pi i}{\lambda} p\right)\right],$$

where * denotes the complex conjugates. Let us write this expression as

$$I(\hat{\mathbf{r}}) = \frac{1}{2}\left\{\left(\frac{|A_Q|}{q}\right)^2 + \left(\frac{|A_P|}{p}\right)^2 + 2\frac{|A_Q|}{q}\frac{|A_P|}{p} \cos\left[\frac{2\pi}{\lambda}(p - q) + \phi - \psi\right]\right\}.$$

This form makes clearly visible that the variation of the phase of the object wave has been changed into a variation of intensity by the reference wave.

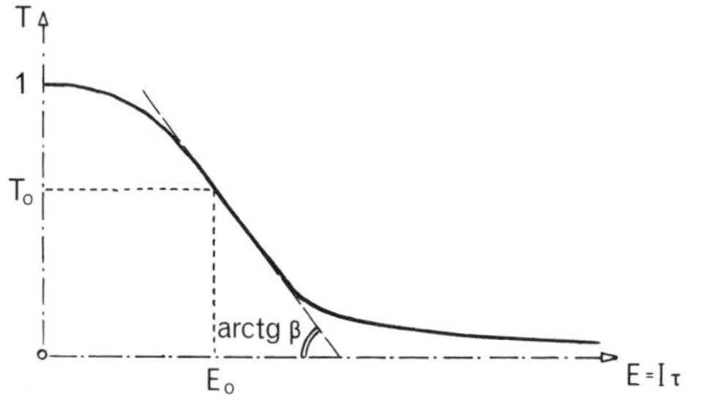

Fig. 3.2. Amplitude transmittance of the photographic plate versus received energy

During the time of exposure τ, the photographic plate receives the energy or exposure: $E = I\tau$. After development, the *amplitude transmittance* T of the plate is proportional to E, provided that the linear range of the curve $T = f(E)$ (Fig. 3.2) is used, i.e. provided that E does not vary too much around a mean value E_0. This may be achieved by giving the two wave fronts different amplitudes: the intensity of the reference wave has to be approximately constant over the whole plate whereas that of the object wave has to be much less, i.e.

$$\left(\frac{|A_Q|}{q}\right)^2 \simeq \text{const}, \qquad \left(\frac{|A_P|}{p}\right)^2 \ll \left(\frac{|A_Q|}{q}\right)^2.$$

Under these circumstances, the amplitude transmittance of the photographic plate is

$$T(\hat{r}) \simeq T_0 - \frac{1}{2}\beta\tau\left\{\frac{A_Q^*}{q}\frac{A_P}{p}\exp\left[\frac{2\pi i}{\lambda}(p-q)\right] + \frac{A_Q}{q}\frac{A_P^*}{p}\exp\left[\frac{2\pi i}{\lambda}(q-p)\right]\right\},$$

where T_0 is the mean transmittance due to the mean energy $E = \frac{1}{2}(|A_Q|/q)^2\tau$ and β is the slope of the curve $T = f(E)$. In this way, the whole information pertaining to the object wave is recorded in the photographic plate, which is therefore called a *hologram* [Ref. 3.3, p. 456].

3.1.2 Reconstruction of the Image of a Point Source

Let us now illuminate the so processed plate with the wave coming from a point source $\tilde{Q}$

$$\frac{A_{\tilde{Q}}}{\tilde{q}}\exp\left(\frac{2\pi i}{\tilde{\lambda}}\tilde{q}\right),$$

where, as in the foregoing, $\tilde{q} = \tilde{q}(\hat{r})$ is the distance from $\tilde{\mathrm{Q}}$ to a point of the hologram and $\tilde{\lambda}$ is the wavelength of the reconstructing wave front. Immediately behind the hologram, the transmitted wave is

$$\begin{aligned} T\frac{A_{\tilde{\mathrm{Q}}}}{\tilde{q}}\exp\left(\frac{2\pi\mathrm{i}}{\tilde{\lambda}}\tilde{q}\right) \simeq T_0\frac{A_{\tilde{\mathrm{Q}}}}{\tilde{q}}\exp\left(\frac{2\pi\mathrm{i}}{\tilde{\lambda}}\tilde{q}\right) \\ -\frac{1}{2}\beta\tau\frac{A_{\tilde{\mathrm{Q}}}}{\tilde{q}}\frac{A_{\mathrm{Q}}^*}{q}\frac{A_{\mathrm{P}}}{p}\exp\left[\frac{2\pi\mathrm{i}}{\tilde{\lambda}}\tilde{q}+\frac{2\pi\mathrm{i}}{\lambda}(p-q)\right] \\ -\frac{1}{2}\beta\tau\frac{A_{\tilde{\mathrm{Q}}}}{\tilde{q}}\frac{A_{\mathrm{Q}}}{q}\frac{A_{\mathrm{P}}^*}{p}\exp\left[\frac{2\pi\mathrm{i}}{\tilde{\lambda}}\tilde{q}+\frac{2\pi\mathrm{i}}{\lambda}(q-p)\right]. \end{aligned} \tag{3.2}$$

The first term represents the part of the reconstruction wave that is simply transmitted, without being diffracted. The other two contain, respectively, the object wave and its conjugate, they are therefore called the *primary or direct-image wave field*, and the *conjugate-image wave field.*

Before showing how the object wave itself may be obtained, let us explain what the conjugate of the object wave represents. It is

$$\mathrm{Re}\left\{\frac{A_{\mathrm{P}}^*}{p}\exp\left[2\pi\mathrm{i}\left(-\frac{p}{\lambda}-\nu t\right)\right]\right\} = \mathrm{Re}\left\{\frac{A_{\mathrm{P}}}{p}\exp\left[2\pi\mathrm{i}\left(\frac{p}{\lambda}+\nu t\right)\right]\right\}^*,$$

which is thus a wave identical to the original wave emitted by the point source P, except that it propagates in a direction opposite to the original direction. Therefore, because the original wave was a diverging wave front from P, its conjugate is a converging wave front, which converges to a focus. In other words, we may imagine that time has been reversed, as in a motion-picture film shown backwards [Ref. 3.26, p. 1087, or Ref. 3.27, p. 73].

Let us consider two special cases of reconstruction.

a) *The reconstructing wave front is identical with the reference wave*: ($A_{\tilde{\mathrm{Q}}} = A_{\mathrm{Q}}$, $\tilde{q} = q$, $\tilde{\lambda} = \lambda$) the primary-image wave field is then

$$-\frac{1}{2}\beta\tau\left(\frac{|A_{\mathrm{Q}}|}{q}\right)^2\frac{A_{\mathrm{P}}}{p}\exp\left(\frac{2\pi\mathrm{i}}{\lambda}p\right). \tag{3.3}$$

Except for a constant factor, we then have immediately behind the hologram the same complex amplitude as that produced by the source P [see (3.1)]. An observer looking through the hologram thus sees a *virtual image* of P in its original position. In other terms, the hologram acts as a window through which the object may be seen (Fig. 3.3).

b) *The reconstructing wave front is identical with the conjugate of the reference wave*: the conjugate-image wave field of (3.2) is then

$$-\frac{1}{2}\beta\tau\left(\frac{|A_{\mathrm{Q}}|}{q}\right)^2\frac{A_{\mathrm{P}}^*}{p}\exp\left(-\frac{2\pi\mathrm{i}}{\lambda}p\right). \tag{3.4}$$

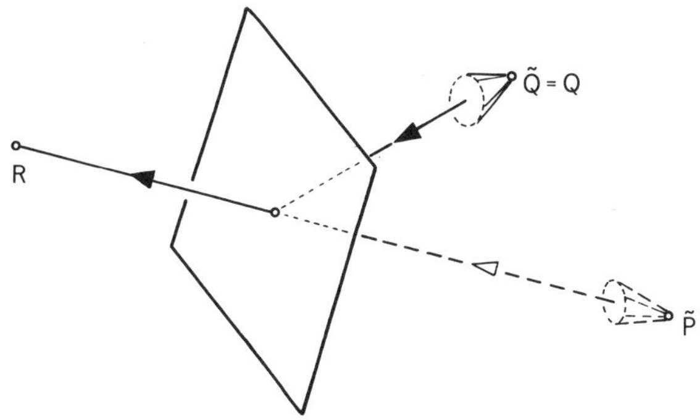

Fig. 3.3. Reconstructing an ideal virtual image P̃. R: position of the observer

The conjugate of the object wave is thereby produced. That is to say, a *real image* of the point source is formed in exactly the position the object had during the recording (Fig. 3.4).

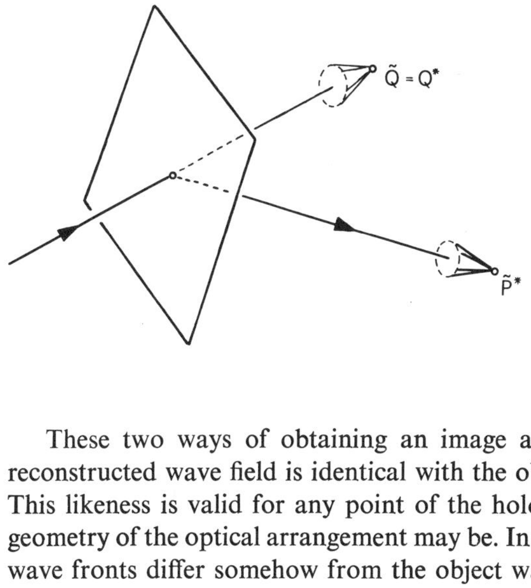

Fig. 3.4. Reconstructing an ideal real image P̃*

These two ways of obtaining an image are the only ones by which the reconstructed wave field is identical with the object wave field or its conjugate. This likeness is valid for any point of the hologram surface and whatever the geometry of the optical arrangement may be. In all other cases, the reconstructed wave fronts differ somehow from the object waves, as may be seen from (3.2). In Sect. 3.2, we shall deal more in detail with this *aberration* problem.

Leith and *Upatnieks* [3.28–30] were the first to separate in space the three wave fields of (3.2) by giving the object and the reference beams different directions, so that each image could be observed without being confused by the other waves. They also first took advantage of the great coherence length of lasers, which provides good interference between the object and the reference wave during the recording.

Let us also point out the premises that we have used so far. The linearity of the $T = f(E)$ curve is an approximation which in certain cases may prove insufficient (refer to the mentioned books or to [3.31–38]). Second, although the hologram has been presumed to be thin, in fact, the thickness of the emulsion, its shrinkage and its change of refractive index during photographic development may cause the reconstructed waves to be not exactly as described by (3.2) (in addition to the books, see e.g. [3.39–41]).

3.1.3 Holography of a Whole Object and Double Exposure

Instead of spherical waves produced by point sources, the object as well as the reference wave generally have fronts of any shape. Particularly, the object wave may consist of the light diffusely reflected by the rough surface of an opaque body. Each surface element dA_n (with vector coordinate $\boldsymbol{r}$) acts then as an elemental source with the complex amplitude $a(\boldsymbol{r})dA_\mathrm{n}$. The total amplitude of the vibration falling on the point H of the hologram plate is

$$U(\hat{\boldsymbol{r}}) = \iint_{A_\mathrm{n}} \frac{a}{p} \exp\left(\frac{2\pi\mathrm{i}}{\lambda} p\right) dA_\mathrm{n}, \tag{3.5}$$

where $p(\boldsymbol{r}, \hat{\boldsymbol{r}}) = |\hat{\boldsymbol{r}} - \boldsymbol{r}|$. Replacing now the spherical wave (3.1) by the wave front (3.5), we might repeat the same train of thought as in the preceding sections. Everything we said about the object wave would now apply to the wave $U(\hat{\boldsymbol{r}})$ reflected by the whole object surface. In particular, a virtual or real *three-dimensional image of the body* may be obtained.

Practically, the recording may take place as indicated schematically in Fig. 3.5: the light emitted by a laser is divided by means of a beam splitter into two

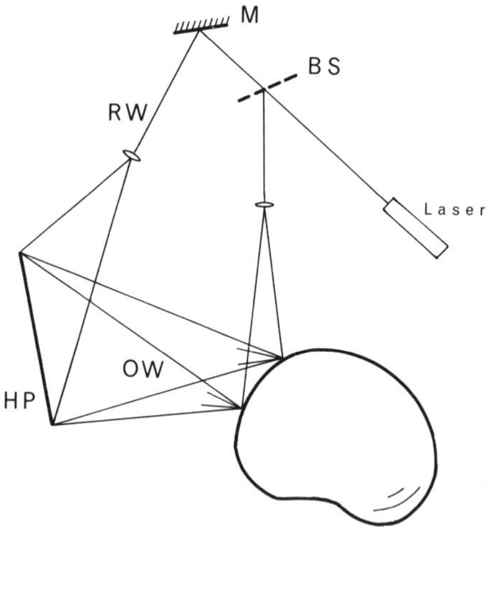

Fig. 3.5. Sketch of a recording arrangement. HP: hologram plate. BS: beam-splitter. M: mirrors. RW: reference wave. OW: object wave

narrow beams that are thereafter expanded by lenses. One of the beam serves as the reference wave whereas the other illuminates the object. For the reconstruction with a beam identical to the reference beam, the same set-up may be used (Fig. 3.6).

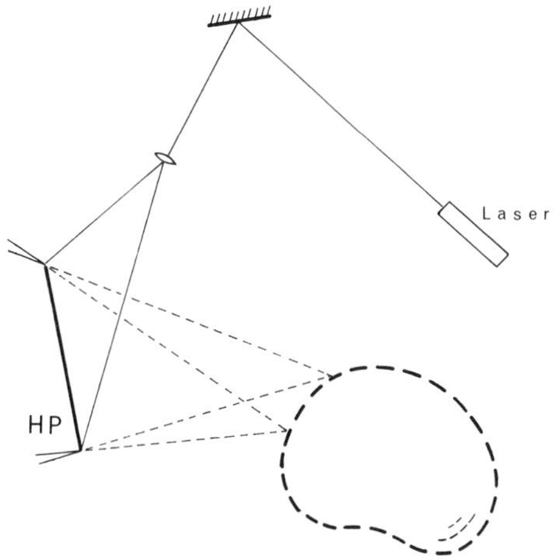

Fig. 3.6. Reconstruction of an ideal virtual image of the object by means of a wave identical to the reference wave of Fig. 3.5

An important point should be mentioned here, namely the *rigidity of the equipment*. In order to form sharp interference fringes on the hologram during the recording, the phase of each wave must stay constant. Therefore, the elements of the optical arrangement should not move more than a fraction of the wavelength with respect to each other. This requires often special precautions, such as the use of damped tables and of heavy and stable components (see also [3.42, 43]). Moreover, the optical paths traversed by the reference and the object beams between the beam splitter and the hologram should be nearly equal. Their difference should not exceed the coherence length of the laser. These are two of the most critical practical measures required to obtain good images of an object.

Let us now see how it is possible to record successively two or more wave fields on the same photographic plate [3.44, 45]. Let us suppose that we desire to record two wave fronts that produce the amplitudes $U(\hat{r})$ and $U'(\hat{r})$ in the hologram plane. U and U' are to be recorded with reference waves having amplitudes $V(\hat{r})$ and $V'(\hat{r})$, respectively. (As the reader may have noticed, we speak here of only the amplitudes that result in the hologram plane and do not explain how these four vibrations are formed.)

During a first exposure, the hologram is exposed to the lights U and V, that is the intensity

$$I = \frac{1}{2}(|V|^2 + |U|^2 + V^*U + VU^*). \tag{3.6}$$

Without developing, we then expose the plate a second time with U' and V', which gives the intensity

$$I' = \frac{1}{2}(|V'|^2 + |U'|^2 + V'^*U' + V'U'^*).$$

The total energy received by the plate is thus $E = I\tau + I'\tau'$, where τ and τ' are the respective exposure times. As previously, if the energy varies only a little around a mean value, i.e. if

$$|V|^2 \simeq \text{const}, \quad |V'|^2 \simeq \text{const and } |U|^2, |U'|^2 \ll |V|^2, |V'|^2, \tag{3.7}$$

the resulting transmittance of the developed emulsion will be proportional to the energy variation, i.e.

$$T(\hat{r}) \simeq T_0 - \frac{1}{2}\beta(\tau V^*U + \tau' V'^*U' + \tau VU^* + \tau' V'U'^*). \tag{3.8}$$

For the reconstruction, let us illuminate the hologram with two wave fields that produce on it the amplitudes $\tilde{V}(\hat{r})$ and $\tilde{V}'(\hat{r})$. These two waves are diffracted and the wave front immediately behind the hologram is

$$(\tilde{V} + \tilde{V}')T_0 - \frac{1}{2}\beta(\tau\tilde{V}V^*U + \tau'\tilde{V}'V'^*U' + \tau\tilde{V}'V^*U + \tau'\tilde{V}V'^*U' \\ + \tau\tilde{V}VU^* + \tau'\tilde{V}'V'U'^* + \tau\tilde{V}'VU^* + \tau'\tilde{V}V'U'^*). \tag{3.9}$$

We may distinguish here again between the *primary-image wave fields* that contain the object waves and *conjugate-image wave fields* that contain the conjugates of the object waves. As previously, only when the reconstruction wave is identical with the reference wave or its conjugate, is a wave identical with the object, or its conjugate, formed. Let us therefore look at the important particular case in which $\tilde{V} \equiv V \equiv \tilde{V}' \equiv V'$ and $\tau = \tau'$, that is where both objects are recorded and reconstructed with the same wave. The primary-image wave field then becomes

$$-\beta\tau|V|^2(U + U'). \tag{3.10}$$

This means that the images of U and of U' are reconstructed.

Thus, the important result of this development is that two wave fronts, *although they were recorded at different times, are reconstructed at the same time* and hence may interfere. In this book, we shall make use of this interesting property in the following way: the waves U and U' originate from the same object, before and after a mechanical deformation takes place. Both images are then reconstructed together, so that the wave fronts interfere and fringes are formed.

Chapter 4 will deal with the interpretation of the fringes in order to find out what distinguishes U from U', which means measurement of the displacement, the strain and the rotation undergone by each point of the object surface.

Let us yet add here that the wave fields of (3.9) generally must be studied by use of the theory explained in the next section.

3.2 Image Formation in the Case of Modification of the Optical Arrangement at the Reconstruction

In Sect. 3.1, we described in general terms first the recording of a hologram and then the wave field created when this hologram is illuminated by a reconstruction source (3.2, 9). We also noted that, with an arbitrary optical arrangement, the wave produced is identical with the object wave or its conjugate only when the reconstruction wave is identical with the reference wave or its conjugate. The problem now is to describe the images formed when that condition is not fulfilled.

We shall consider as an elementary object a point source P, because a whole finite object may be thought of as a set of points {P}. The first modification of the optical arrangement to be considered will be a shift of the reconstruction source with respect to the reference source and a change of wavelength. To investigate the image, we shall compare the wave front that would produce a point image with the actual wave front; we shall see that the actual image suffers aberrations. A second modification of the optical arrangement will be when the hologram is moved relative to its position during the recording. Finally, we shall calculate the difference of the displacements of two image points reconstructed with two movable sandwiched holograms.

In these developments, we shall distinguish between large and small changes, similar to the analysis of the deformation of a body in Sect. 2.2. Small changes are especially interesting for two reasons. On the one hand, if we reconstruct at the same time the images of an undeformed and of the corresponding deformed object, the two images must overlap so that fringes can be formed (see later Chap. 4). On the other hand, linearized relations are easier to apply.

3.2.1 Large Modification of the Reference Source: Position of an Image Point, Aberrations

As mentioned in the introduction, we shall now assume that the reconstruction source $\tilde{Q}$ differs from the recording source Q as for its position and as for the wavelength and the amplitude of the light it emits (Fig. 3.7). Besides, we suppose that the hologram has not moved and that it has been developed in ideal conditions.

Under these circumstances, the wave fronts produced are those given by the general formula (3.2). By analogy with the special cases of reconstruction that

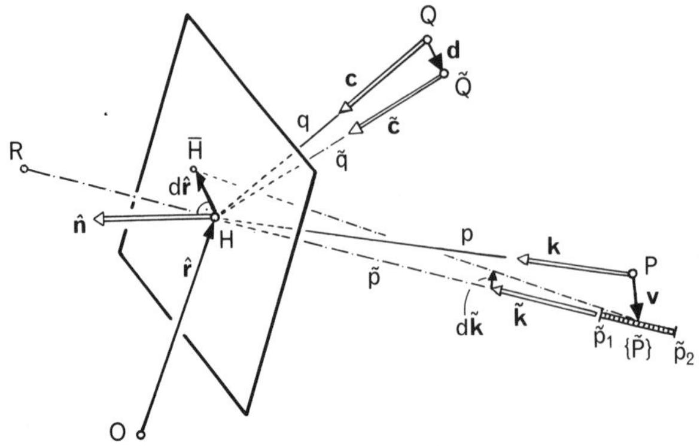

Fig. 3.7. Position of the set $\{\tilde{\mathrm{P}}\}$ of the images of the object point source P when the reconstruction source $\tilde{\mathrm{Q}}$ differs from the reference source Q. R: location of the observer. O: origin of the position vectors. $\hat{\boldsymbol{n}}$, $\boldsymbol{c}$, $\tilde{\boldsymbol{c}}$, $\boldsymbol{k}$, $\tilde{\boldsymbol{k}}$: unit direction vectors

we described in Sect. 3.1.2, we expect the primary-image wave front to form a virtual image point at a place $\tilde{\mathrm{P}}$ that still must be determined. Such a source would produce on the hologram the amplitude

$$\tilde{U}(\hat{\boldsymbol{r}}) = \frac{A_{\tilde{\mathrm{P}}}}{\tilde{p}} \exp\left(\frac{2\pi\mathrm{i}}{\tilde{\lambda}}\tilde{p}\right). \tag{3.11}$$

As previously, $A_{\tilde{\mathrm{P}}} = |A_{\tilde{\mathrm{P}}}| \exp(\mathrm{i}\tilde{\phi})$ represents the complex amplitude of the image source, $\tilde{p}(\hat{\boldsymbol{r}})$ is the distance between the point $\tilde{\mathrm{P}}$ and a variable point H on the hologram, finally $\tilde{\lambda}$ is the wavelength of the reconstruction light. It is worth while noting here that this method, in which the expected wave front is compared to the actual one (see for instance [Ref. 3.9, p. 67; Ref. 3.10, p. 35; 3.46–66]) is similar to the calculus in the geometrical aberration theory of rotationally symmetrical optical systems [Ref. 3.5, p. 203; 3.67; Ref. 3.68, p. 102]. The difference is, however, that here the reference image, i.e. the expected image, is not known *a priori*.

If the statement concerning the image "point" were correct, we could find this point $\tilde{\mathrm{P}}$ such that the primary-image wave field of (3.2) could be replaced by (3.11) in each point of the hologram. The amplitudes should be the same, as well as the phase terms. For the latter, the following geometrical condition should be fulfilled in any point H on the hologram

$$\frac{2\pi}{\tilde{\lambda}}\tilde{p} + \tilde{\phi} = \frac{2\pi}{\lambda}p + \phi - \frac{2\pi}{\lambda}q - \psi + \frac{2\pi}{\tilde{\lambda}}\tilde{q} + \tilde{\psi},$$

or

$$\frac{2\pi}{\tilde{\lambda}}(\tilde{p} - \tilde{q}) + \tilde{\phi} - \tilde{\psi} = \frac{2\pi}{\lambda}(p - q) + \phi - \psi. \tag{3.12}$$

This relation states that the "interference" in H produced by $\tilde{\mathrm{Q}}$ and $\tilde{\mathrm{P}}$ during reconstruction is the same as the interference produced by Q and P during recording; therefore, we call it the *condition of interference identity* in H. $\tilde{p}(\hat{\boldsymbol{r}})$, $\tilde{q}(\hat{\boldsymbol{r}})$, $p(\hat{\boldsymbol{r}})$ and $q(\hat{\boldsymbol{r}})$ are all functions of the vector $\hat{\boldsymbol{r}}$, i.e. of the position of H. Grouping these lengths, we may define a *phase-difference function*

$$\theta(\hat{\boldsymbol{r}}) = \frac{2\pi}{\tilde{\lambda}}(\tilde{p} - \tilde{q}) - \frac{2\pi}{\lambda}(p - q). \tag{3.13}$$

If $\tilde{p}(\hat{\boldsymbol{r}})$ does not depend on any other variables besides $\hat{\boldsymbol{r}}$, that is to say if it has to be the distance between a *fixed* image point $\tilde{\mathrm{P}}$ and the variable point H, the condition (3.12) cannot be fulfilled over the entire hologram surface. Therefore, we must conclude that the *whole hologram no longer reconstructs one definite image point*. Yet, let us see if at least a *small* region around the point H produces such a point; this region may be thought of as the part of the hologram through which an observer sees the image when he uses a diaphragm of small diameter. Therefore, we now develop the functions of $\hat{\boldsymbol{r}}$ in (3.12) into series around the axes $\tilde{\mathrm{P}}$H, $\tilde{\mathrm{Q}}$H, PH and QH. Thereupon, we shall try to *match* the development of $\tilde{p}$ with that of the known functions.

We thus develop the phase difference $\theta(\hat{\boldsymbol{r}})$ into a Taylor series around H (see (2.58), with $d^2\hat{\boldsymbol{r}} = 0$, $d^3\hat{\boldsymbol{r}} = 0, \ldots$)

$$\begin{aligned}\bar{\theta} &= \theta + d\theta + \frac{1}{2!}d^2\theta + \frac{1}{3!}d^3\theta + \cdots \\ &= \theta + d\hat{\boldsymbol{r}}\cdot\boldsymbol{\nabla}_{\hat{\mathrm{n}}}\theta + \frac{1}{2}d\hat{\boldsymbol{r}}\cdot(\boldsymbol{\nabla}_{\hat{\mathrm{n}}} \otimes \boldsymbol{\nabla}_{\hat{\mathrm{n}}}\theta)d\hat{\boldsymbol{r}} \\ &\qquad + \frac{1}{6}d\hat{\boldsymbol{r}}\cdot[d\hat{\boldsymbol{r}}(\boldsymbol{\nabla}_{\hat{\mathrm{n}}} \otimes \boldsymbol{\nabla}_{\hat{\mathrm{n}}} \otimes \boldsymbol{\nabla}_{\hat{\mathrm{n}}}\theta)d\hat{\boldsymbol{r}}] + \cdots .\end{aligned} \tag{3.14}$$

θ is a scalar, its derivatives are vectors and tensors, all being considered in H, whereas $\bar{\theta}$ denotes the value of the scalar function in a neighboring point $\overline{\mathrm{H}}$ separated from H by the vector increment $d\hat{\boldsymbol{r}}$ (Fig. 3.7). Similar to (2.49), the formal vector $\boldsymbol{\nabla}_{\hat{\mathrm{n}}} = \hat{\boldsymbol{N}}\boldsymbol{\nabla}$ is the derivative operator in the hologram plane, whose unit normal is $\hat{\boldsymbol{n}}$. The normal projection $\hat{\boldsymbol{N}} = \boldsymbol{I} - \hat{\boldsymbol{n}} \otimes \hat{\boldsymbol{n}}$ is thus present. Because the hologram has no curvature, the derivative of $\hat{\boldsymbol{n}}$ is zero, so that we may take out of the derivatives the projections $\hat{\boldsymbol{N}}$

$$\begin{aligned}\bar{\theta} &= \theta + d\hat{\boldsymbol{r}}\cdot\hat{\boldsymbol{N}}\boldsymbol{\nabla}\theta + \frac{1}{2}d\hat{\boldsymbol{r}}\cdot\hat{\boldsymbol{N}}(\boldsymbol{\nabla} \otimes \boldsymbol{\nabla}\theta)\hat{\boldsymbol{N}}d\hat{\boldsymbol{r}} \\ &\qquad + \frac{1}{6}d\hat{\boldsymbol{r}}\cdot\hat{\boldsymbol{N}}[d\hat{\boldsymbol{r}}\hat{\boldsymbol{N}}(\boldsymbol{\nabla} \otimes \boldsymbol{\nabla} \otimes \boldsymbol{\nabla}\theta)\hat{\boldsymbol{N}}d\hat{\boldsymbol{r}}] + \cdots .\end{aligned} \tag{3.15}$$

The derivatives apply to the four lengths between the fixed points P, $\tilde{\mathrm{P}}$, Q, $\tilde{\mathrm{Q}}$

and the variable point H. We can thus make use of the results of Sect. 2.1.2 and find: a) $\nabla p = \boldsymbol{k}$ [see (2.35)] where $\boldsymbol{k}$ is a *unit direction vector* along PH (Fig. 3.7), b) $\nabla \otimes \boldsymbol{k} = \boldsymbol{K}/p$ [see (2.36)] where $\boldsymbol{K}$ is a *normal projection* onto the plane perpendicular to $\boldsymbol{k}$, c) $\nabla \otimes (\boldsymbol{K}/p) = -\mathscr{K}/p^2$ [see (2.37)] with the *superprojection* $\mathscr{K}$. Similarly, the derivatives of $\tilde{p}$, q, $\tilde{q}$ give the unit direction vectors $\tilde{\boldsymbol{k}}$, $\boldsymbol{c}$, $\tilde{\boldsymbol{c}}$, the normal projections $\tilde{\boldsymbol{K}}$, $\boldsymbol{C}$, $\tilde{\boldsymbol{C}}$ and the superprojections $\tilde{\mathscr{K}}$, $\mathscr{C}$, $\tilde{\mathscr{C}}$.

The zero-order term in (3.15) has already been shown explicitly by (3.12), which defines the phase of the light ray that comes from $\tilde{\mathrm{P}}$ and goes through the point H.

Next, the *first-order term* must be zero for any $d\hat{\boldsymbol{r}}$ in the hologram plane. The following condition of *stationary behavior* must thus be fulfilled

$$\nabla_{\hat{\mathrm{n}}}\theta = \hat{\boldsymbol{N}}\nabla\theta = 0, \tag{3.16}$$

that is to say with (2.35)

$$\hat{\boldsymbol{N}}\left[\frac{1}{\tilde{\lambda}}(\tilde{\boldsymbol{k}} - \tilde{\boldsymbol{c}}) - \frac{1}{\lambda}(\boldsymbol{k} - \boldsymbol{c})\right] = 0. \tag{3.17}$$

Because of the projection $\hat{\boldsymbol{N}}$, this expresses only the interior part of $\tilde{\boldsymbol{k}}$, i.e. the component situated in the hologram plane. However, the supplementary condition $|\tilde{\boldsymbol{k}}| = 1$ enables us to determine *the direction $\tilde{\boldsymbol{k}}$ of the image point* $\tilde{\mathrm{P}}$. It must be noted here that the latter condition can be fulfilled only if

$$|\hat{\boldsymbol{N}}\tilde{\boldsymbol{k}}| = \left|\hat{\boldsymbol{N}}\left[\tilde{\boldsymbol{c}} + \frac{\tilde{\lambda}}{\lambda}(\boldsymbol{k} - \boldsymbol{c})\right]\right| < 1. \tag{3.18}$$

If, on the contrary, this value is greater than 1, the exterior part $\hat{\boldsymbol{n}}\cdot\tilde{\boldsymbol{k}}$ will be imaginary; the reconstructed wave is then an *evanescent wave*, that propagates along the hologram (see also [3.69, 70]). Let us add also that

$$\operatorname{sgn}(\hat{\boldsymbol{n}}\cdot\tilde{\boldsymbol{k}}) = \operatorname{sgn}(\hat{\boldsymbol{n}}\cdot\tilde{\boldsymbol{c}}), \tag{3.19}$$

in order for the light to propagate on both paths in the same direction.

As for the distance $\tilde{p}$ from the hologram, it can be determined only from the higher-order terms in the series (3.15), since it did not appear in (3.17). To a first approximation, we may expect it to be given by the *second-order term*, i.e. by

$$d^2\theta = d\hat{\boldsymbol{r}}\cdot\hat{\boldsymbol{N}}(\nabla \otimes \nabla\theta)\hat{\boldsymbol{N}}d\hat{\boldsymbol{r}} = 0. \tag{3.20}$$

With (2.36) we then find (see also [Ref. 3.71, Eqs. (5.5) and (A.14)], where the same tensors appear)

$$d\hat{r}\cdot\hat{N}\left[\frac{1}{\tilde{\lambda}}\left(\frac{1}{\tilde{p}}\tilde{K} - \frac{1}{\tilde{q}}\tilde{C}\right) - \frac{1}{\lambda}\left(\frac{1}{p}K - \frac{1}{q}C\right)\right]\hat{N}d\hat{r} = 0. \tag{3.21}$$

The $\hat{N}$'s, which, in fact, could be omitted here, indicate that only the full projection of the bracket can come into account in (3.21). Because the only remaining undetermined quantity is $\tilde{p}$, it will generally not be possible to assert the vanishing of $\hat{N}(\nabla \otimes \nabla\theta)\hat{N}$. Nevertheless, in order to isolate the unknown $\tilde{p}$, we change from the increment $d\hat{r}$ in the hologram plane to an increment $d\tilde{k} = \tilde{m}d\alpha$, which describes the angular variation $d\alpha$ between $\tilde{\mathrm{P}}\mathrm{H}$ and $\tilde{\mathrm{P}}\overline{\mathrm{H}}$ by use of a unit vector $\tilde{m}$ perpendicular to $\tilde{k}$. Therefore, we use a connection between two surfaces, here the hologram and a unit sphere, which we discussed in Sect. 2.1.5. So we use the *oblique projection* (I is still the identity)

$$\tilde{M} = I - \frac{1}{\hat{n}\cdot\tilde{k}}\hat{n}\otimes\tilde{k}.$$

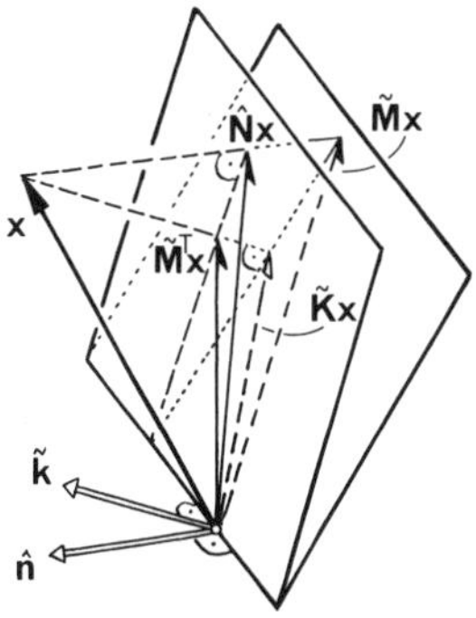

Fig. 3.8. Normal and oblique projections of an arbitrary vector x onto the hologram plane (with normal $\hat{n}$) and the plane perpendicular to $\tilde{k}$

When it is applied to a vector to the right, this linear transformation acts as a projection along $\hat{n}$ onto the plane normal to $\tilde{k}$ (see Fig. 3.8). Using its transpose $\tilde{M}^{\mathrm{T}} = I - \tilde{k}\otimes\hat{n}/\tilde{k}\cdot\hat{n}$, we may write a relation similar to (2.74) between the two increments

$$d\hat{r} = \tilde{p}\tilde{M}^{\mathrm{T}}d\tilde{k} = \tilde{p}d\alpha\tilde{M}^{\mathrm{T}}\tilde{m}. \tag{3.22}$$

With this relation and $\tilde{M}\hat{N} = \tilde{M}$, $\hat{N}\tilde{M}^{\mathrm{T}} = \tilde{M}^{\mathrm{T}}$ similar to (2.71), (3.21) becomes

$$\tilde{p}^2(d\alpha)^2\tilde{m}\cdot\tilde{M}\left[\frac{1}{\tilde{\lambda}\tilde{p}}\tilde{K}\right]\tilde{M}^{\mathrm{T}}\tilde{m} = \tilde{p}^2(d\alpha)^2\tilde{m}\cdot\tilde{M}\left[\frac{1}{\lambda p}K - \frac{1}{\lambda q}C + \frac{1}{\tilde{\lambda}\tilde{q}}\tilde{C}\right]\tilde{M}^{\mathrm{T}}\tilde{m}.$$

Moreover, because of $\tilde{M}\tilde{K} = \tilde{K}$, $\tilde{K}\tilde{M}^{\mathrm{T}} = \tilde{K}$ and $\tilde{m}\cdot\tilde{K}\tilde{m} = 1$ we obtain, after division by $\tilde{p}^2(d\alpha)^2$,

$$\frac{1}{\tilde{p}} = \tilde{\boldsymbol{m}} \cdot \tilde{\boldsymbol{T}} \tilde{\boldsymbol{m}}, \tag{3.23}$$

where

$$\tilde{\boldsymbol{T}} = \tilde{\lambda} \tilde{\boldsymbol{M}} \left(\frac{1}{\lambda p} \boldsymbol{K} - \frac{1}{\lambda q} \boldsymbol{C} + \frac{1}{\tilde{\lambda} \tilde{q}} \tilde{\boldsymbol{C}} \right) \tilde{\boldsymbol{M}}^{\mathrm{T}}. \tag{3.24}$$

$\tilde{\boldsymbol{T}}$ is a given symmetric tensor; because of the full projection with $\tilde{\boldsymbol{M}}$ and $\tilde{\boldsymbol{M}}^{\mathrm{T}}$, it is situated in the plane perpendicular to $\tilde{\boldsymbol{k}}$. So, the second-order terms lead us to the following conclusion: in the general case, to each direction $\tilde{\boldsymbol{m}}$ (or to each corresponding $d\hat{\boldsymbol{r}}/|d\hat{\boldsymbol{r}}|$) belongs a distance $\tilde{p}$. The maximum and the minimum of $\tilde{p}$, when $\tilde{\boldsymbol{m}}$ rotates in its plane, is found in the following classical manner (see for instance [Ref. 2.11, p. 30]): considering $|\tilde{\boldsymbol{m}}| = 1$ as an auxiliary condition, $1/\tilde{p}$ as a Lagrange multiplier and $\tilde{\boldsymbol{K}}$ as the two-dimensional identity tensor in the plane of interest, the extreme values of $\tilde{p}$ are given by

$$\frac{\partial}{\partial \tilde{\boldsymbol{m}}} \left[\tilde{\boldsymbol{m}} \cdot \left(\tilde{\boldsymbol{T}} - \frac{1}{\tilde{p}} \tilde{\boldsymbol{K}} \right) \tilde{\boldsymbol{m}} \right] = 0,$$

that is to say by the characteristic eigenvalue equation

$$\left| \tilde{\boldsymbol{T}} - \frac{1}{\tilde{p}} \tilde{\boldsymbol{K}} \right| = 0. \tag{3.25}$$

Let us add that, according to our convention about $\tilde{\mathrm{P}}$ (see Fig. 3.7), the image formed is *virtual* if $\tilde{p} > 0$ and *real* if $\tilde{p} < 0$.

Thus, our assumption concerning the existence of an image point $\tilde{\mathrm{P}}$ is only partly confirmed: by matching the first-order terms, we were able to find the direction of the image [see (3.17)] but matching of the second-order terms led us to an *interval* along this direction, whose limits are given by the two eigenvalues $\tilde{p}_1$, $\tilde{p}_2$ of (3.25). This interval shows the *astigmatism of the image* $\tilde{\mathrm{P}}$. In other words, the primary-image wave field is generally an astigmatic pencil of rays whose foci are located on the interval $\{\tilde{\mathrm{P}}\}$. Nevertheless, there is a particular case in which we can immediately see that the interval $\{\tilde{\mathrm{P}}\}$ reduces to one point. In fact, if $\tilde{\boldsymbol{k}} = \boldsymbol{k} = \tilde{\boldsymbol{c}} = \boldsymbol{c}$, (3.25) becomes

$$\left| \left[\frac{\tilde{\lambda}}{\lambda} \left(\frac{1}{p} - \frac{1}{q} \right) + \frac{1}{\tilde{q}} - \frac{1}{\tilde{p}} \right] \tilde{\boldsymbol{K}} \right| = 0,$$

so that we have the unique eigenvalue

$$\frac{1}{\tilde{\lambda} \tilde{p}_1} = \frac{1}{\tilde{\lambda} \tilde{p}_2} = \frac{1}{\lambda p} - \frac{1}{\lambda q} + \frac{1}{\tilde{\lambda} \tilde{q}}. \tag{3.26}$$

In this particular case, in which the illumination or observation directions are the same (but not necessarily the same as the normal $\hat{\boldsymbol{n}}$ of the hologram), the image is thus stigmatic and is given by a formula similar to that of Gaussian optics for lens imagery (see also [3.26, 72]). In this case, we can speak of an *axis* and of a paraxial approximation with respect to the oblique, and common, illumination and observation direction $\tilde{\boldsymbol{k}} = \boldsymbol{k} = \tilde{\boldsymbol{c}} = \boldsymbol{c} \neq \hat{\boldsymbol{n}}$.

Concerning again the general case $\tilde{\boldsymbol{c}} \neq \boldsymbol{c}$, it must be added that it is also possible to keep close to the geometric-optics aberration theory of rotationally symmetrical optical systems (refer in particular to *Champagne* [3.52], *Kiemle* and *Röss* [Ref. 3.10, pp. 35, 62], or *Smith* [Ref. 3.20, p. 172]). Indeed, the tensor that appears in the expression (3.21) of $d^2\theta$ may be decomposed into the part

$$\hat{\boldsymbol{N}}\left[\frac{1}{\tilde{\lambda}}\left(\frac{1}{\tilde{p}} - \frac{1}{\tilde{q}}\right)\boldsymbol{I} - \frac{1}{\lambda}\left(\frac{1}{p} - \frac{1}{q}\right)\boldsymbol{I}\right]\hat{\boldsymbol{N}} = \left[\frac{1}{\tilde{\lambda}\tilde{p}} - \frac{1}{\lambda p} + \frac{1}{\lambda q} - \frac{1}{\tilde{\lambda}\tilde{q}}\right]\hat{\boldsymbol{N}} \tag{3.27}$$

and the remainder

$$-\hat{\boldsymbol{N}}\left[\frac{1}{\tilde{\lambda}}\left(\frac{1}{\tilde{p}}\tilde{\boldsymbol{k}} \otimes \tilde{\boldsymbol{k}} - \frac{1}{\tilde{q}}\tilde{\boldsymbol{c}} \otimes \tilde{\boldsymbol{c}}\right) - \frac{1}{\lambda}\left(\frac{1}{p}\boldsymbol{k} \otimes \boldsymbol{k} - \frac{1}{q}\boldsymbol{c} \otimes \boldsymbol{c}\right)\right]\hat{\boldsymbol{N}}. \tag{3.28}$$

If we state now that the bracket of (3.27) is zero, we obtain a relation similar to (3.26) which, when it is joined to (3.12, 17), defines the distance $\tilde{p}_0$, the phase $\tilde{\phi}_0 \equiv \tilde{\phi}$ and the direction $\tilde{\boldsymbol{k}}_0 \equiv \tilde{\boldsymbol{k}}$ of a "source" $\tilde{\mathrm{P}}_0$. This point is called the "*Gaussian image*" and may be used as a reference point. Relative to it, we can then calculate the *wave aberration*, i.e. the difference between the phase of this reference wave front and that of the actual wave front,

$$\begin{aligned} \Delta(\hat{\boldsymbol{r}}) &= \left(\frac{2\pi}{\tilde{\lambda}}\tilde{p}_0 + \tilde{\phi}_0\right) - \left(\frac{2\pi}{\lambda}p + \phi - \frac{2\pi}{\lambda}q - \psi + \frac{2\pi}{\tilde{\lambda}}\tilde{q} + \tilde{\psi}\right) \\ &= \theta_0(\hat{\boldsymbol{r}}) + \tilde{\phi}_0 - \phi - \tilde{\psi} + \psi. \end{aligned} \tag{3.29}$$

The zero-order term as well as the first-order term in the series development of Δ around H vanish because of the definition of $\tilde{\mathrm{P}}_0$. As for the second-order term $d^2\Delta = d^2\theta_0$, the part that contains (3.27) is equal to zero also by definition; on the contrary, the part (3.28) does not generally disappear. Still by analogy with geometric-optics aberration theory, one may call

$$-\hat{\boldsymbol{N}}\left[\frac{1}{\tilde{\lambda}}\left(\frac{1}{\tilde{p}_0}\tilde{\boldsymbol{k}}_0 \otimes \tilde{\boldsymbol{k}}_0 - \frac{1}{\tilde{q}}\tilde{\boldsymbol{c}} \otimes \tilde{\boldsymbol{c}}\right) - \frac{1}{\lambda}\left(\frac{1}{p}\boldsymbol{k} \otimes \boldsymbol{k} - \frac{1}{q}\boldsymbol{c} \otimes \boldsymbol{c}\right)\right]\hat{\boldsymbol{N}}$$

the tensor characterizing the *astigmatism*.

Let us now briefly pass to the higher-order derivatives in the development of θ. We have first with (2.37)

$$d^3\theta = -2\pi d\hat{\mathbf{r}}\cdot\hat{\mathbf{N}}\left\{d\hat{\mathbf{r}}\hat{\mathbf{N}}\left[\frac{1}{\tilde{\lambda}}\left(\frac{1}{\tilde{p}^2}\tilde{\mathscr{K}} - \frac{1}{\tilde{q}^2}\tilde{\mathscr{C}}\right) - \frac{1}{\lambda}\left(\frac{1}{p^2}\mathscr{K} - \frac{1}{q^2}\mathscr{C}\right)\right]\hat{\mathbf{N}}d\hat{\mathbf{r}}\right\}. \tag{3.30}$$

The position of the expected image $\tilde{\mathrm{P}}$ is already "determined" by the foregoing investigation of the first- and second-order terms. If we want to take into account the preceding results, we must put into (3.30) the vector $\tilde{\boldsymbol{k}}$ given by (3.17) and also one value of $\tilde{p}$, either that given by (3.23), corresponding to $d\hat{\mathbf{r}}$, or the reference value $\tilde{p}_0$ of the "Gaussian image". In both cases, $d^3\theta$ is usually not zero. Again by analogy with the geometric-optics theory of aberrations, we could call the sum of the four superprojections in (3.30) the third-order tensor characterizing the *coma*. In the same way, we could also calculate $d^4\theta$. The corresponding fourth-order tensor which appears there should characterize the so-called *spherical aberration.*

So far, we have dealt with the primary-image wave field; we have expected the reconstructed image of a point source to be an image point and we have seen the limitations of this assumption. Nevertheless, the same reasoning also proves useful to describe the image produced by the *conjugate-image wave field.* Indeed, let us suppose that this latter [such as shown in (3.2)] can be replaced by the following wave, emitted from a point source $\tilde{\mathrm{P}}^*$,

$$\tilde{U}^*(\hat{\mathbf{r}}) = \frac{\overset{*}{A_{\tilde{\mathrm{P}}}}}{\tilde{p}^*}\exp\left(\frac{2\pi i}{\tilde{\lambda}}\tilde{p}^*\right). \tag{3.31}$$

Instead of (3.13) we have now first

$$\theta^*(\hat{\mathbf{r}}) = \frac{2\pi}{\tilde{\lambda}}(\tilde{p}^* - \tilde{q}) + \frac{2\pi}{\lambda}(p - q). \tag{3.32}$$

Then, the series development of θ^* around H, instead of (3.17), gives,

$$\nabla_{\hat{\mathrm{n}}}\theta^* = 2\pi\hat{\mathbf{N}}\left[\frac{1}{\tilde{\lambda}}(\tilde{\boldsymbol{k}}^* - \tilde{\boldsymbol{c}}) + \frac{1}{\lambda}(\boldsymbol{k} - \boldsymbol{c})\right] = 0, \tag{3.33}$$

and, instead of (3.20), $d^2\theta^* = 0$, which implies, similar to (3.23),

$$\frac{1}{\tilde{p}^*} = \tilde{\boldsymbol{m}}^*\cdot\tilde{\boldsymbol{T}}^*\tilde{\boldsymbol{m}}^*, \tag{3.34}$$

where

$$\tilde{\boldsymbol{T}}^* = \tilde{\lambda}\tilde{\boldsymbol{M}}^*\left(-\frac{1}{\lambda p}\boldsymbol{K} + \frac{1}{\lambda q}\boldsymbol{C} + \frac{1}{\tilde{\lambda}\tilde{q}}\tilde{\boldsymbol{C}}\right)\tilde{\boldsymbol{M}}^{*\mathrm{T}}. \tag{3.35}$$

Again, the stationary behavior (3.33) defines the direction $\tilde{\boldsymbol{k}}^*$ of the image and (3.34) enables us to find a set of values $\{\tilde{p}^*\}$ along this direction; $\tilde{\boldsymbol{m}}^*$ is here a unit vector perpendicular to $\tilde{\boldsymbol{k}}^*$ and $\tilde{\boldsymbol{M}}^*$ projects along $\hat{\boldsymbol{n}}$ onto the plane perpendicular to $\tilde{\boldsymbol{k}}^*$.

Let us now examine some particular cases of reconstruction. First, suppose that the reconstruction source $\tilde{\mathrm{Q}}$ is identical to the reference source Q, that is to say $\tilde{\boldsymbol{c}} = \boldsymbol{c}$, $\tilde{q} = q$, $\tilde{\lambda} = \lambda$. Then (3.17, 21) concerning the primary-image wave field show that $\tilde{\boldsymbol{k}} = \boldsymbol{k}$ and $\tilde{p} = p$, which is precisely the result we had in Sect. 3.1.2, i.e. $\tilde{\mathrm{P}} \equiv \mathrm{P}$. As for the conjugate wave field in this case, it forms an image described by (3.33, 34), which become

$$\hat{\boldsymbol{N}}\tilde{\boldsymbol{k}}^* = \hat{\boldsymbol{N}}(2\boldsymbol{c} - \boldsymbol{k}), \tag{3.36}$$

$$\frac{1}{\tilde{p}^*} = \tilde{\boldsymbol{m}}^* \cdot \tilde{\boldsymbol{M}}^*\left(\frac{2}{q}\boldsymbol{C} - \frac{1}{p}\boldsymbol{K}\right)\tilde{\boldsymbol{M}}^{*\mathrm{T}}\tilde{\boldsymbol{m}}^*. \tag{3.37}$$

So, according to the geometry of the optical arrangement, the image $\tilde{\mathrm{P}}^*$ may exist ($|\hat{\boldsymbol{N}}\tilde{\boldsymbol{k}}^*| < 1$) or not ($|\hat{\boldsymbol{N}}\tilde{\boldsymbol{k}}^*| > 1$) and, if it exists, be virtual ($\tilde{p}^* > 0$) or real ($\tilde{p}^* < 0$) and suffer more or less astigmatism. Figures 3.9, 10 represent two examples of such a reconstruction. If we now suppose that the reconstruction wave front is identical with the conjugate of the reference wave front, i.e. $\tilde{\boldsymbol{c}} =$

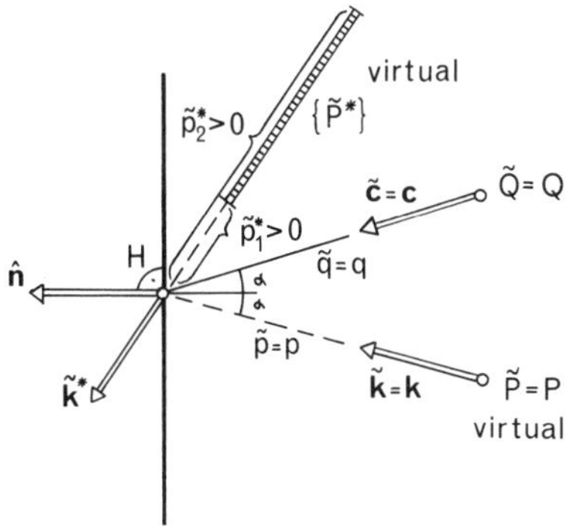

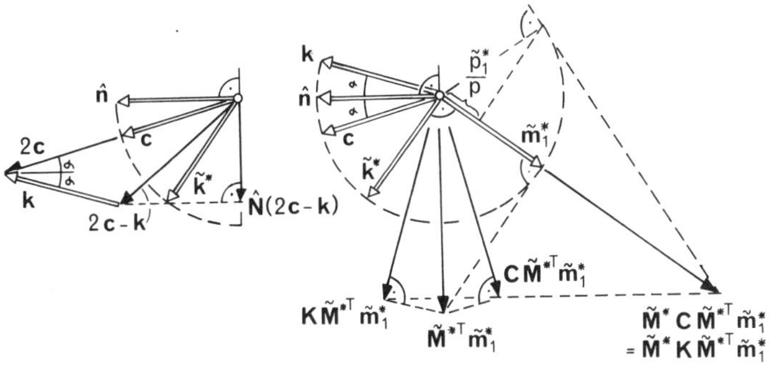

Fig. 3.9. Example of a reconstruction with $\tilde{\mathrm{Q}} = \mathrm{Q}$. All direction vectors lie in one plane and $q = p$. The lower drawings represent geometrically (3.36) and (3.37), respectively

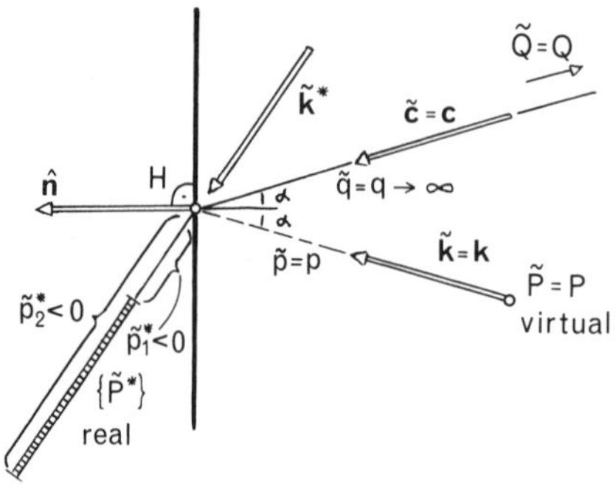

Fig. 3.10. Example of a reconstruction with $\tilde{\mathrm{Q}} = \mathrm{Q}$. Except that $q, \tilde{q} \to \infty$ (collimated reference and reconstruction beams), the arrangement is the same as in Fig. 3.9

$-\boldsymbol{c}$, $\tilde{q} = -q$, $\tilde{\lambda} = \lambda$, then (3.33, 34) yield the result of Sect. 3.1.2, $\tilde{\boldsymbol{k}}^* = -\boldsymbol{k}$, $\tilde{p}^* = -p$. The image $\tilde{\mathrm{P}}^*$ is thus real and is situated at exactly the same place as P. (3.17, 23), which describe the primary-image wave field, become on the other hand

$$\hat{\boldsymbol{N}}\tilde{\boldsymbol{k}} = \hat{\boldsymbol{N}}(\boldsymbol{k} - 2\boldsymbol{c}), \tag{3.38}$$

$$\frac{1}{\tilde{p}} = \tilde{\boldsymbol{m}} \cdot \tilde{\boldsymbol{M}}\left(\frac{1}{p}\boldsymbol{K} - \frac{2}{q}\boldsymbol{C}\right)\tilde{\boldsymbol{M}}^{\mathrm{T}}\tilde{\boldsymbol{m}}. \tag{3.39}$$

What we have just said about $\tilde{\mathrm{P}}^*$ in the preceding case is thus now valid for $\tilde{\mathrm{P}}$. The example in which $q, \tilde{q} \to \infty$ is presented in Fig. 3.11.

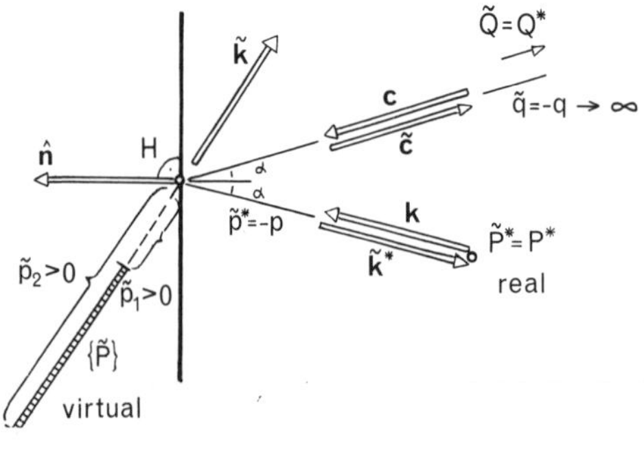

Fig. 3.11. Reconstruction with the hologram of Fig. 3.10 and with $\tilde{\mathrm{Q}} = \mathrm{Q}^*$

3.2.2 Small Modification of the Reference Source: Position of an Image Point

In this section, we shall look at two cases of reconstruction similar to the ideal ones we have described in Sect. 3.1.2. Firstly, if the position and the wavelength of the reconstruction source $\tilde{\mathrm{Q}}$ differs *little* from those of the reference source Q, the primary-image wave field forms a virtual image $\tilde{\mathrm{P}}$ very *near* to the object point source P. Second, if $\tilde{\mathrm{Q}}$ differs little from the conjugate of Q, the real image $\tilde{\mathrm{P}}^*$ formed by the conjugate-image wave field is again near to P. In fact, it is not necessary to treat these two cases separately because we can change, without risk of confusion, the signs of $\tilde{\boldsymbol{c}}$, $\tilde{\boldsymbol{k}}^*$, $\tilde{q}$ and $\tilde{p}^*$ for the second case so that the equations become those of the first case. Thus, both cases may be described by Fig. 3.7.

Similar to what we did in Sect. 2.2 in connection with the *small* deformation of a body, we shall express here the optical phenomena relative to a *reference configuration*, which we choose to be the recording arrangement. The changes are then characterized by (see Fig. 3.7) the *displacement vectors* $\boldsymbol{d}$ from Q to $\tilde{\mathrm{Q}}$, and $\boldsymbol{v}$ from P to $\tilde{\mathrm{P}}$, or more precisely to the set $\{\tilde{\mathrm{P}}\}$. Further, the assumptions $|\boldsymbol{d}| = d \ll q$ and $|\boldsymbol{v}| = v \ll p$ allow us to obtain *linear* relations [3.64]. In addition, we take into account that $|\lambda - \tilde{\lambda}| = |\Delta\lambda| \ll \lambda$ and, in order to relate the magnitude of the two small quantities, we presume that

$$\frac{\Delta\lambda}{\lambda} = O\left(\frac{d}{q}\right)\left[= O\left(\frac{v}{p}\right)\right]. \tag{3.40}$$

As previously mentioned, small shifts are of interest for holographic interferometry: the reader may anticipate that the "optical displacement" will be combined later with the displacement due to the mechanical deformation.

Let us then calculate all quantities as functions of the reference quantities and their small changes. First, we have the distances

$$\begin{aligned} \tilde{q} &= |q\boldsymbol{c} - \boldsymbol{d}| = \sqrt{q^2 - 2q\boldsymbol{c}\cdot\boldsymbol{d} + d^2} \simeq q\left(1 - \frac{1}{q}\boldsymbol{c}\cdot\boldsymbol{d}\right) = q - \boldsymbol{c}\cdot\boldsymbol{d}, \\ \tilde{p} &= |p\boldsymbol{k} - \boldsymbol{v}| \simeq p - \boldsymbol{k}\cdot\boldsymbol{v}. \end{aligned} \tag{3.41}$$

The unit vectors become, thereupon,

$$\tilde{\boldsymbol{c}} = \frac{q\boldsymbol{c} - \boldsymbol{d}}{|q\boldsymbol{c} - \boldsymbol{d}|} \simeq \frac{q\boldsymbol{c} - \boldsymbol{d}}{q\left(1 - \frac{1}{q}\boldsymbol{c}\cdot\boldsymbol{d}\right)} \simeq \boldsymbol{c} - \frac{1}{q}[\boldsymbol{d} - \boldsymbol{c}(\boldsymbol{c}\cdot\boldsymbol{d})],$$

or, by means of the normal projections $\boldsymbol{C} = \boldsymbol{I} - \boldsymbol{c} \otimes \boldsymbol{c}$, and also $\boldsymbol{K} = \boldsymbol{I} - \boldsymbol{k} \otimes \boldsymbol{k}$ [see (2.36)]

$$\tilde{\boldsymbol{c}} \simeq \boldsymbol{c} - \frac{1}{q}\boldsymbol{C}\boldsymbol{d}, \qquad \tilde{\boldsymbol{k}} \simeq \boldsymbol{k} - \frac{1}{p}\boldsymbol{K}\boldsymbol{v}. \tag{3.42}$$

With these results, we may also modify the normal projections

$$\begin{aligned} \frac{1}{\tilde{q}}\tilde{\boldsymbol{C}} &\simeq \frac{1}{q\left(1 - \frac{1}{q}\boldsymbol{c}\cdot\boldsymbol{d}\right)}\left[\boldsymbol{I} - \left(\boldsymbol{c} - \frac{1}{q}\boldsymbol{C}\boldsymbol{d}\right) \otimes \left(\boldsymbol{c} - \frac{1}{q}\boldsymbol{C}\boldsymbol{d}\right)\right] \\ &\simeq \frac{1}{q}\boldsymbol{C} + \frac{1}{q^2}[\boldsymbol{c} \otimes \boldsymbol{C}\boldsymbol{d} + \boldsymbol{C}(\boldsymbol{c}\cdot\boldsymbol{d}) + \boldsymbol{C}\boldsymbol{d} \otimes \boldsymbol{c}], \end{aligned}$$

or, more briefly, with the superprojection $\mathscr{C} = \boldsymbol{c} \otimes \boldsymbol{C} + \boldsymbol{C} \otimes \boldsymbol{c} + \boldsymbol{C} \otimes \boldsymbol{c})^{\mathrm{T}}$ and also $\mathscr{K} = \boldsymbol{k} \otimes \boldsymbol{K} + \boldsymbol{K} \otimes \boldsymbol{k} + \boldsymbol{K} \otimes \boldsymbol{k})^{\mathrm{T}}$ [see (2.37)]

$$\frac{1}{\tilde{q}}\tilde{\boldsymbol{C}} \simeq \frac{1}{q}\boldsymbol{C} + \frac{1}{q^2}\mathscr{C}\boldsymbol{d}, \qquad \frac{1}{\tilde{p}}\tilde{\boldsymbol{K}} \simeq \frac{1}{p}\boldsymbol{K} + \frac{1}{p^2}\mathscr{K}\boldsymbol{v}. \tag{3.43}$$

The approximations we have just written are actually series developments in function of the variable positions of Q or P. Instead of an increment $d\boldsymbol{r}$ we have here the small vectors $\boldsymbol{d}$ or $\boldsymbol{v}$. This explains the likeness of these results to those of Sect. 2.1.2.

So then, let us rewrite the principal relations of the preceding section. First, we get with $\lambda - \tilde{\lambda} = \Delta\lambda$ for the function θ, defined by (3.13) at a point H of the hologram,

$$\theta(\hat{\boldsymbol{r}}) \simeq \frac{2\pi}{\lambda}\left[\frac{\Delta\lambda}{\lambda}(p - q) - \boldsymbol{k}\cdot\boldsymbol{v} + \boldsymbol{c}\cdot\boldsymbol{d}\right]. \tag{3.44}$$

Second, the *stationarity condition* $d\theta = 0$, i.e. (3.17), becomes

$$\hat{\boldsymbol{N}}\left[\frac{\Delta\lambda}{\lambda}(\boldsymbol{k} - \boldsymbol{c}) - \frac{1}{p}\boldsymbol{K}\boldsymbol{v} + \frac{1}{q}\boldsymbol{C}\boldsymbol{d}\right] = 0. \tag{3.45}$$

By means of the oblique projection

$$\hat{\boldsymbol{M}} = \boldsymbol{I} - \frac{1}{\hat{\boldsymbol{n}}\cdot\boldsymbol{k}}\hat{\boldsymbol{n}} \otimes \boldsymbol{k},$$

which projects along $\hat{\boldsymbol{n}}$ onto the plane normal to the reference direction $\boldsymbol{k}$ (note that $\hat{\boldsymbol{M}} \neq \tilde{\boldsymbol{M}}$) and the relations (2.71) $\hat{\boldsymbol{M}}\hat{\boldsymbol{N}} = \hat{\boldsymbol{M}}$, $\hat{\boldsymbol{M}}\boldsymbol{K} = \boldsymbol{K}$, we may isolate the unknown part

$$\boldsymbol{K}\boldsymbol{v} = \frac{p\Delta\lambda}{\lambda}\hat{\boldsymbol{M}}(\boldsymbol{k} - \boldsymbol{c}) + \frac{p}{q}\hat{\boldsymbol{M}}\boldsymbol{C}\boldsymbol{d}. \tag{3.46}$$

Thus, the first-order term in the series development of θ around H leads to the *lateral part* $\boldsymbol{K}\boldsymbol{v}$ *of the displacement* of P, which is another way of determining the *direction* of the image $\tilde{\mathrm{P}}$ than that used in Sect. 3.2.1. Finally, let us give the equivalent of (3.21) stating $d^2\theta = 0$

$$d\hat{\boldsymbol{r}}\cdot\hat{\boldsymbol{N}}\left[\frac{\Delta\lambda}{\lambda}\left(\frac{1}{p}\boldsymbol{K} - \frac{1}{q}\boldsymbol{C}\right) + \frac{1}{p^2}\mathscr{K}\boldsymbol{v} - \frac{1}{q^2}\mathscr{C}\boldsymbol{d}\right]\hat{\boldsymbol{N}}d\hat{\boldsymbol{r}} = 0. \tag{3.47}$$

In this expression, the tensor $\mathscr{K}\boldsymbol{v} = \boldsymbol{k} \otimes \boldsymbol{K}\boldsymbol{v} + \boldsymbol{K}(\boldsymbol{k}\cdot\boldsymbol{v}) + \boldsymbol{K}\boldsymbol{v} \otimes \boldsymbol{k}$ contains both the known lateral part $\boldsymbol{Kv}$ and the unknown longitudinal part $\boldsymbol{k}\cdot\boldsymbol{v}$ of the displacement of P. In order to isolate the unknown scalar, we use again a connection between surfaces similar to (2.74), now in the form

$$d\hat{\boldsymbol{r}} = p\hat{\boldsymbol{M}}^{\mathrm{T}}d\boldsymbol{k} = pd\alpha\hat{\boldsymbol{M}}^{\mathrm{T}}\boldsymbol{m}. \tag{3.48}$$

This replaces (3.22) and relates the increment $d\hat{\boldsymbol{r}}$ on the hologram surface to an increment $d\boldsymbol{k} = \boldsymbol{m}d\alpha$ that is perpendicular to $\boldsymbol{k}$. In place of (3.23, 24), we have therefore

$$\boldsymbol{k}\cdot\boldsymbol{v} = \boldsymbol{m}\cdot\boldsymbol{Tm}, \tag{3.49}$$

where

$$\boldsymbol{T} = \hat{\boldsymbol{M}}\left[\frac{-p^2\Delta\lambda}{\lambda}\left(\frac{1}{p}\boldsymbol{K} - \frac{1}{q}\boldsymbol{C}\right) + \frac{p^2}{q^2}\mathscr{C}\boldsymbol{d} - (\boldsymbol{k}\otimes\boldsymbol{Kv} + \boldsymbol{Kv}\otimes\boldsymbol{k})\right]\hat{\boldsymbol{M}}^{\mathrm{T}}. \tag{3.50}$$

As previously, the second-order term determines the distance from the hologram to the image $\tilde{\mathrm{P}}$, now by means of the *longitudinal part* $\boldsymbol{k}\cdot\boldsymbol{v}$ *of the displacement* $\boldsymbol{v}$ (Fig. 3.7). Again, astigmatism is present: indeed, for each direction $\boldsymbol{m}$, or for each corresponding $d\hat{\boldsymbol{r}}$, there is one $\boldsymbol{k}\cdot\boldsymbol{v}$. The extreme values of the interval are given by the following characteristic eigenvalue equation, which is similar to (3.25),

$$|\boldsymbol{T} - (\boldsymbol{k}\cdot\boldsymbol{v})\boldsymbol{K}| = 0. \tag{3.51}$$

Instead of linearizing each relation of the preceding section, we could also have derived directly the linearized expression of θ (3.44). Of course, we would have found the same results. The discussion of the second- and higher-order terms could be continued, with results that would be similar to those obtained in Sect. 3.2.1. But because we shall use merely the linearized first-order terms, such an extended discussion may be dispensed with.

3.2.3 Modification of the Reference Source: Analysis in Terms of Transverse Ray Aberration

As we saw in Sect. 3.2.1, all of the light rays that come from the vicinity of a point H on the hologram do not converge to one image point. To describe this phenomenon, the geometrical theory of aberration of optical systems makes use, on the one hand, of the concept of wave aberration of which we have already spoken, and on the other hand of the concept of *transverse ray aberration* [Ref. 3.5, p. 190; 3.67; Ref. 3.68, p. 110]: the latter is defined as the vector between a reference image, which a ray should ideally go to, and the intersection of the ray with a reference plane. The reference plane contains the reference image and is perpendicular to the axis of the system. In designing optical instruments constituted of lenses, the reference image is of course the Gaussian image and the rays are calculated, i.e. traced, by successive application of the laws of refraction or reflection. In holography, similar considerations may be used (see e.g. [3.39, 59, 60, 71, 73–78]), as we shall see in the present section.

First we must define how the image-forming rays have to be traced. Recalling Sect. 3.2.1, we ascribe to *each* point $\overline{\mathrm{H}}$ of the hologram a light ray whose phase at $\overline{\mathrm{H}}$ is again given by the *condition of interference identity* (3.12), which in this case is

$$\frac{2\pi}{\tilde{\lambda}}\tilde{\bar{\bar{p}}} + \tilde{\bar{\bar{\phi}}} = \frac{2\pi}{\lambda}\bar{p} + \phi - \frac{2\pi}{\lambda}\bar{q} - \psi + \frac{2\pi}{\tilde{\lambda}}\tilde{\bar{\bar{q}}} + \tilde{\psi}. \tag{3.52}$$

Similar to (2.13), we define the function

$$\bar{\bar{\theta}} = \frac{2\pi}{\tilde{\lambda}}(\tilde{\bar{\bar{p}}} - \tilde{\bar{\bar{q}}}) - \frac{2\pi}{\lambda}(\bar{p} - \bar{q}), \tag{3.53}$$

not to be confused with $\bar{\theta}$, because it is not related to the point $\tilde{\mathrm{P}}$ but rather to some point $\tilde{\bar{\bar{\mathrm{P}}}}$. Then, the direction of the ray through $\overline{\mathrm{H}}$ is determined by the *stationary behavior* of $\bar{\bar{\theta}}$ there, i.e. by

$$\nabla_{\hat{\mathrm{n}}}\bar{\bar{\theta}} = 0. \tag{3.54}$$

Similar to (3.17) this equation becomes explicitly the system

$$\hat{\boldsymbol{N}}\left[\frac{1}{\tilde{\lambda}}(\tilde{\bar{\bar{\boldsymbol{k}}}} - \tilde{\bar{\bar{\boldsymbol{c}}}}) - \frac{1}{\lambda}(\bar{\boldsymbol{k}} - \bar{\boldsymbol{c}})\right] = 0, \qquad |\tilde{\bar{\bar{\boldsymbol{k}}}}| = 1. \tag{3.55}$$

The relations (3.52, 55) are called the *ray-tracing equations*. Let us recall that $\bar{p}$, $\bar{q}$, $\tilde{\bar{\bar{q}}}$ are the distances between the known fixed point sources P, Q, $\tilde{\mathrm{Q}}$, respectively, and the variable point $\overline{\mathrm{H}}$ on the hologram (Figs. 3.12, 7); $\bar{\boldsymbol{k}}$, $\bar{\boldsymbol{c}}$, $\tilde{\bar{\bar{\boldsymbol{c}}}}$ are the corresponding unit direction vectors, and ϕ, ψ, $\tilde{\psi}$ represent the phases of the sources. The quantity $\tilde{\bar{\bar{\phi}}}$ may also be interpreted as the phase of an image point $\tilde{\bar{\bar{\mathrm{P}}}}$ in the direction $\tilde{\bar{\bar{\boldsymbol{k}}}}$, which can be situated at an arbitrary distance $\tilde{\bar{\bar{p}}}$ from $\overline{\mathrm{H}}$ and which is supposed to produce a unique light ray through $\overline{\mathrm{H}}$.

Further, we choose as a *reference axis* the ray that goes through one definite point H and propagates in a direction $\tilde{\boldsymbol{k}}$. This ray may, for example, be the axis of the optical system used to observe the image. The transverse ray aberration $\boldsymbol{\Delta y}$ at an arbitrary point Y on the reference ray is then defined as the vector perpendicular to $\tilde{\boldsymbol{k}}$ that links Y to the ray that goes through one point $\overline{\mathrm{H}}$ (Fig. 3.12). $\boldsymbol{\Delta y}$ depends thus on the distance y from H to Y and also on the increment $d\hat{\boldsymbol{r}} = \Delta\hat{\boldsymbol{r}}$ for the position of $\overline{\mathrm{H}}$ relative to H. Fig. 3.12 shows that

$$\boldsymbol{\Delta y} = y\tilde{\boldsymbol{k}} + \Delta\hat{\boldsymbol{r}} - \bar{y}\tilde{\bar{\bar{\boldsymbol{k}}}},$$

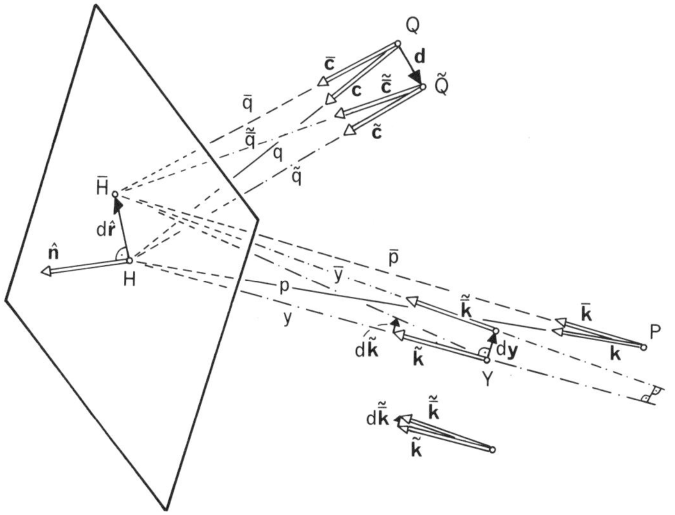

Fig. 3.12. Transverse ray aberration $d\boldsymbol{y}$ between image-forming rays through H and $\overline{\mathrm{H}}$. The hologram containing the information about the object point source P has been recorded with the reference source Q and is illuminated by the reconstruction source $\tilde{\mathrm{Q}}$

where the distance $\bar{y}$ is determined by the condition that $\Delta\boldsymbol{y}$ is perpendicular to $\tilde{\boldsymbol{k}}$. Thus, let us first project onto the plane normal to the latter direction

$$\Delta\boldsymbol{y} = \tilde{\boldsymbol{K}}(\Delta\hat{\boldsymbol{r}} - \bar{y}\tilde{\tilde{\boldsymbol{k}}}),$$

and, second, form the component along $\tilde{\boldsymbol{k}}$

$$0 = y + \Delta\hat{\boldsymbol{r}}\cdot\tilde{\boldsymbol{k}} - \bar{y}\tilde{\tilde{\boldsymbol{k}}}\cdot\tilde{\boldsymbol{k}}.$$

Eliminating the quantity $\bar{y}$ from these two equations, we find a first result for the transverse ray aberration

$$\Delta\boldsymbol{y} = \tilde{\boldsymbol{K}}\left(\Delta\hat{\boldsymbol{r}} - \frac{y + \Delta\hat{\boldsymbol{r}}\cdot\tilde{\boldsymbol{k}}}{\tilde{\tilde{\boldsymbol{k}}}\cdot\tilde{\boldsymbol{k}}}\tilde{\tilde{\boldsymbol{k}}}\right). \tag{3.56}$$

This is the relation used by *Miles* [Ref. 3.60, Eqs. (21, 22)].

Moreover, we may also deduce a linearized form, whence appear interesting features. If $|\Delta\hat{\boldsymbol{r}}|$ is *small* compared with the distances p, q, $\tilde{q}$ the unit direction vectors related to $\overline{\mathrm{H}}$ may be approximated by their values relative to H and their first differentials, so for example [see again (2.36)]

$$\bar{\boldsymbol{k}} \simeq \boldsymbol{k} + d\boldsymbol{k} = \boldsymbol{k} + d\hat{\boldsymbol{r}}(\nabla_{\hat{\mathrm{n}}} \otimes \boldsymbol{k}) = \boldsymbol{k} + \frac{1}{p}d\hat{\boldsymbol{r}}\hat{\boldsymbol{N}}\boldsymbol{K},$$

$$\tilde{\tilde{\boldsymbol{k}}} \simeq \tilde{\boldsymbol{k}} + d\tilde{\tilde{\boldsymbol{k}}},$$

where the unknown differential $d\tilde{\tilde{k}}$ is perpendicular to $\tilde{k}$ because both $\tilde{k}$ and $\tilde{\tilde{k}}$ are unit vectors. The condition of stationarity (3.55) becomes thus

$$\hat{N}\left[\frac{1}{\tilde{\lambda}}\left(\tilde{k} + d\tilde{\tilde{k}} - \tilde{c} - \frac{1}{\tilde{q}} d\hat{r}\hat{N}\tilde{C}\right) - \frac{1}{\lambda}\left(k + \frac{1}{p} d\hat{r}\hat{N}K - c - \frac{1}{q} d\hat{r}\hat{N}C\right)\right] = 0,$$

or, because (3.55) is also valid on the reference axis,

$$\hat{N} d\tilde{\tilde{k}} = \tilde{\lambda} d\hat{r}\hat{N}\left(\frac{1}{\lambda p}K - \frac{1}{\lambda q}C + \frac{1}{\tilde{\lambda}\tilde{q}}\tilde{C}\right)\hat{N}.$$

To isolate the unknown $d\tilde{\tilde{k}}$, we again utilize the connection between the hologram plane and a unit sphere, this time with center in Y [see (2.74)]

$$d\hat{r} = y\tilde{M}^{\mathrm{T}} d\tilde{k} = y d\tilde{k}\tilde{M}, \tag{3.57}$$

with the oblique projection $\tilde{M} = I - \hat{n} \otimes \tilde{k}/\hat{n}\cdot\tilde{k}$ (see Fig. 3.8). Note that $d\tilde{k}$, which is different from $d\tilde{\tilde{k}}$, is the increment to be added to $\tilde{k}$ in order to get the unit vector along $\mathrm{Y}\overline{\mathrm{H}}$ (Fig. 3.12). With this relation and with an additional projection $\tilde{M}^{\mathrm{T}}$ from the right, we obtain

$$d\tilde{\tilde{k}} = \tilde{\lambda} y d\tilde{k}\tilde{M}\left(\frac{1}{\lambda p}K - \frac{1}{\lambda q}C + \frac{1}{\tilde{\lambda}\tilde{q}}\tilde{C}\right)\tilde{M}^{\mathrm{T}},$$

which, with the definition (3.24), may also be written briefly

$$d\tilde{\tilde{k}} = y d\tilde{k}\tilde{T}. \tag{3.58}$$

If in (3.56) we replace $\tilde{\tilde{k}}$ by $\tilde{k} + d\tilde{\tilde{k}}$ and $\Delta\hat{r}$ by $y d\tilde{k}\tilde{M}$, we get

$$dy = \tilde{K}[y d\tilde{k}\tilde{M} - (y + y d\tilde{k}\tilde{M}\cdot\tilde{k})(\tilde{k} + y d\tilde{k}\tilde{T})].$$

Observing that $\tilde{K}(d\tilde{k}\tilde{M}) = d\tilde{k}\tilde{M}\tilde{K} = d\tilde{k}\tilde{K}$, $\tilde{K}\tilde{k} = 0$ and $\tilde{K}(d\tilde{k}\tilde{T}) = d\tilde{k}\tilde{T}\tilde{K} = d\tilde{k}\tilde{T}$, we arrive at the *linearized* form

$$dy = y d\tilde{k}(\tilde{K} - y\tilde{T}). \tag{3.59}$$

At first, we remark that the tensor $\tilde{T}$, which is symmetric and situated in the plane normal to $\tilde{k}$ (see also [Ref. 3.71, Eqs. (5.5, A.14)]), already appeared in (3.23) which we deduced from the second-order terms in the development of the function $\theta(\hat{r})$ around H. To obtain (3.59), we have indeed approximated the first derivative (3.55) in $\overline{\mathrm{H}}$ by means of the first and second derivative in H. Both relations (3.23, 59) describe, in different ways, the same fact, i.e. the astigmatism of the reconstructed image. We saw in Sect. 3.1.1 that (3.23) leads to an interval for a set of images $\{\tilde{\mathrm{P}}\}$. As for (3.59), it shows the following. Let us suppose a *circular aperture* located at H in the plane normal to the reference axis $\tilde{\boldsymbol{k}}$, and determine the whole set of $\boldsymbol{dy}$ related to the rays that go through the edge of this diaphragm, i.e. with $y|d\tilde{\boldsymbol{k}}|$ constant. The result is generally an *ellipse* in the reference plane, whose equation is found by eliminating $d\tilde{\boldsymbol{k}}/|d\tilde{\boldsymbol{k}}|$

$$\frac{1}{y^2|d\tilde{\boldsymbol{k}}|^2}\boldsymbol{dy}\cdot(\tilde{\boldsymbol{K}} - y\tilde{\boldsymbol{T}})^{-2}\boldsymbol{dy} = 1. \tag{3.60}$$

The center of this ellipse lies in Y; its axes are parallel to the eigenvectors of $\tilde{\boldsymbol{T}}$. By varying y, we obtain the set of ellipses represented in Fig. 3.13, i.e. the *astigmatic pencil of rays* that go through the chosen small aperture. This result is identical to that obtained in the case of optical imaging by lenses with small extra-axial circular apertures (see particularly [Ref. 3.5, Fig. 4.20; Ref. 3.68, Fig. 114]).

Let us yet look at further properties of this beam. First, if Y lies in one end of the interval of the images $\{\tilde{\mathrm{P}}\}$, we have $y = \tilde{p}_1$, or $= \tilde{p}_2$ and, because these are the reciprocals of the eigenvalues of $\tilde{\boldsymbol{T}}$, one axis of the ellipse has zero length; the ellipse turns thus into a line segment, which is called a *focal line*.

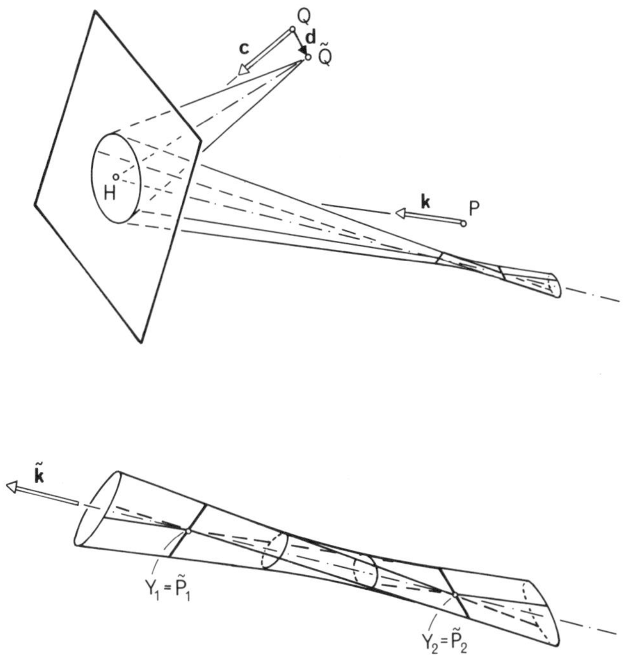

Fig. 3.13. Astigmatic pencil of rays calculated by the ray-tracing method

Second, let us examine one of the rays that corresponds to some given $\overline{\mathrm{H}}$, i.e. to some given $y\, d\tilde{\boldsymbol{k}}$. The related transverse aberration is minimal (shortest distance between the rays through H and $\overline{\mathrm{H}}$) at the point Y where $d\boldsymbol{y}$ is perpendicular not only to $\tilde{\boldsymbol{k}}$ but also to $\tilde{\bar{\boldsymbol{k}}}$, and thus to $d\tilde{\boldsymbol{k}}$. The condition to be fulfilled is therefore with (3.58, 59)

$$0 = d\boldsymbol{y}\cdot d\tilde{\bar{\boldsymbol{k}}} = y^2 d\tilde{\boldsymbol{k}}\cdot(\tilde{\boldsymbol{T}} - y\tilde{\boldsymbol{T}}^2)d\tilde{\boldsymbol{k}}. \tag{3.61}$$

Now, if the aberrations are small, i.e. if $|\tilde{p}_1 - \tilde{p}_2| \ll p$, then $\tilde{\boldsymbol{T}}$ differs little from $\tilde{\boldsymbol{K}}/\tilde{p}$, where $\tilde{p}$ is chosen to be the value given by (3.23), i.e. by

$$\frac{1}{\tilde{p}}|d\tilde{\boldsymbol{k}}|^2 = d\tilde{\boldsymbol{k}}\cdot\tilde{\boldsymbol{T}}d\tilde{\boldsymbol{k}}.$$

Because this value lies in the interval $\langle\tilde{p}_1, \tilde{p}_2\rangle$, we may write the approximation

$$\tilde{\boldsymbol{T}} \simeq \frac{1}{\tilde{p}}\tilde{\boldsymbol{K}} + \frac{\varepsilon}{\tilde{p}}\tilde{\boldsymbol{U}},$$

with a small parameter $\varepsilon = |(\tilde{p}_1 - \tilde{p}_2)/(\tilde{p}_1 + \tilde{p}_2)| \ll 1$. Equation (3.61) becomes then

$$d\tilde{\boldsymbol{k}}\cdot\left[\left(1 - \frac{y}{\tilde{p}}\right)\tilde{\boldsymbol{K}} + \left(1 - \frac{2y}{\tilde{p}}\right)\varepsilon\tilde{\boldsymbol{U}}\right]d\tilde{\boldsymbol{k}} = 0, \tag{3.62}$$

whereas (3.23) changes to

$$\varepsilon d\tilde{\boldsymbol{k}}\cdot\tilde{\boldsymbol{U}}d\tilde{\boldsymbol{k}} = 0.$$

We thus see that, if $y = \tilde{p}$, the condition (3.62) is fulfilled because of the latter equation. In other words, the transverse ray aberration between the rays through H and $\overline{\mathrm{H}}$ is *minimal* in the point Y which is precisely the image $\tilde{\mathrm{P}}$ produced through a slit along $\mathrm{H}\overline{\mathrm{H}}$, as described by (3.23).

Third, if Y is chosen in the *middle of the interval* of the images $\{\tilde{\mathrm{P}}\}$, the length of the transverse ray aberration is, with the same approximation as previously,

$$|d\boldsymbol{y}|^2 = y^2 d\tilde{\boldsymbol{k}}\cdot(\tilde{\boldsymbol{K}} - y\tilde{\boldsymbol{T}})^2 d\tilde{\boldsymbol{k}} \simeq y^2\varepsilon^2 d\tilde{\boldsymbol{k}}\cdot\tilde{\boldsymbol{U}}^2 d\tilde{\boldsymbol{k}} = y^2\varepsilon^2|d\tilde{\boldsymbol{k}}|^2,$$

because, if y is equal to $(\tilde{p}_1 + \tilde{p}_2)/2$, $\tilde{\boldsymbol{U}}^2 = \tilde{\boldsymbol{K}}$. We may assert that $|d\boldsymbol{y}|$ is then independent of the direction of $d\tilde{\boldsymbol{k}}$, and thus the ellipse becomes a *circle* with a radius of order ε.

In this section, we have spoken of the transverse ray aberration and we have seen how it is related to the wave-front matching described in Sect. 3.2.1. Note here that, besides these two methods of the geometrical theory of optical imaging, we could also use the diffraction theory of aberrations (see e.g. [Ref. 3.5, p.

459; 3.79; Ref. 3.80, p. 317]). We might then calculate isophotes in some reference plane. The obtained curves or photographs would then be analogous to those presented in the mentioned references (see particularly [Ref. 3.79, Fig. 45–48, 50]).

3.2.4 Movement of the Hologram: Position of an Image Point

Up to now, we have investigated the image produced when a hologram is illuminated by a reconstruction source that differs from the reference source, the position of the hologram itself being not altered. If now we assume, on the contrary, that the positions of the sources remain unchanged but that the hologram is moved between the recording and the reconstruction, the same optical phenomena occur. We could describe them by the foregoing results if we took into account the relative motion of the reconstruction source with respect to the hologram. However, because later, in interferometry, we shall have to consider what happens relative to a reference configuration (e.g. the position of the undeformed object or that of a fixed hologram, the other being movable) it is worth while mentioning again briefly the principal considerations concerning the primary-image wave field. Also, this permits further investigations of the effects of a hologram deformation. As we are particularly interested in the position of the image, we use the method of wave-front matching, in the cases of both large and small movements [3.81, 82].

As previously, we denote by q the distance between the reference source Q and a variable point H on the hologram in the recording position (vector $\hat{\boldsymbol{r}}$, see Fig. 3.14) and by p the distance between the object-point source P and again H. For the reconstruction, the hologram point H has moved to the point $\tilde{\mathrm{H}}$ (position vector $\tilde{\boldsymbol{r}}$); the displacement vector between these points will be denoted

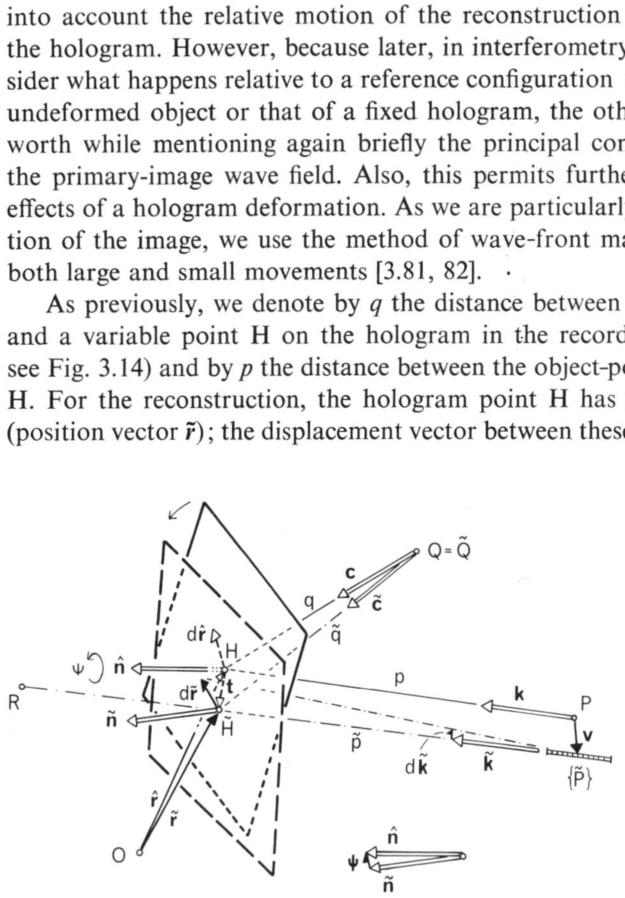

Fig. 3.14. Position of the set $\{\tilde{\mathrm{P}}\}$ of the images of the object-point source P when the hologram is shifted, the point H being displaced to $\tilde{\mathrm{H}}$ (displacement $\boldsymbol{t}$), and rotated (pivot rotation Ψ and inclination ψ). R: location of the observer. $\hat{\boldsymbol{n}}$, $\tilde{\boldsymbol{n}}$, $\boldsymbol{k}$, $\tilde{\boldsymbol{k}}$, $\boldsymbol{c}$, $\tilde{\boldsymbol{c}}$: unit direction vectors

$\boldsymbol{t}$. Further, $\tilde{q}$ is now the distance from Q ($=\tilde{\mathrm{Q}}$) to $\tilde{\mathrm{H}}$ and $\tilde{p}$, that from the expected image $\tilde{\mathrm{P}}$ to $\tilde{\mathrm{H}}$. The *condition of interference identity* concerns now the related points H and $\tilde{\mathrm{H}}$; it reads exactly like (3.12). Similar to (3.13), we define the phase difference

$$\theta(\hat{\boldsymbol{r}}, \tilde{\boldsymbol{r}}) = \frac{2\pi}{\tilde{\lambda}}(\tilde{p} - \tilde{q}) - \frac{2\pi}{\lambda}(p - q), \tag{3.63}$$

which is at this moment a function of the *two* variables $\hat{\boldsymbol{r}}$ and $\tilde{\boldsymbol{r}}$. Again, this function must be *stationary*, now around H *and* $\tilde{\mathrm{H}}$, which lie in different planes. We have therefore

$$d\theta = \frac{2\pi}{\tilde{\lambda}} d\tilde{\boldsymbol{r}} \cdot \boldsymbol{\nabla}_{\tilde{\mathrm{n}}}(\tilde{p} - \tilde{q}) - \frac{2\pi}{\lambda} d\hat{\boldsymbol{r}} \cdot \boldsymbol{\nabla}_{\hat{\mathrm{n}}}(p - q) = 0. \tag{3.64}$$

In addition to the already encountered two-dimensional derivative operator $\boldsymbol{\nabla}_{\hat{\mathrm{n}}} = \hat{\boldsymbol{N}}\boldsymbol{\nabla}$ in the hologram plane during the recording, the operator $\boldsymbol{\nabla}_{\tilde{\mathrm{n}}} = \tilde{\boldsymbol{N}}\boldsymbol{\nabla}$ appears here; it concerns the hologram plane during the reconstruction, whose unit normal is $\tilde{\boldsymbol{n}}$; $d\tilde{\boldsymbol{r}}$ and $d\hat{\boldsymbol{r}}$ are corresponding increments in these planes; they are connected with a linear transformation similar to (2.98)

$$d\tilde{\boldsymbol{r}} = \tilde{\boldsymbol{N}}\hat{\boldsymbol{F}}\hat{\boldsymbol{N}} d\hat{\boldsymbol{r}}, \tag{3.65}$$

where $\hat{\boldsymbol{F}}$ represents the deformation gradient of the hologram. Because we assume here that the hologram remains a rigid body, we have instead of $\hat{\boldsymbol{F}}$ an *orthogonal rotation tensor* $\boldsymbol{R}$, so that

$$d\tilde{\boldsymbol{r}} = \tilde{\boldsymbol{N}}\boldsymbol{R}\hat{\boldsymbol{N}} d\hat{\boldsymbol{r}}, \qquad d\hat{\boldsymbol{r}} = \hat{\boldsymbol{N}}\boldsymbol{R}^{\mathrm{T}}\tilde{\boldsymbol{N}} d\tilde{\boldsymbol{r}}. \tag{3.66}$$

On the other hand, the derivatives of the distances still consist of unit direction vectors [see (2.35)] so that we conclude from (3.64) and the fact that either $d\hat{\boldsymbol{r}}$ or $d\tilde{\boldsymbol{r}}$ is arbitrary

$$\hat{\boldsymbol{N}}\left[\frac{1}{\tilde{\lambda}}\boldsymbol{R}^{\mathrm{T}}\tilde{\boldsymbol{N}}(\tilde{\boldsymbol{k}} - \tilde{\boldsymbol{c}}) - \frac{1}{\lambda}(\boldsymbol{k} - \boldsymbol{c})\right] = 0,$$

$$\tilde{\boldsymbol{N}}\left[\frac{1}{\tilde{\lambda}}(\tilde{\boldsymbol{k}} - \tilde{\boldsymbol{c}}) - \frac{1}{\lambda}\boldsymbol{R}\hat{\boldsymbol{N}}(\boldsymbol{k} - \boldsymbol{c})\right] = 0.$$

Let us note that, the hologram being supposed rigid, the rotation of its normal is also described by $\boldsymbol{R}$: $\tilde{\boldsymbol{n}} = \boldsymbol{R}\hat{\boldsymbol{n}}$. We can thus write, with $\boldsymbol{R}\boldsymbol{R}^{\mathrm{T}} = \boldsymbol{I}$,

$$\tilde{\boldsymbol{N}} = \boldsymbol{I} - \boldsymbol{R}\hat{\boldsymbol{n}} \otimes \hat{\boldsymbol{n}}\boldsymbol{R}^{\mathrm{T}} = \boldsymbol{R}\hat{\boldsymbol{N}}\boldsymbol{R}^{\mathrm{T}},$$

and further

$$\boldsymbol{R}^{\mathrm{T}}\tilde{\boldsymbol{N}} = \hat{\boldsymbol{N}}\boldsymbol{R}^{\mathrm{T}}, \qquad \tilde{\boldsymbol{N}}\boldsymbol{R} = \boldsymbol{R}\hat{\boldsymbol{N}}. \tag{3.67}$$

The foregoing relations may thus be simplified as follows (this would not be possible if $\hat{\boldsymbol{F}}$ replaced $\boldsymbol{R}$!)

$$\hat{\boldsymbol{N}}\left[\frac{1}{\tilde{\lambda}}\boldsymbol{R}^{\mathrm{T}}(\tilde{\boldsymbol{k}} - \tilde{\boldsymbol{c}}) - \frac{1}{\lambda}(\boldsymbol{k} - \boldsymbol{c})\right] = 0,$$
$$\tilde{\boldsymbol{N}}\left[\frac{1}{\tilde{\lambda}}(\tilde{\boldsymbol{k}} - \tilde{\boldsymbol{c}}) - \frac{1}{\lambda}\boldsymbol{R}(\boldsymbol{k} - \boldsymbol{c})\right] = 0. \tag{3.68}$$

Either of these equations, in addition to the condition $|\tilde{\boldsymbol{k}}| = 1$, enables us to determine the direction of the image $\tilde{\mathrm{P}}$.

Next, we get for the *second-order* terms

$$d^2\theta = \frac{2\pi}{\tilde{\lambda}}d\tilde{\boldsymbol{r}}\cdot[\boldsymbol{\nabla}_{\tilde{\mathrm{n}}} \otimes \boldsymbol{\nabla}_{\tilde{\mathrm{n}}}(\tilde{p} - \tilde{q})]d\tilde{\boldsymbol{r}} - \frac{2\pi}{\lambda}d\hat{\boldsymbol{r}}\cdot[\boldsymbol{\nabla}_{\hat{\mathrm{n}}} \otimes \boldsymbol{\nabla}_{\hat{\mathrm{n}}}(p - q)]d\hat{\boldsymbol{r}} = 0. \tag{3.69}$$

Using again (2.36), which states that the second derivatives of distances consist of projections, and considering (3.66, 67), we obtain

$$d\hat{\boldsymbol{r}}\cdot\hat{\boldsymbol{N}}\left[\frac{1}{\tilde{\lambda}}\boldsymbol{R}^{\mathrm{T}}\left(\frac{1}{\tilde{p}}\tilde{\boldsymbol{K}} - \frac{1}{\tilde{q}}\tilde{\boldsymbol{C}}\right)\boldsymbol{R} - \frac{1}{\lambda}\left(\frac{1}{p}\boldsymbol{K} - \frac{1}{q}\boldsymbol{C}\right)\right]\hat{\boldsymbol{N}}d\hat{\boldsymbol{r}} = 0,$$
$$d\tilde{\boldsymbol{r}}\cdot\tilde{\boldsymbol{N}}\left[\frac{1}{\tilde{\lambda}}\left(\frac{1}{\tilde{p}}\tilde{\boldsymbol{K}} - \frac{1}{\tilde{q}}\tilde{\boldsymbol{C}}\right) - \frac{1}{\lambda}\boldsymbol{R}\left(\frac{1}{p}\boldsymbol{K} - \frac{1}{q}\boldsymbol{C}\right)\boldsymbol{R}^{\mathrm{T}}\right]\tilde{\boldsymbol{N}}d\tilde{\boldsymbol{r}} = 0. \tag{3.70}$$

To isolate the unknown $\tilde{p}$, we utilize here another oblique projection

$$\hat{\boldsymbol{M}} = \boldsymbol{I} - \frac{1}{\tilde{\boldsymbol{n}}\cdot\tilde{\boldsymbol{k}}}\tilde{\boldsymbol{n}} \otimes \tilde{\boldsymbol{k}},$$

and the corresponding connection between surfaces [similar to (2.74)]

$$d\tilde{\boldsymbol{r}} = \tilde{p}\hat{\boldsymbol{M}}^{\mathrm{T}}d\tilde{\boldsymbol{k}} = \tilde{p}d\alpha\hat{\boldsymbol{M}}^{\mathrm{T}}\tilde{\boldsymbol{m}}, \tag{3.71}$$

where $\tilde{\boldsymbol{m}} = d\tilde{\boldsymbol{k}}/|d\tilde{\boldsymbol{k}}|$ and where $d\alpha$ is an angular increment. Because of $\hat{\boldsymbol{M}}\tilde{\boldsymbol{N}} = \hat{\boldsymbol{M}}$, $\tilde{\boldsymbol{N}}\hat{\boldsymbol{M}}^{\mathrm{T}} = \hat{\boldsymbol{M}}^{\mathrm{T}}$, $\hat{\boldsymbol{M}}\tilde{\boldsymbol{K}} = \tilde{\boldsymbol{K}}$, $\tilde{\boldsymbol{K}}\hat{\boldsymbol{M}}^{\mathrm{T}} = \tilde{\boldsymbol{K}}$ [similar to (2.71)], we obtain from the second equation (3.70) the counterparts of (3.23, 24)

$$\frac{1}{\tilde{p}} = \tilde{\boldsymbol{m}}\cdot\tilde{\boldsymbol{T}}_{\mathrm{H}}\tilde{\boldsymbol{m}}, \tag{3.72}$$

where

$$\tilde{\boldsymbol{T}}_{\mathrm{H}} = \tilde{\lambda}\hat{\boldsymbol{M}}\left[\frac{1}{\tilde{\lambda}\tilde{q}}\tilde{\boldsymbol{C}} + \frac{1}{\lambda}\boldsymbol{R}\left(\frac{1}{p}\boldsymbol{K} - \frac{1}{q}\boldsymbol{C}\right)\boldsymbol{R}^{\mathrm{T}}\right]\hat{\boldsymbol{M}}^{\mathrm{T}}. \tag{3.73}$$

The discussion of these results would be the same as in Sect. 3.2.1, especially as far as the direction of $\tilde{\boldsymbol{k}}$ is concerned, the sign of $\tilde{p}$ or the astigmatism. So we do not repeat it here, but proceed to the case of small shifts.

As in Sect. 3.2.2, we choose as a reference the recording configuration and suppose the *displacement vectors* $\boldsymbol{t}$ from H to $\tilde{\mathrm{H}}$ and $\boldsymbol{v}$ from P to $\tilde{\mathrm{P}}$ to be *small*: $t \ll p, q$, and $v \ll p$. In the same way, the hologram rotation $\boldsymbol{R}$ differs little from the identity $\boldsymbol{I}$ and finally the change of wavelength is small: $|\lambda - \tilde{\lambda}| = |\Delta\lambda| \ll \lambda$. Moreover, all small quantities are assumed to be of the same order of magnitude. Similar to (3.41), the distances become

$$\tilde{q} = |q\boldsymbol{c} + \boldsymbol{t}| \simeq q + \boldsymbol{c}\cdot\boldsymbol{t}, \qquad \tilde{p} \simeq p + \boldsymbol{k}\cdot(\boldsymbol{t} - \boldsymbol{v}). \tag{3.74}$$

Further, the unit vectors are approximated by [see (3.42)]

$$\tilde{\boldsymbol{c}} \simeq \boldsymbol{c} + \frac{1}{q}\boldsymbol{C}\boldsymbol{t}, \qquad \tilde{\boldsymbol{k}} \simeq \boldsymbol{k} + \frac{1}{p}\boldsymbol{K}(\boldsymbol{t} - \boldsymbol{v}). \tag{3.75}$$

Third, the projections are written like (3.43)

$$\frac{1}{\tilde{q}}\tilde{\boldsymbol{C}} \simeq \frac{1}{q}\boldsymbol{C} - \frac{1}{q^2}\mathscr{C}\boldsymbol{t}, \qquad \frac{1}{\tilde{p}}\tilde{\boldsymbol{K}} \simeq \frac{1}{p}\boldsymbol{K} - \frac{1}{p^2}\mathscr{K}(\boldsymbol{t} - \boldsymbol{v}). \tag{3.76}$$

Finally, for the hologram rotation, we take again the result (2.92), which in the present case is

$$\boldsymbol{R} \simeq \boldsymbol{I} - \boldsymbol{\Psi}, \qquad \boldsymbol{R}^{\mathrm{T}} - \boldsymbol{I} = \boldsymbol{\nabla} \otimes \boldsymbol{t} \simeq \boldsymbol{\Psi}, \tag{3.77}$$

$\boldsymbol{\Psi}$ denoting the *skew-symmetric* tensor describing, instead of $\boldsymbol{R}$, the *small* hologram rotation.

The phase difference θ takes now the linearized form

$$\theta(\hat{\boldsymbol{r}}) = \frac{2\pi}{\lambda}\left[\frac{\Delta\lambda}{\lambda}(p - q) + \boldsymbol{k}\cdot(\boldsymbol{t} - \boldsymbol{v}) - \boldsymbol{c}\cdot\boldsymbol{t}\right], \tag{3.78}$$

and the stationarity condition, in the first form of (3.68), becomes thereafter

$$\hat{\boldsymbol{N}}\left[\frac{\Delta\lambda}{\lambda}(\boldsymbol{k} - \boldsymbol{c}) + \frac{1}{p}\boldsymbol{K}(\boldsymbol{t} - \boldsymbol{v}) - \frac{1}{q}\boldsymbol{C}\boldsymbol{t} + \boldsymbol{\Psi}(\boldsymbol{k} - \boldsymbol{c})\right] = 0. \tag{3.79}$$

The same oblique projection $\hat{\boldsymbol{M}} = \boldsymbol{I} - \hat{\boldsymbol{n}} \otimes \boldsymbol{k}/\hat{\boldsymbol{n}}\cdot\boldsymbol{k}$ as in Sect. 3.2.2 allows us to isolate the unknown *lateral part of the displacement*

$$\boldsymbol{K}\boldsymbol{v} = \frac{p\Delta\lambda}{\lambda}\hat{\boldsymbol{M}}(\boldsymbol{k} - \boldsymbol{c}) + \boldsymbol{K}\boldsymbol{t} - \frac{p}{q}\hat{\boldsymbol{M}}\boldsymbol{C}\boldsymbol{t} + p\hat{\boldsymbol{M}}\boldsymbol{\Psi}(\boldsymbol{k} - \boldsymbol{c}). \tag{3.80}$$

As for the second-order term, the linearized form of the first relation (3.70) is

$$d\hat{\boldsymbol{r}}\cdot\hat{\boldsymbol{N}}\left[\frac{\Delta\lambda}{\lambda}\left(\frac{1}{p}\boldsymbol{K}-\frac{1}{q}\boldsymbol{C}\right)-\frac{1}{p^2}\mathscr{K}(\boldsymbol{t}-\boldsymbol{v})+\frac{1}{q^2}\mathscr{C}\boldsymbol{t}\right.$$
$$\left.+\frac{1}{p}(\boldsymbol{\Psi}\boldsymbol{K}-\boldsymbol{K}\boldsymbol{\Psi})-\frac{1}{q}(\boldsymbol{\Psi}\boldsymbol{C}-\boldsymbol{C}\boldsymbol{\Psi})\right]\hat{\boldsymbol{N}}d\hat{\boldsymbol{r}}=0. \tag{3.81}$$

Again, in order to isolate the unknown, here the *longitudinal part* $\boldsymbol{k}\cdot\boldsymbol{v}$ of the displacement, we proceed as in Sect. 3.2.2 with the help of (3.48) and obtain

$$\boldsymbol{k}\cdot\boldsymbol{v}=\boldsymbol{m}\cdot\boldsymbol{T}_{\mathrm{H}}\boldsymbol{m}, \tag{3.82}$$

where

$$\boldsymbol{T}_{\mathrm{H}}=\hat{\boldsymbol{M}}\left[-\frac{p^2\Delta\lambda}{\lambda}\left(\frac{1}{p}\boldsymbol{K}-\frac{1}{q}\boldsymbol{C}\right)+\mathscr{K}\boldsymbol{t}-\frac{p^2}{q^2}\mathscr{C}\boldsymbol{t}\right.$$
$$\left.-(\boldsymbol{k}\otimes\boldsymbol{K}\boldsymbol{v}+\boldsymbol{K}\boldsymbol{v}\otimes\boldsymbol{k})-p(\boldsymbol{\Psi}\boldsymbol{K}-\boldsymbol{K}\boldsymbol{\Psi})+\frac{p^2}{q}(\boldsymbol{\Psi}\boldsymbol{C}-\boldsymbol{C}\boldsymbol{\Psi})\right]\hat{\boldsymbol{M}}^{\mathrm{T}}. \tag{3.83}$$

In the first part of this section, we established the relations valid for large movements of the hologram, which replace those of Sect. 3.2.1, whereas in the second part we deduced the counterparts of those in Sect. 3.2.2, which are valid only for small movements. In the following, equations (3.46, 80) particularly will become of importance for the lateral shift between the object point P and its near image $\tilde{\mathrm{P}}$.

3.2.5 Movement of Two Sandwiched Holograms: Difference Between the Lateral Displacements of Two Image Points

At the end of the present section devoted to the images formed in the case of a modification of the optical arrangement during reconstruction, we must add the description of a particular problem related to the so-called *sandwich holography* [3.83]. This technique, which we shall deal with more in detail in Chap. 4, requires two holograms (Fig. 3.15). On the first, which we suppose to be on the side of the observer, the wave field emitted by a point source P is recorded by means of a reference source Q. A second hologram, parallel to the first, is recorded with the same reference source Q but with a slightly different point source P′. P and P′ are in fact one point of an object before and after a deformation and are thus separated by a small displacement vector $\boldsymbol{u}$. We shall examine here the peculiar problem of the relative motion of P and P′ when the two holograms are displaced together. However, we shall proceed in such a way as will be needed later. Thus, we do not consider the effects of any change of wavelength. We also neglect the influence of the glass in the photographic plates;

this is warrantable provided that the hologram on the observer's side has been recorded with a glass plate in front of it to replace the other hologram and that the emulsions face each other, i.e. that both glass plates are located "out of the interferometer". Moreover we are only interested in the difference of the *lateral parts of the movements* of P and P′ to P̃ and P̃′ in the case of *small shifts* of the holograms.

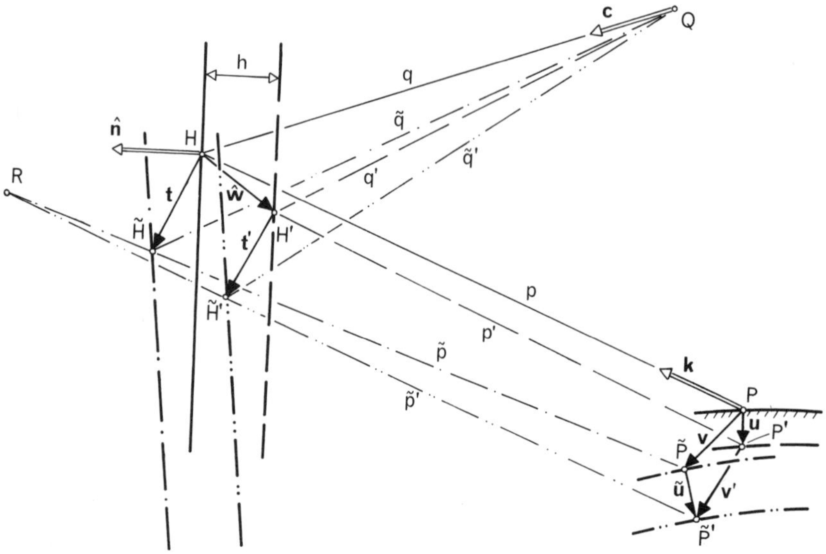

Fig. 3.15. Position of the images P̃ and P̃′ of the object points P and P′ when two sandwiched holograms are moved

We should actually write two equations of the type (3.80) related to the same reference axis and take the difference $\boldsymbol{K v}' - \boldsymbol{K v}$. However, to establish (3.80) we made the series developments taking into account only the order of magnitude of the quantities t/q and v/p, whereas we must now pay attention to three orders of magnitude. Indeed, we can first compare the *distance h between the two holograms* with a length l typical for the optical arrangement, as for instance p, and write

$$\frac{h}{l} = \delta,$$

where δ is a small dimensionless parameter. Second, with another small dimensionless parameter η, we have for the *movements of the holograms*

$$\frac{\boldsymbol{t} \cdot \boldsymbol{e}}{l} = O(\eta), \qquad \boldsymbol{e} \cdot \boldsymbol{\Psi} \boldsymbol{e} = O(\eta),$$

where $\boldsymbol{e}$ is an arbitrary unit vector. Third, a small dimensionless parameter ε may characterize the deformation of the object

$$\frac{\boldsymbol{u}\cdot\boldsymbol{e}}{l} = O(\varepsilon), \qquad \boldsymbol{e}\cdot(\boldsymbol{\nabla}\otimes\boldsymbol{u})\boldsymbol{e} = O(\varepsilon);$$

$\boldsymbol{u}$ represents the displacement from P to P′ and $\boldsymbol{\nabla}\otimes\boldsymbol{u}$, its derivative (see Sect. 2.2). The three quantities δ, η and ε are, so far, independent. But, because in practice the magnitude of the displacement $\boldsymbol{u}$ is very small (e.g. $\sim 10^{-2}$ mm), the distance h and the shift $\boldsymbol{t}$ are moderately small (e.g. $\sim 10^{0}$mm) and the lengths l, p, q . . . , finite (e.g. $\sim 10^2$ mm), it seems reasonable to presume that

$$\eta = O(\delta), \qquad \varepsilon = O(\delta^2). \tag{3.84}$$

Let us now indicate the order of magnitude of the relevant quantities. An observer located in R looks at the image $\tilde{\mathrm{P}}$ reconstructed by the vicinity of a point $\tilde{\mathrm{H}}$ on one hologram and at the image $\tilde{\mathrm{P}}'$ reconstructed by the vicinity of $\tilde{\mathrm{H}}'$ on the other hologram. During the recording, $\tilde{\mathrm{H}}$ and $\tilde{\mathrm{H}}'$ lay in H and H′, respectively, which are separated by the vector $\hat{\boldsymbol{w}}$. This vector consists of a part $-kh/\hat{\boldsymbol{n}}\cdot\boldsymbol{k}$ parallel to $\boldsymbol{k}$ and of a part $\hat{\boldsymbol{M}}^{\mathrm{T}}\hat{\boldsymbol{w}}$ situated in the hologram plane and of the order of u. From this, we conclude that $\hat{w}/l = O(\delta)$ and that $|\boldsymbol{K}\hat{\boldsymbol{w}}|/l = O(\delta^2)$. Moreover, if we suppose that the two holograms do not move with respect to each other, we may write a kinematic relation similar to (2.94)

$$\boldsymbol{t}' - \boldsymbol{t} \simeq \boldsymbol{\psi}^{\mathrm{e}} \times \hat{\boldsymbol{w}} = \hat{\boldsymbol{w}}\boldsymbol{\Psi}, \tag{3.85}$$

where $\boldsymbol{\psi}^{\mathrm{e}}$ represents the vector of small rotation of the holograms. Consequently, we know that $|\boldsymbol{t}' - \boldsymbol{t}|/l = O(\delta^2)$. Next, (3.80, 82) show that v/l and also v'/l are of the same order of magnitude δ as t/l and t'/l. From the preceding, we may finally argue that $|\boldsymbol{v}' - \boldsymbol{v}|/l = O(\delta^2)$. The following table sums up these orders of magnitude.

$O(1)$	$O(\delta)$	$O(\delta^2)$
$\frac{p}{l}, \frac{q}{l}, \ldots$	$\frac{h}{l}, \frac{\hat{w}}{l}, \frac{t}{l}, \frac{v}{l}, \ldots$	$\frac{u}{l}, \frac{\lvert\boldsymbol{t}'-\boldsymbol{t}\rvert}{l}, \frac{\lvert\boldsymbol{v}'-\boldsymbol{v}\rvert}{l}, \frac{\lvert\boldsymbol{K}\hat{\boldsymbol{w}}\rvert}{l},$
	$\boldsymbol{e}\cdot\boldsymbol{\Psi}\boldsymbol{e}$	$\boldsymbol{e}\cdot(\boldsymbol{\nabla}\otimes\boldsymbol{u})\boldsymbol{e}$
lengths in the optical arrangement	sandwich thickness and motion of the holograms	mechanical object deformation and relative motions

We can now proceed to the calculation of the difference of the lateral parts of the displacements of P and P′. As a reference configuration, we choose the recording position and the undeformed object, i.e. the lines HQ and HP. Moreover, instead of writing both the stationarity conditions around H and $\tilde{\mathrm{H}}$ and around H′ and $\tilde{\mathrm{H}}'$, it is advisable, because some terms vanish, to consider at the very start the difference

$$\Theta' - \Theta = (\tilde{p}' - p') - (\tilde{p} - p) - (\tilde{q}' - q') + (\tilde{q} - q). \tag{3.86}$$

The *optical-path differences* Θ' and Θ replace here the phase differences θ' and θ defined by (3.63) because we do not take into account any wavelength change.

Similar to what we did in Sect. 3.2.2, 4, let us express all lengths as functions of the reference lengths. First of all, we develop $\tilde{q} = |q\boldsymbol{c} + \boldsymbol{t}| = q|\boldsymbol{c} + \boldsymbol{t}/q|$ up to the second-order terms with regard to the small quantity $t/q = O(\delta)$. We apply thus a MacLaurin series of the type

$$\sqrt{1 + 2ax + x^2} = 1 + ax + \frac{1 - a^2}{2}x^2 + O(x^3).$$

Noting that x is replaced by t/q, we get, in our case,

$$\begin{aligned}
\tilde{q} &= q\sqrt{\left|\boldsymbol{c} + \frac{\boldsymbol{t}}{q}\right|^2} = q\sqrt{1 + 2\frac{\boldsymbol{c}\cdot\boldsymbol{t}}{t}\frac{t}{q} + \frac{t^2}{q^2}} \\
&= q\left\{1 + \frac{\boldsymbol{c}\cdot\boldsymbol{t}}{q} + \frac{1}{2}\left[1 - \frac{(\boldsymbol{c}\cdot\boldsymbol{t})^2}{t^2}\right]\frac{t^2}{q^2} + O(\delta^3)\right\} \\
&= q + \boldsymbol{c}\cdot\boldsymbol{t} + \frac{1}{2q}\boldsymbol{t}\cdot(\boldsymbol{I} - \boldsymbol{c}\otimes\boldsymbol{c})\boldsymbol{t} + O(\delta^3),
\end{aligned}$$

$$\tilde{q} = q + \boldsymbol{c}\cdot\boldsymbol{t} + \frac{1}{2q}\boldsymbol{t}\cdot\boldsymbol{C}\boldsymbol{t} + O(\delta^3). \tag{3.87}$$

Let us remark that this result, just like (3.74) [see also (3.41)], is a series development around H; here, because we take account of second-order terms, the second derivative of q, i.e. the normal projection $\boldsymbol{C}$, appears. In the same way, we have

$$q' = |q\boldsymbol{c} + \hat{\boldsymbol{w}}| = q + \boldsymbol{c}\cdot\hat{\boldsymbol{w}} + \frac{1}{2q}\hat{\boldsymbol{w}}\cdot\boldsymbol{C}\hat{\boldsymbol{w}} + O(\delta^3),$$

$$\tilde{q}' = |q\boldsymbol{c} + \hat{\boldsymbol{w}} + \boldsymbol{t}'| = q + \boldsymbol{c}\cdot(\hat{\boldsymbol{w}} + \boldsymbol{t}') + \frac{1}{2q}(\hat{\boldsymbol{w}} + \boldsymbol{t}')\cdot\boldsymbol{C}(\hat{\boldsymbol{w}} + \boldsymbol{t}') + O(\delta^3).$$

With these developments, we obtain for the second part of $\Theta' - \Theta$

$$(\tilde{q}' - q') - (\tilde{q} - q) \simeq (\boldsymbol{t}' - \boldsymbol{t})\cdot\boldsymbol{c} + \frac{1}{2q}(2\hat{\boldsymbol{w}}\cdot\boldsymbol{C}\boldsymbol{t}' + \boldsymbol{t}'\cdot\boldsymbol{C}\boldsymbol{t}' - \boldsymbol{t}\cdot\boldsymbol{C}\boldsymbol{t}).$$

Because $\boldsymbol{t}'\cdot\boldsymbol{C}\boldsymbol{t}' - \boldsymbol{t}\cdot\boldsymbol{C}\boldsymbol{t} = (\boldsymbol{t}' - \boldsymbol{t})\cdot\boldsymbol{C}(\boldsymbol{t}' + \boldsymbol{t}) = O(\delta^3)$, we may neglect it, so that

$$(\tilde{q}' - q') - (\tilde{q} - q) \simeq (\boldsymbol{t}' - \boldsymbol{t})\cdot\boldsymbol{c} + \frac{1}{q}\hat{\boldsymbol{w}}\cdot\boldsymbol{C}\boldsymbol{t}',$$

where we may observe that the unexpected term $\hat{\boldsymbol{w}}\cdot\boldsymbol{C}\boldsymbol{t}'/q$ is of order δ^2 and thus comparable to $(\boldsymbol{t}' - \boldsymbol{t})\cdot\boldsymbol{c}$.

In the same way, we may develop the "p's", where we have simply to replace q by p, $\boldsymbol{c}$ by $\boldsymbol{k}$, $\boldsymbol{t}$ by $\boldsymbol{t} - \boldsymbol{v}$, $\hat{\boldsymbol{w}}$ by $\hat{\boldsymbol{w}} - \boldsymbol{u}$ and $\boldsymbol{t}'$ by $\boldsymbol{t}' - \boldsymbol{v}'$. So we get for the first part of $\Theta' - \Theta$

$$(\tilde{p}' - p') - (\tilde{p} - p) \simeq (\boldsymbol{t}' - \boldsymbol{t})\cdot\boldsymbol{k} - (\boldsymbol{v}' - \boldsymbol{v})\cdot\boldsymbol{k} + \frac{1}{2p}[2(\hat{\boldsymbol{w}} - \boldsymbol{u})\cdot\boldsymbol{K}(\boldsymbol{t}' - \boldsymbol{v}') + (\boldsymbol{t}' - \boldsymbol{v}')\cdot\boldsymbol{K}(\boldsymbol{t}' - \boldsymbol{v}') - (\boldsymbol{t} - \boldsymbol{v})\cdot\boldsymbol{K}(\boldsymbol{t} - \boldsymbol{v})].$$

Contrary to the foregoing, the whole bracket is here of order δ^3 because $|\boldsymbol{K}\hat{\boldsymbol{w}}|/l = O(\delta^2)$ and $u/l = O(\delta^2)$. Thus

$$(\tilde{p}' - p') - (\tilde{p} - p) \simeq (\boldsymbol{t}' - \boldsymbol{t})\cdot\boldsymbol{k} - (\boldsymbol{v}' - \boldsymbol{v})\cdot\boldsymbol{k}.$$

Furthermore, as we already mentioned, $\hat{\boldsymbol{w}}$ is almost parallel to $\boldsymbol{k}$, so that

$$\hat{\boldsymbol{w}} = -\frac{h}{\hat{\boldsymbol{n}}\cdot\boldsymbol{k}}\boldsymbol{k} + O(\delta^2).$$

Consequently, (3.85) becomes

$$\boldsymbol{t}' - \boldsymbol{t} \simeq -\frac{h}{\hat{\boldsymbol{n}}\cdot\boldsymbol{k}}\boldsymbol{k}\boldsymbol{\Psi}. \tag{3.88}$$

Observing on the one hand that this vector is perpendicular to $\boldsymbol{k}$, and on the other hand that $\boldsymbol{t}'$, which appears alone, may be replaced by $\boldsymbol{t}$, we write the optical-path difference in the form

$$\Theta' - \Theta \simeq (\boldsymbol{v} - \boldsymbol{v}')\cdot\boldsymbol{k} + \frac{h}{\hat{\boldsymbol{n}}\cdot\boldsymbol{k}}\boldsymbol{k}\cdot\left(\boldsymbol{\Psi}\boldsymbol{c} + \frac{1}{q}\boldsymbol{C}\boldsymbol{t}\right). \tag{3.89}$$

To write explicitly the stationary behavior of this function, i.e.

$$\boldsymbol{\nabla}_{\hat{\mathrm{n}}}(\Theta' - \Theta) = \hat{\boldsymbol{N}}\boldsymbol{\nabla}(\Theta' - \Theta) = 0, \tag{3.90}$$

we need some known derivatives (see Sect. 3.2.4), namely

$$\boldsymbol{\nabla} \otimes \boldsymbol{c} = \frac{1}{q}\boldsymbol{C}, \quad \boldsymbol{\nabla} \otimes \left(\frac{1}{q}\boldsymbol{C}\right) = -\frac{1}{q^2}\mathscr{C}, \quad \boldsymbol{\nabla} \otimes \boldsymbol{k} = \frac{1}{p}\boldsymbol{K}, \quad \boldsymbol{\nabla} \otimes \boldsymbol{t} = \boldsymbol{\Psi},$$

and, in addition, the following derivative

$$\boldsymbol{\nabla} \otimes \left(\frac{\boldsymbol{k}}{\hat{\boldsymbol{n}}\cdot\boldsymbol{k}}\right) = \frac{1}{(\hat{\boldsymbol{n}}\cdot\boldsymbol{k})^2}[(\boldsymbol{\nabla} \otimes \boldsymbol{k})\hat{\boldsymbol{n}}\cdot\boldsymbol{k} - (\boldsymbol{\nabla} \otimes \boldsymbol{k})\hat{\boldsymbol{n}} \otimes \boldsymbol{k}] = \frac{1}{p\hat{\boldsymbol{n}}\cdot\boldsymbol{k}}\boldsymbol{K}\hat{\boldsymbol{M}},$$

$$\boldsymbol{\nabla} \otimes \left(\frac{\boldsymbol{k}}{\hat{\boldsymbol{n}}\cdot\boldsymbol{k}}\right) = \frac{1}{p\hat{\boldsymbol{n}}\cdot\boldsymbol{k}}\hat{\boldsymbol{M}}. \tag{3.91}$$

Thus, (3.90) becomes, with $\hat{\boldsymbol{N}}\hat{\boldsymbol{M}} = \hat{\boldsymbol{N}}$,

$$\frac{1}{p}\hat{\boldsymbol{N}}\boldsymbol{K}(\boldsymbol{v}' - \boldsymbol{v}) = \frac{h}{p\hat{\boldsymbol{n}}\cdot\boldsymbol{k}}\hat{\boldsymbol{N}}\left(\boldsymbol{\Psi}\boldsymbol{c} + \frac{1}{q}\boldsymbol{C}\boldsymbol{t}\right) + \frac{h}{\hat{\boldsymbol{n}}\cdot\boldsymbol{k}}\hat{\boldsymbol{N}}\left[\frac{1}{q}\boldsymbol{C}\boldsymbol{\Psi}^{\mathrm{T}} + \frac{1}{q}\boldsymbol{\Psi}\boldsymbol{C} - \frac{1}{q^2}\mathscr{C}\boldsymbol{t}\right]\boldsymbol{k}.$$

Taking into account $\boldsymbol{\Psi}^{\mathrm{T}} = -\boldsymbol{\Psi}$ and the symmetry of $\mathscr{C}$, we may write

$$\frac{1}{p}\hat{\boldsymbol{N}}\boldsymbol{K}(\boldsymbol{v}' - \boldsymbol{v}) = \frac{h}{p\hat{\boldsymbol{n}}\cdot\boldsymbol{k}}\hat{\boldsymbol{N}}\left(\boldsymbol{\Psi}\boldsymbol{c} + \frac{1}{q}\boldsymbol{C}\boldsymbol{t}\right) + \frac{h}{\hat{\boldsymbol{n}}\cdot\boldsymbol{k}}\hat{\boldsymbol{N}}\left[\frac{1}{q}(\boldsymbol{\Psi}\boldsymbol{C} - \boldsymbol{C}\boldsymbol{\Psi})\boldsymbol{k} - \frac{1}{q^2}\boldsymbol{k}\mathscr{C}\boldsymbol{t}\right].$$

To isolate the lateral part of the displacements, we use again the oblique projection $\hat{\boldsymbol{M}}$ and obtain the sought result [3.81]

$$\boldsymbol{K}(\boldsymbol{v}' - \boldsymbol{v}) = \frac{h}{\hat{\boldsymbol{n}}\cdot\boldsymbol{k}}\hat{\boldsymbol{M}}\left[\frac{p}{q}(\boldsymbol{\Psi}\boldsymbol{C} - \boldsymbol{C}\boldsymbol{\Psi})\boldsymbol{k} + \left(\boldsymbol{\Psi}\boldsymbol{c} + \frac{1}{q}\boldsymbol{C}\boldsymbol{t}\right) - \frac{p}{q^2}\boldsymbol{k}\mathscr{C}\boldsymbol{t}\right]. \tag{3.92}$$

As we expected, this expression is of the order of magnitude δ^2.

3.3 Conclusion

In the present chapter, we turned our attention first to the image formed in standard holography and saw that it is identical with the object if the reconstruction wave field is the same as the reference one. When the latter condition is not fulfilled, the image generally is astigmatic, as we saw in the second part of this chapter. We then had to take into consideration only a small region of the hologram; consequently we had to use series developments of phases, or rather of optical paths. Every time, we encountered the same features: successive differentiations of the distance between a fixed and a variable point made appear unit vectors, projections and superprojections.

Having in this way explained the properties of a holographically reconstructed wave field, we may proceed to the problem of the superposition of two wave fields and thus of the interference fringes.

Chapter 4 Fringe Interpretation in Holographic Interferometry

Up to now, we have investigated, on the one hand, the kinematics of the deformation of a body (Sect. 2.2) and, on the other hand, the formation of holographic images (Chap. 3). In this chapter, these two subjects will be combined, because we are going to explain how holography can be used to measure deformations.

As we have seen, holography enables us to store wave fields and to reconstruct them at a later time. Thanks to this striking property, wave fields that do not exist simultaneously can be caused to interfere. In the same way as in conventional interferometry, the observation of the formed fringes permits us to compare the wave fields and consequently also the two objects from which they originate. The first experiments in interferometry with holographically produced wave fields were made and published about 1965 (refer, e.g., to [4.1–14]). The basic techniques described therein, and still utilized, are the following.

First, a reconstructed wave field and an "actually existing" field can be superimposed; this is called the *real-time or live-fringe technique*. Provided that the hologram is exactly repositioned and reconstructed with the same wave as that used as the reference wave during the recording, a wave field is produced that differs only by a constant factor from the recorded object wave field [see (3.3)]. If, moreover, the superimposed wave field has undergone no change from the recorded wave field, except that its amplitude has been adjusted, the waves cancel each other because of the negative sign in (3.3); thus no image can be seen. If the "actually existing" wave is slightly modified, both wave fields become visible. In fact, they still are so similar that they cannot be seen separately, but macroscopic interference fringes are formed. Thus, at the same time the modifications are made, their consequences can be observed by looking at the changes of the fringes. Instead of being a stationary wave field, the non-reconstructed wave field can also be produced by *stroboscopic* illumination of a vibratory process (e.g. [4.15–16]). However, apart from the advantage of immediate observation, drawbacks do exist: inexact reconstruction as well as shrinkage of the emulsion often produce spurious fringes.

Second, as we have seen in Sect. 3.1.3, both wave fields can be recorded successively on the same hologram and reconstructed simultaneously; this is called the *double-exposure or frozen-fringe technique*. With this method, small differences between the arrangement of reconstruction and of recording are much less critical than with the preceding technique, because both wave fields undergo the same changes. On the other hand, once the two wave fields are stored, they cannot be influenced any more. However, as we shall see, by using two reference sources or two holograms, the fringes may nevertheless be modified. Furthermore, by dividing the hologram plate into several parts, by use of masks, or by using several reference waves, it is possible to record many pairs of wave fields on the same plate; this is often called *multiplexing* of double exposures (see e.g. [4.17–21]). Wave fields produced by use of *pulsed lasers* can also be recorded (e.g. [4.22–23]). In such cases, the exposure times are very short, which overcomes problems caused by vibrations of the optical arrangement and is most often used to store the shape, at a given instant, of a wave field that varies quite rapidly in time.

Third, the *time-average* technique consists of a single recording of all wave fields reflected or transmitted by an oscillatory process, the vibration period of which is short compared with the exposure time. During reconstruction, an ensemble of wave fields is produced; roughly said, this integrated effect reproduces the most frequent shape of the wave fields during the exposure time.

The three methods we have briefly described can be practiced with wave fields produced in many ways and applied to a variety of cases (see e.g. the surveys [4.24–31]). For instance, we may investigate waves transmitted by transparent media or reflected by mirror surfaces; optical methods of measuring the deformation such as photoelasticity, contouring, speckle or moiré techniques can draw advantages from the holographic storage of waves. In this treatise however, we shall concentrate on the problem of determining the deformation between two configurations of a nontransparent body that reflects light diffusely. Thus, the wave fields we shall deal with are formed in the following manner: a light source illuminates the object to be investigated; owing to the roughness of the surface, the specular reflection is negligible and the light is reflected almost uniformly in all directions. This behavior is typical for the machined surfaces of metallic parts; and precisely there lies one of the great advantages of holographic nondestructive testing: the very component of a structure can be tested.

Although the holographic process makes no difference between the two compared object states, we shall take one as a reference configuration and call it the undeformed object. The other one, which visually differs undistinguishably from the first, will be named the deformed object.

In the first two sections, we assume that the wave fields are exactly such as they are produced by the illuminated object surfaces. This means that the holographic reconstruction is expected to be ideal (the pertinent conditions were described in Sect. 3.1.2) and that the hologram simply acts as a window through which the waves can be observed (This is what we call *standard* holographic interferometry). Thus, it is not necessary to specify whether only one or both wave fields are holographically reconstructed; nor shall we say whether the two configurations are due to a static deformation or whether they are two transient states during movement. In Sect. 4.1, we use a simple description of the interference phenomenon in terms of the *optical-path difference* between two rays. This quantity depends on the *displacement* vector between the pair of points to which these rays go. This vector may then be measured by interpreting the fringes. In Sect. 4.2, we look further into the interference phenomenon, considering small regions around the points on the object surface and the ensemble of rays they reflect. The main feature here is the presence of first *derivatives of the optical-path difference* and consequently of the derivative of the displacement, i.e. the *strain* and rotation tensors, in which the engineer is mostly interested. The results relate then the latter quantities to the direction, the spacing, the visibility (contrast) and the localization of the fringes.

In Sect. 4.3, we show the counterparts of the main results established in the preceding sections for cases in which the optical arrangement used during the reconstruction is *modified* with respect to that used during the recording. In this way, when both wave fields are reconstructed ones, the fringes can be adjusted and the measurements become more flexible.

4.1 Optical-Path Difference in Standard Holographic Interferometry

The order of an interference fringe is determined by the phase difference, i.e. the optical-path difference, of two superimposed beams. To calculate the latter quantity, we take into account in this section only the two light rays that are reflected by the same material point in the undeformed and in the deformed configuration. This simplified interpretation, which ignores whether and how much the fringes are visible, will be justified and completed in the next section. For the time being, anticipating somewhat, we can only say that, because of the roughness of the surface microprofile, arbitrary points on the surfaces represent mutually incoherent sources, so that only rays reflected by the same material point can cause visible interference.

The mechanical quantity that appears here is the displacement vector. After demonstrating the basic relation, we shall explain how this information can be obtained, and then conclude by showing how the displacement can be "differentiated" by means of finite differences in order to calculate approximately the components of the strain tensor.

4.1.1 Basic Relation

Because the assumption is that the holographic reconstruction is ideal, the fringe formation can be described by the simple Fig. 4.1. The two surfaces of the body are lit by a point source S placed in front of them and reflect light diffusely in all directions, in particular to an arbitrary point K where the interference is examined. As in Sect. 2.2, we denote by $\boldsymbol{r}$ the position vector of a point P on the undeformed surface and by $\boldsymbol{r}'$ that of the corresponding point P′ on the deformed surface. The displacement vector is then $\boldsymbol{u} = \boldsymbol{r}' - \boldsymbol{r}$. As a reference configuration, we choose the set {P}, so that $\boldsymbol{r}$ becomes the independent variable.

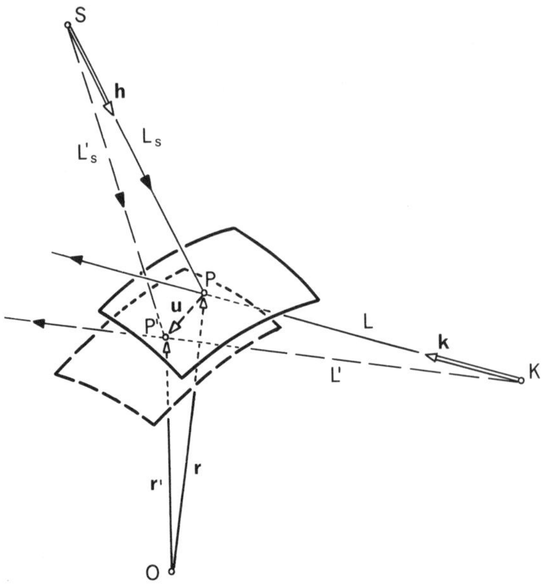

Fig. 4.1. Simplified model of the fringe formation at a point K. S: light source. O: origin of the position vectors. P, P′: object point before and after deformation

From S, light travels the distances L_S to P and L_S' to P′. After reflection, it covers further the distances L or L' to K. L is assumed here to be a virtual path, i.e. K is supposed to lie behind the object surfaces. In case K lies in front of them, L has to be taken negative. The *optical-path difference* at K is then expressed as

$$D = (L_S - L) - (L_S' - L'). \tag{4.1}$$

For the wavelength λ, the *fringe-order* n is

$$n = \frac{D}{\lambda}. \tag{4.2}$$

If both wave fields are reconstructed ones, the fringes that correspond to integer values of n are bright, whereas they are dark if only one wave field is holographically produced, as in the real-time technique. This results from the sign change caused by the reconstruction [see (3.3)].

Let us now express D as a function of the reference position, with the direction unit vectors $\boldsymbol{k}$ along KP and $\boldsymbol{h}$ along SP, and of the displacement $\boldsymbol{u}$. We write for instance (as in [4.32])

$$L' = |L\boldsymbol{k} + \boldsymbol{u}| = L\sqrt{1 + 2\frac{\boldsymbol{u}\cdot\boldsymbol{k}}{L} + \frac{u^2}{L^2}},$$

with $u \equiv |\boldsymbol{u}|$, so that

$$D = L\left(\sqrt{1 + 2\frac{\boldsymbol{u}\cdot\boldsymbol{k}}{L} + \frac{u^2}{L^2}} - 1\right) - L_S\left(\sqrt{1 + 2\frac{\boldsymbol{u}\cdot\boldsymbol{h}}{L_S} + \frac{u^2}{L_S^2}} - 1\right). \tag{4.3}$$

In order for the two wave fields to superimpose well and to form good interference fringes (see also Sect. 4.2.2), the *deformation* must be *small* (this is fortunately the case in most practical problems), so that we may assume

$$u \ll L_S, \qquad u \ll |L|. \tag{4.4}$$

Consequently, we are able to linearize (4.3). Using a MacLaurin series of the type

$$\sqrt{1 + 2ax + x^2} = 1 + ax + \frac{1 - a^2}{2}x^2 + O(x^3),$$

we obtain for instance for L', x being replaced by u/L,

$$L' = L\left\{1 + \frac{\boldsymbol{u}\cdot\boldsymbol{k}}{L} + \frac{1}{2}\left[1 - \frac{(\boldsymbol{u}\cdot\boldsymbol{k})^2}{u^2}\right]\frac{u^2}{L^2} + O\left(\frac{u^3}{L^3}\right)\right\}$$

$$= L + \boldsymbol{u}\cdot\boldsymbol{k} + \frac{1}{2L}\boldsymbol{u}\cdot(\boldsymbol{I} - \boldsymbol{k}\otimes\boldsymbol{k})\boldsymbol{u} + \cdots$$

$$L' = L + \boldsymbol{u}\cdot\boldsymbol{k} + \frac{1}{2L}\boldsymbol{u}\cdot\boldsymbol{K}\boldsymbol{u} + \cdots. \tag{4.5}$$

This constitutes of course the series development around P, up to the second-order terms, of the length L between the fixed center K and the variable point P (We could also have written it in the form (2.31), $\boldsymbol{u}$ taking the place of the increment $d\boldsymbol{r}$). That is why we recognize the derivatives we calculated at the end of Sect. 2.1.2: the first derivative of L is the unit direction vector $\boldsymbol{k}$, and the second derivative of L contains the normal projection $\boldsymbol{K} = \boldsymbol{I} - \boldsymbol{k}\otimes\boldsymbol{k}$ onto the plane perpendicular to $\boldsymbol{k}$ (we recall that $\boldsymbol{K}\boldsymbol{u} = \boldsymbol{u} - \boldsymbol{k}(\boldsymbol{k}\cdot\boldsymbol{u})$). Taking account of the first-order terms only, we find the following *linear expression* of the optical-path difference

$$D = \boldsymbol{u}\cdot(\boldsymbol{k} - \boldsymbol{h}) = \boldsymbol{u}\cdot\boldsymbol{g}. \tag{4.6}$$

$\boldsymbol{g} = \boldsymbol{k} - \boldsymbol{h}$ expresses how a given vector $\boldsymbol{u}$ makes itself felt and is therefore called the *sensitivity vector*. One essential feature of this equation, in contradistinction to (4.1, 3), is that D does not depend completely on the position of K but only on the direction $\boldsymbol{k}$ in which the point P is observed. The other important fact is that the displacement vector appears and can thus be measured by means of this basic relation, as we shall presently see.

Let us also remark here that the fringe-order, and thus D, is a function, on the one hand, of the position $\boldsymbol{r}$ of P, because $\boldsymbol{u}$ and $\boldsymbol{h}$ depend on this variable, and, on the other hand, of the observation direction $\boldsymbol{k}$; thus,

$$D = D(\boldsymbol{r}, \boldsymbol{k}). \tag{4.7}$$

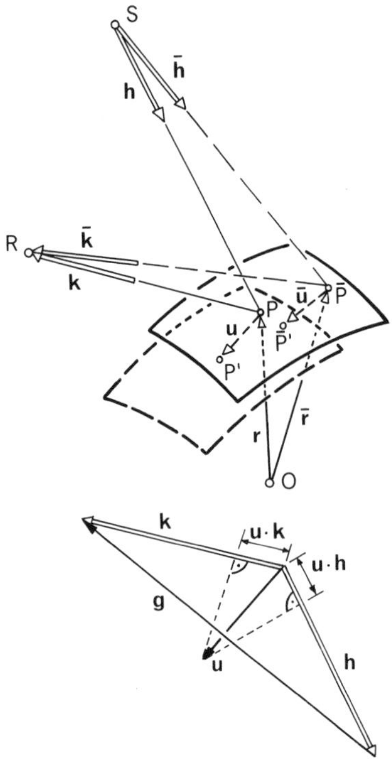

Fig. 4.2. Variation of $D_R(\boldsymbol{r})$ or $D_R(\boldsymbol{k})$ for a fixed position R of the observer and a variable object point P

In particular, when an observer looks at a fringe pattern from a point R (Fig. 4.2), or when the fringes are photographed from this point (see for instance Figs. 4.27–29), what is seen is the ensemble of fringes related to the fixed point R and to the variable point P: the change from D to $\bar{D}$ is then determined both by the changes from $\boldsymbol{u}$ and $\boldsymbol{h}$ in P to $\bar{\boldsymbol{u}}$ and $\bar{\boldsymbol{h}}$ in $\bar{\mathrm{P}}$ and by the change from $\boldsymbol{k}$ along PR to $\bar{\boldsymbol{k}}$ along $\bar{\mathrm{P}}$R. In contrast to this, as we shall see in Sect. 4.1.3, it is also possible to display a fringe pattern concerning one given point P and the ensemble of the points R, i.e. of the directions $\boldsymbol{k}$: the change of D is then due solely to the change of $\boldsymbol{k}$.

4.1.2 Determination of the Displacement by Means of the Optical-Path Difference

First, let us stress that, although the whole fringe field contains the information related to the whole visible surface of the object, the displacement vector must be determined, by means of (4.6), for each interesting point P of the surface. As proposed by *Sollid* [4.33–35], we shall distinguish two fundamental procedures.

a) *The Absolute Fringe-Order Method, also Called the Zero-Order Fringe Method or the Static Method*

This method, first described by *Ennos* [4.36], consists in measuring the order n (not necessarily an integer) of the fringe seen when looking at the point P in some direction $\boldsymbol{k}$. This direction as well as the illumination direction $\boldsymbol{h}$ must be known too. Taking three such measurements, with three linearly independent sensibility vectors $\boldsymbol{g}_i$, we obtain

$$\lambda n_i = D_i = \boldsymbol{u}\cdot\boldsymbol{g}_i \qquad (i = 1, 2, 3). \tag{4.8}$$

This system of three linear equations allows us to determine all three components (these being related to any chosen coordinate system) of the displacement vector.

Recalling Sect. 2.1.1, we can also interpret the D_i's as the covariant components of the vector $\boldsymbol{u}$ with respect to the covariant base vectors $\boldsymbol{g}_i$ (see also [4.33, 37]). The displacement vector is thus obtained as

$$\boldsymbol{u} = D_k\boldsymbol{g}^k = \lambda n_k\boldsymbol{g}^k, \tag{4.9}$$

where the repetition of the index means summation from 1 to 3 and where the contravariant base vectors $\boldsymbol{g}^k$ are given by the orthogonality relations (2.2), i.e. by

$$\boldsymbol{g}_i\cdot\boldsymbol{g}^k = \delta_i^k = \begin{cases} 1, & \text{if } i = k, \\ 0, & \text{if } i \neq k. \end{cases} \tag{4.10}$$

In usual terms of components with respect to a cartesian coordinate system, the latter nine relations are written as the product of two matrices, the result of which is the unit matrix. The first matrix is formed by the components of the vectors $\boldsymbol{g}_i$

$$G = \begin{bmatrix} g_{1x} & g_{1y} & g_{1z} \\ g_{2x} & g_{2y} & g_{2z} \\ g_{3x} & g_{3y} & g_{3z} \end{bmatrix};$$

the second, which is the inverse of the preceding matrix, contains the components of the vectors $\boldsymbol{g}^k$

$$G^{-1} = \begin{bmatrix} g^1{}_x & g^2{}_x & g^3{}_x \\ g^1{}_y & g^2{}_y & g^3{}_y \\ g^1{}_z & g^2{}_z & g^3{}_z \end{bmatrix}.$$

In the language of linear algebra, the passage from (4.8) to (4.9) may thus be considered as the determination of the three unknowns u, v, w, that are the components of $\boldsymbol{u}$ in the chosen system, by inverting a matrix [4.27, 38]. Now, this general calculation is interesting only if the same base vectors can be used for many vectors $\boldsymbol{u}$, such as those that correspond to several points P if S and R are at infinity. In most practical cases, it will however be simpler to solve the system (4.8) directly.

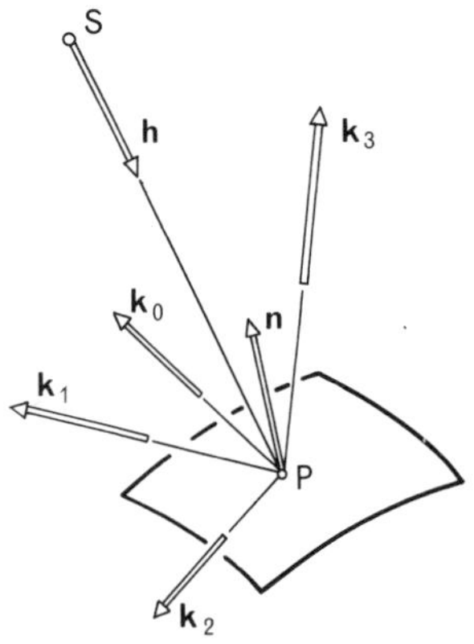

Fig. 4.3. Getting three base vectors $\boldsymbol{g}_i$ or $\boldsymbol{f}_i$ by means of several observation directions $\boldsymbol{k}_i$

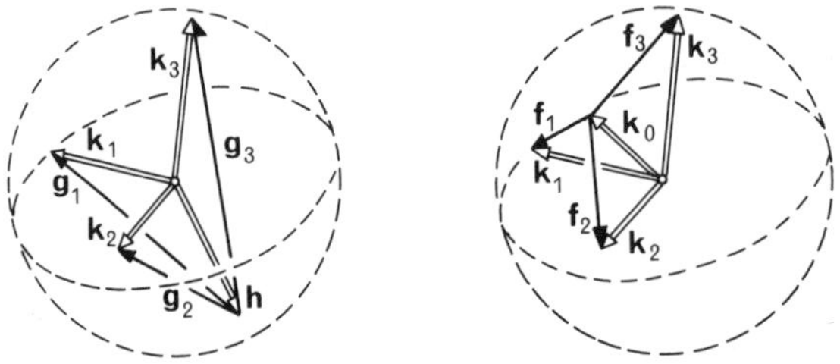

So, one major problem is to find three *sufficiently different vectors* $\boldsymbol{g}_i$ to permit a complete and accurate calculation of $\boldsymbol{u}$. First, these vectors can be obtained by looking at the point P in three directions $\boldsymbol{k}_i$ (Fig. 4.3). For that purpose, because the limits of one hologram do not provide very different directions of view, unless the plate is very large or located close to P, it is often necessary to record three holograms [4.36, 39], at different loci, or to place mirrors near the object. Second, $\boldsymbol{g}$ can be changed by illuminating the object from different directions $\boldsymbol{h}_i$ [4.40]. This also requires three recordings, which however can be made at the same place or can even be recorded on the same plate, but then with three different reference waves.

The second important problem to be solved is the determination of the fringe-order. Generally, this is achieved by counting the fringes, starting from a *zero-*

order fringe. As seen from (4.6), $D = 0$ either if $\boldsymbol{u}$ is perpendicular to $\boldsymbol{g}$ or if $\boldsymbol{u} = 0$. In practice, only the points where the latter, more restrictive, condition is fulfilled are of interest. Often, it is known from the start that a particular point or a region of the object undergoes no displacement. But such points can also be located by observing the fringe system. Indeed, where $\boldsymbol{u} = 0$, the fringe-order remains zero whatever the parameters of (4.6) are. In other words, this part or point of the zero-order fringe does not move if the directions of observation [4.36] or of illumination [4.41] are varied. The latter direction may be changed either by placing several light sources in front of the object, taking care that all $\boldsymbol{h}_i$ do not lie on a cone $D = \text{constant} \neq 0$, or by putting a ground-glass plate in front of the region of the object where no displacement is expected. This diffuse illumination is equivalent to an infinity of sources so that a fringe is seen only where the fringe systems for all of them coincide, i.e. where $\boldsymbol{u} = 0$. Finally, it is also possible to let $\boldsymbol{u}$ vary, or more exactly only its length [4.41, 42]; this is done by recording more than two deformation states in the frozen-fringe technique, or by looking at the fringes while the object is variously loaded in the real-time technique. If no fixed object point can be found or exists, the fringe-order can also be counted on a rubber strip, one end of which is fixed to the object whereas the other end is attached to a (visible) part of the apparatus, which surely does not move while the object is deformed [4.40, 43–46].

b) The Relative Fringe-Order Method, also Called the Fringe-Order-Difference Method or the Dynamic Method

This method, which was proposed by *Aleksandrov* and *Bonch-Bruevich* [4.47], eludes the problem of finding the zero-order fringe. The measured quantity is the variation Δn of the fringe-order when passing continuously from one observation direction $\boldsymbol{k}_{0i}$ to another $\boldsymbol{k}_i$ (hence the name dynamic method). With three such measurements, we obtain from (4.6)

$$\lambda \Delta n_i = \Delta D_i = D_i - D_{0i} = \boldsymbol{u} \cdot (\boldsymbol{k}_i - \boldsymbol{k}_{0i}) = \boldsymbol{u} \cdot \boldsymbol{f}_i \quad (i = 1, 2, 3). \tag{4.11}$$

The vector $\boldsymbol{u}$ can be determined in exactly the same way as in the case described previously, the base vectors being now the three $\boldsymbol{f}_i$. For this method, fringe-order differences and observation directions must be known; furthermore, the relative signs of the ΔD_i are important, but neither the absolute fringe-order nor the illumination direction need be known.

To obtain very different base vectors $\boldsymbol{f}_i$, several holograms [4.48] or mirrors can be used. In the general case, six observation directions are required. In order to simplify the measurements, some observation directions may be used several times; especially, if all observations are made through the same hologram plate, there can be only one common $\boldsymbol{k}_0$ (Fig. 4.3) [4.49, 50]. Another particular case of (4.11) occurs when the observation directions are chosen so that $\Delta n_i = 0$ (see, e.g., the discussion following [4.41] and also [4.51–54]); then, the observation directions lie on a circular cone, the axis of which is the vector

$\boldsymbol{u}$. Consequently, with three directions on such a cone, we can write

$$\boldsymbol{u} \cdot \boldsymbol{k}_1 = \boldsymbol{u} \cdot \boldsymbol{k}_2 = \boldsymbol{u} \cdot \boldsymbol{k}_3, \tag{4.12}$$

from which the direction of $\boldsymbol{u}$ can be calculated. This direction can also be determined by another property of the fringes. Indeed, the fringe-order goes through a maximum or minimum when $\boldsymbol{k}$ is parallel to $\boldsymbol{u}$, as can be seen from (4.6). If the direction of $\boldsymbol{u}$ intersects the hologram, it is possible to aim at P in a variable direction until the fringe-order remains stationary and thus $\boldsymbol{k} = \pm \boldsymbol{u}/u$ [4.55]. After the direction of $\boldsymbol{u}$ is known, its length is obtained by one measurement of the absolute or of the relative fringe-order [4.52–55].

As numerous examples (among others [4.56–107]) show, these two techniques are efficient in many kinds of practical problems. However, their use is not without difficulties. We have already mentioned that three very different base vectors $\boldsymbol{g}_i$ or $\boldsymbol{f}_i$ must be used. As a matter of fact, the possibility of error in the measurement of $\boldsymbol{u}$ is greatly increased if the base vectors are close to each other [4.45, 108–110]. Generally, because only one of the two unit vectors that constitute $\boldsymbol{g}$ or $\boldsymbol{f}$ can be varied, a change of this unit vector brings only a small change of the base vector. Furthermore, for practical reasons, the vectors $\boldsymbol{k}$ and $\boldsymbol{h}$ do not get far away from the normal to the plane tangent to the object surface. Then, as shown by Fig. 4.3, the vectors $\boldsymbol{g}_i$ will be nearly parallel to this normal whereas the vectors $\boldsymbol{f}_i$ will be more or less tangent to the object. This suggests that it can be useful to combine the two methods or, as proposed by *Dhir* and *Sikora* [4.111, 112] (see also [4.44] for method *a* or [4.50] for *b*), to calculate $\boldsymbol{u}$ from more than three equations with the least-squares method.

Usually, the unit direction vectors are measured conventionally, i.e. mechanically. Nevertheless, it may be useful to determine the base vectors by "inverting" the methods we have described, i.e. by measuring one $\boldsymbol{g}_i$ by means of three linearly independent, known displacements $\boldsymbol{u}_j$ [4.113].

Sometimes, the measurements can be simplified because some information about the displacement vector is given [4.114]. Particularly, if the direction of this vector is known, as in the case of the deflection of beams or plates, of if the plane in which it lies is known, only one or two base vectors are required [4.115–119].

The sign of the displacement cannot be determined from the fringe pattern because the interference process makes no distinction between the two wave fields. It must thus be determined from external knowledge, as for example the direction of the applied force, or by a mechanical measurement, or by the additional optical means that we shall describe in the next section.

Let us note that localization of the fringes does not come into account when the basic relation (4.6) is used. But, as we shall see in Sects. 4.2.2, 3, the visibility of the fringes is not the same everywhere. However, if the aperture of the observation system is small, this phenomenon does not make itself felt.

4.1.3 Determination of the Displacement with Additional Equipment

So far, we have assumed that the observer was looking at the two configurations of the object and at the fringe pattern through the hologram plate. The interfering wave fields were thus supposed to be strictly identical to those produced by the illuminated object surfaces. In this section, we describe some possible variations or complements to this basic procedure.

Instead of examining a virtual image of the object, we can also make measurements on a *real image* of it. This image can be obtained by a lens with a large aperture placed in front of the hologram or can be produced by illuminating the hologram with a reconstruction wave conjugate to the reference wave (see Sect. 3.1.2). Of course the basic relation (4.6) (and the relations to be developed in Sect. 4.2) still apply to the interference phenomenon, the problem being however the selection of a convenient fringe pattern among the whole possible set.

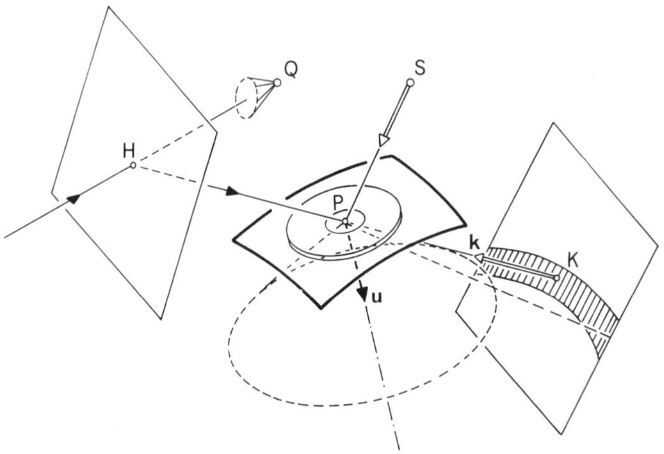

Fig. 4.4. Fringe observed at K, on a screen located behind the object surface, when a small aperture is placed on the real images of the object

In a first manner, proposed by *Tsujiuchi*, *Takeya* and *Matsuda* [4.120, 121], and by *Gates* [4.49, 122] (see also e.g. [4.123–126]), a small aperture is put around the images of P and P′ (Fig. 4.4), thus isolating the light rays that come from these areas. Fringes can first be observed by looking through the aperture, in which case the points K where these fringes are examined lie in front of the object surface (i.e. on the same side as the source S), for instance on the hologram plate. Second, we can place a screen behind the aperture (Fig. 4.4), i.e. behind the object surface. The ensemble of fringes seen on this screen is thus related to one position vector $\boldsymbol{r}$, i.e. one point P and to many observation directions $\boldsymbol{k}$ [remembering (4.7)]; in other words, the fringes mark loci of constant D with the only variable $\boldsymbol{k}$. So, the ensemble of the observations which were made in the preceding section by changing the direction $\boldsymbol{k}$ is now displayed simultaneously on the screen. As we already said [see (4.12)] keeping D constant while changing the direction in which P is observed, means moving on a circular cone, the axis of which lies along $\boldsymbol{u}$. The fringes, which represent the intersection of such cones with the screen plane, are thus circles, ellipses, parabolas or hyperbolas [4.51, 52]. The displacement vector can be measured by either of the two methods described in Sect. 4.1.2. However, because, for the determination of the absolute fringe-order, it is necessary to move the aperture over the object surface, starting from a point that has undergone no displacement (zero-order fringe) and bringing it to the envisaged point P while counting the fringes that pass K, it would be preferable to use the relative fringe-order method. We should remember here that the ensemble of possible observation directions has the same limitations, due to the dimensions of the hologram plate [4.127], as in the case in which a virtual image is observed through this plate. Therefore, for an accurate determination of $\boldsymbol{u}$, it may again be advisable to use several holograms.

A second manner of using the real image of the object surface consists in reconstructing this image by use of only a small region around a point H of the hologram, in the way suggested by *Matsumoto*, *Iwata* and *Nagata* [4.128], or by *Fossati-Bellani* and *Sona* [4.129, 130] (see also [4.109, 110]). For that purpose, the hologram can be illuminated simply by an unexpanded laser beam, acting as part of a wave conjugate to the reference beam (Fig. 4.5). The whole image of the object surface is reconstructed, but the fringes which can be seen on a screen relate all to H, i.e. to observation directions defined by the lines HP. For simplification, the screen can be located at the position of P itself.

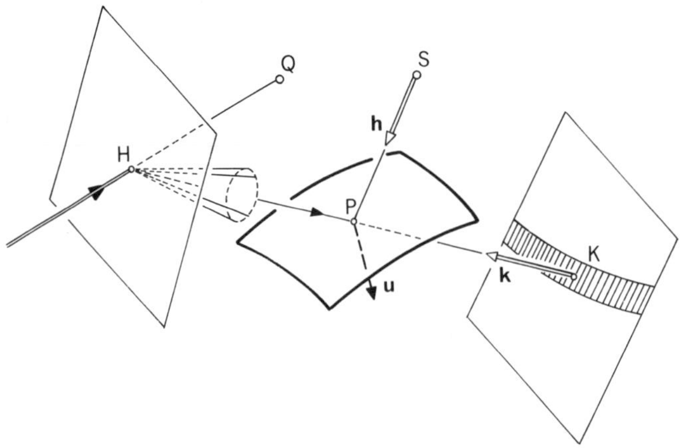

Fig. 4.5. Fringe observed at K, on a screen, when the real images of the object are reconstructed by only one ray

To visualize the relations established in Sect. 4.1.1, we may represent graphically the loci of constant optical path or optical-path difference [4.124, 125, 131–141] by means of what *Abramson* has named *holo-diagrams*. Especially, the nonlinearized equation (4.1) can be interpreted easily: if we choose a fixed point K, for example in front of the object surface, i.e. $L < 0$, the locus of the points P for which the optical-path difference $L_S - L = L_S + |L|$ is a constant forms an ellipsoid with foci S and K; on the other hand, if P and P′ are fixed, the locus of the points K where the optical-path difference $L' - L$ is constant is a hyperboloid with foci at P and P′. These hyperboloids become the already mentioned cones of apex P and of axis $\boldsymbol{u}$ if, as in the preceding, we use the linearized expression (4.6) instead of (4.1).

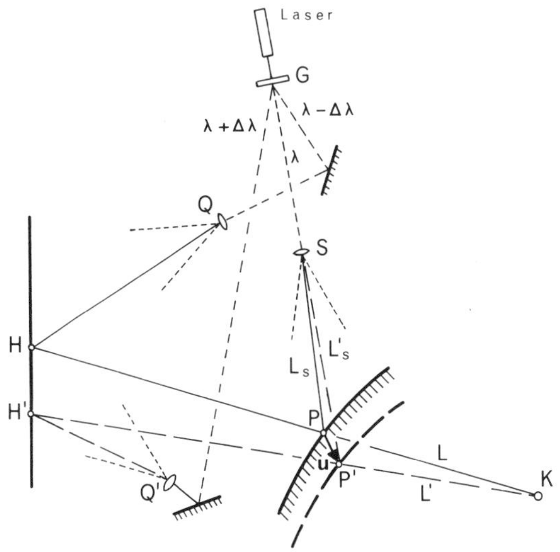

Fig. 4.6. Optical arrangement for heterodyne holographic interferometry. G: for example, a rotating radial grating

To provide faster and more accurate measurements of the displacement vector, some authors make use of an *electro-optical* apparatus [4.129, 130, 142–150]. Because this additional equipment in no way affects the principles of the methods we have presented, we shall not go into its technical details; nevertheless, it may be useful to describe briefly the fundamentals of the interesting proceeding chosen by *Dändliker*, *Ineichen* and *Mottier*, which are somewhat different from those assumed so far. The basic idea of their *heterodyne holographic interferometry* is to have a small frequency shift between the two interfering wave fields. In the case of double exposure, this is achieved by having two reference sources (Fig. 4.6): a source Q is used to record and reconstruct the wave field that emanates from the undeformed object {P} whereas a source Q′ serves for the deformed object {P′}. The difference between the frequencies of the lights emitted by Q and Q′ is produced for example by a rotating radial grating that provides coherent lights with wavelengths λ, $\lambda + \Delta\lambda$ and $\lambda - \Delta\lambda$.

This grating, however, rotates only during the reconstruction so that the recordings are made with wave fields of the same wavelength λ. The presence of two reference sources and a wavelength change between the recording and the reconstruction suggests that this technique ought to be classed among those taking advantage of the modifications of the optical arrangement at the reconstruction, but, because the arrangement keeps the same geometry and because $\Delta\lambda$ is small, the changes of the interference phenomenon may be neglected, as we shall show in Sect. 4.3.1. In point K, with position vector $\boldsymbol{\rho}$, we then write the total amplitude due to the two considered rays as

$$A(\boldsymbol{\rho}) \exp[-2\pi\mathrm{i}(\nu + \Delta\nu)t] + A'(\boldsymbol{\rho}) \exp \frac{-2\pi\mathrm{i}D}{\lambda} \exp[-2\pi\mathrm{i}(\nu - \Delta\nu)t],$$

where $A(\boldsymbol{\rho})$ and $A'(\boldsymbol{\rho})$ are the complex amplitudes of the rays that come from P and P′, respectively (at the moment, we only need to specify that they are almost equal), and where $\Delta\nu$ is the frequency shift. D is the optical-path difference between the two rays and is given by (4.6). Isolating the factor $\exp(-2\pi\mathrm{i}\nu t)$, we may calculate the intensity as in the case of monochromatic light, i.e. as

$$\frac{1}{2}\left[|A|^2 + |A'|^2 + 2|A|\,|A'| \cos\left(\frac{2\pi}{\lambda} D - 4\pi\Delta\nu t\right)\right]. \tag{4.13}$$

The interesting term $2\pi D/\lambda$ appears as the phase of the intensity modulation, which has the beat frequency $2\Delta\nu$. Because $\Delta\nu$ is small, this oscillation can be resolved by a photodetector. By comparison with a reference signal, as for example the light that comes from parts of the object that have undergone no displacement (this plays the role of the zero-order fringe), the phase of the vibration, and thus D, can be measured very accurately: the fringe order $n = D/\lambda$ can be determined up to $\sim 1/1000$. Contrary to most other methods, it does not matter if n is an integer or not, which much improves the absolute fringe-order method.

Another variation of the basic arrangement consists in using *two sources* S and S', separated by a vector $\boldsymbol{s}$ or with different phases, to illuminate the object (Fig. 4.7). If the first source is used to record the undeformed object and the second to record the deformed object [Ref. 3.21, p. 100, 4.151–154], the optical-path difference (4.1) is

$$\begin{aligned} D &= (L_S^r - L) - (L_S' - L') \\ &= \boldsymbol{u}\cdot\boldsymbol{g} + \boldsymbol{s}\cdot\boldsymbol{h}, \end{aligned} \tag{4.14}$$

provided that $|\boldsymbol{s}| \ll L_S^r$, or

$$\begin{aligned} D &= (L' - L) - (L_S' - L_S'^r) + (L_S^r - L_S'^r) \\ &= \boldsymbol{u}\cdot(\boldsymbol{k} - \boldsymbol{h}') + (L_S^r - L_S'^r), \end{aligned} \tag{4.15}$$

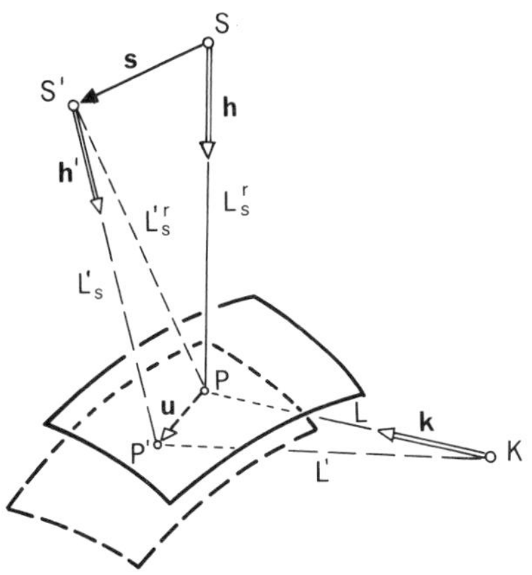

Fig. 4.7. Fringe formation with two light sources S and S′ illuminating the undeformed and deformed objects

if $\boldsymbol{s}$ is arbitrary. The latter shows that the *sign of the displacement* now affects the value of D and not only its sign. Indeed, the first term, that contains $\boldsymbol{u}$, appears as a perturbation (let us recall that $u \ll L_S^r$ and also $u \ll |L_S^r - L_S'^r|$) of the fringe pattern expressed by the second term, which depends on only the position of P and represents the fringes obtained by illuminating the undeformed body, alone, with both S and S′. As shown particularly by (4.14) where $s = O(u)$, using two sources or shifting S to S′ permits modification of the fringe order with respect to that of conventional holographic interferometry; in Sect. 4.3, we shall come back to this point. It is also possible to use the two sources together to illuminate the undeformed and also the deformed object. So, each object surface is covered with a fringe field. Superimposing two object surfaces means adding two fringe families, which thus form *moiré* patterns (e.g. [4.56, 137, 155–160]). Similar results are obtained when the fringe pattern related to two observation directions $\boldsymbol{k}$ are superimposed [4.161]. However, the description of the moiré-pattern method of measuring displacement lies beyond the scope of this treatise.

Finally, let us mention that a Fourier transform of a photograph of the holographic fringes can also be analyzed for determination of the displacement [4.162].

Concluding, we might add that, for clarity's sake, we have grouped current methods, as seen from a theoretical standpoint. In practice however, combinations of these methods either with each other, with those described in the following, or even with other techniques, may prove useful.

4.1.4 Determination of Strain with Finite Differences

In the previous sections, we explained some of the typical methods that, by use of the basic relation (4.6), allow us to determine the displacement field on the surface of an object. However, except for some cases, such as vibration analysis, this is only a first step, the wanted quantities being the components of the strain tensor, which in turn depends on the derivative of the displacement [see (2.105)]. To approximate this derivative, a quite common method in experimental strain analysis consists in forming *finite differences* (see e.g. [4.112, 144, 163]). It must be noted that this second step is possible only if the displacement vectors are known with sufficient accuracy, because the error of the difference of two neighboring quantities is much larger than the errors of the quantities themselves. Therefore, while some of the efforts on holographic interferometry are devoted to improving measurements of displacement, others try to avoid this problem by measuring directly the derivative, as we shall see in Sect. 4.2.5.

At present, we assume that the displacement vector $\boldsymbol{u}$ is explicitly known at a series of neighboring points on the object surface, and that the tensor $\nabla_{\mathrm{n}} \otimes \boldsymbol{u}$ is sought (let us recall that ∇_{n} is the two-dimensional surface-derivative operator). Because the deformation is supposed to be small, the tensor γ of the strain at the surface, the pivot rotation Ω around the unit normal $\boldsymbol{n}$, as well as the vector $\boldsymbol{\omega}$ of the inclination of $\boldsymbol{n}$, can be obtained from the additive decomposition (2.109), i.e.

$$\nabla_{\mathrm{n}} \otimes \boldsymbol{u} = \gamma + \Omega E + \boldsymbol{\omega} \otimes \boldsymbol{n},$$

or conversely from (2.105), i.e.

$$\gamma = \frac{1}{2} N[\nabla_{\mathrm{n}} \otimes \boldsymbol{u} + (\nabla_{\mathrm{n}} \otimes \boldsymbol{u})^{\mathrm{T}}]N,$$

and from (2.111). To calculate $\nabla_{\mathrm{n}} \otimes \boldsymbol{u}$ at a point P, we write the first-order approximations

$$\Delta \boldsymbol{u}_{\alpha} = \boldsymbol{u}_{\alpha} - \boldsymbol{u} \simeq \Delta \boldsymbol{r}_{\alpha}(\nabla_{\mathrm{n}} \otimes \boldsymbol{u}) \qquad (\alpha = 1, 2), \tag{4.16}$$

where $\Delta \boldsymbol{r}_1$ and $\Delta \boldsymbol{r}_2$ are finite increments from P to neighboring points P_1 and P_2 (Fig. 4.8). In components, these two vectorial equations give us six linear relations. To take account of the fact that $\nabla_{\mathrm{n}} \otimes \boldsymbol{u}$ is a semi-interior tensor, because ∇_{n} lies in the plane tangent to the surface at P, the condition

$$\boldsymbol{n}(\nabla_{\mathrm{n}} \otimes \boldsymbol{u}) = 0 \tag{4.17}$$

must also be fulfilled, so that we get nine equations for the nine components of the unknown tensor.

Let us break the two latter relations into their components, first with respect to an arbitrary *global* coordinate system. This system, centered at a point O, with base vectors $\mathring{\boldsymbol{g}}_i$ ($i = 1, 2, 3$, see Sect. 2.1.1), serves for all points of the object surface; for convenience, it could be cartesian (Fig. 4.8). The quantities that

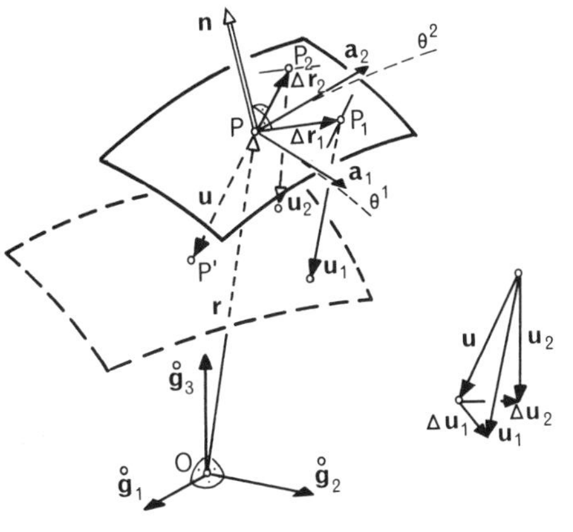

Fig. 4.8. Forming finite differences of the displacement vector $\boldsymbol{u}$

appear in (4.16, 17) are then written

$$\Delta \boldsymbol{r}_\alpha = \Delta x^i_\alpha \mathring{\boldsymbol{g}}_i, \qquad \Delta \boldsymbol{u}_\alpha = \Delta u^j_\alpha \mathring{\boldsymbol{g}}_j,$$

$$\boldsymbol{n} = n^k \mathring{\boldsymbol{g}}_k, \qquad \boldsymbol{\nabla}_{\mathrm{n}} \otimes \boldsymbol{u} = U_l{}^m \mathring{\boldsymbol{g}}^l \otimes \mathring{\boldsymbol{g}}_m,$$

where the latin indices range from 1 to 3, whereas the greek indices take only the values 1, 2. We now obtain for (4.16, 17)

$$\Delta u^m_\alpha = \Delta x^l_\alpha U_l{}^m, \tag{4.18}$$

$$n^l U_l{}^m = 0. \tag{4.19}$$

We may then choose a better-suited system, namely that formed by the base vectors $\boldsymbol{a}_\alpha$ in the tangent plane at P (see Sect. 2.1.3). First, this will mean that the coordinates η^α are *local* and situated in the tangent plane; the components of $\boldsymbol{u}_\alpha$ in P_α must thus be expressed in this system bound to P. Assuming furthermore that the increments $\Delta \boldsymbol{r}_\alpha$ are parallel to the base vectors $\boldsymbol{a}_\alpha$, we have

$$\Delta \boldsymbol{r}_\alpha = \Delta \eta^\alpha \boldsymbol{a}_\alpha \qquad \text{(no sum over } \alpha),$$

$$\Delta \boldsymbol{u}_\alpha = \Delta u^\beta_\alpha \boldsymbol{a}_\beta + \Delta u^3_\alpha \boldsymbol{n},$$

$$\boldsymbol{\nabla}_{\mathrm{n}} \otimes \boldsymbol{u} = U_\lambda{}^\mu \boldsymbol{a}^\lambda \otimes \boldsymbol{a}_\mu + U_\lambda{}^3 \boldsymbol{a}^\lambda \otimes \boldsymbol{n},$$

because $U_3{}^\lambda = U_3{}^3 = 0$ from (4.17). Equation (4.16) is then written (again without summing over α)

$$\Delta u^\mu_\alpha \boldsymbol{a}_\mu + \Delta u^3_\alpha \boldsymbol{n} = \Delta \eta^\alpha U_\alpha{}^\mu \boldsymbol{a}_\mu + \Delta \eta^\alpha U_\alpha{}^3 \boldsymbol{n},$$

that is to say

$$U_\alpha{}^\mu = \frac{\Delta u^\mu_\alpha}{\Delta \eta^\alpha}, \qquad U_\alpha{}^3 = \frac{\Delta u^3_\alpha}{\Delta \eta^\alpha}.$$

The unknown tensor, as such, is thus expressed as (α, μ summed)

$$\nabla_{\mathrm{n}} \otimes \boldsymbol{u} = \frac{\Delta u_{\alpha}^{\mu}}{\Delta \eta^{\alpha}} \boldsymbol{a}^{\alpha} \otimes \boldsymbol{a}_{\mu} + \frac{\Delta u_{\alpha}^{3}}{\Delta \eta^{\alpha}} \boldsymbol{a}^{\alpha} \otimes \boldsymbol{n}. \tag{4.20}$$

Second, if we suppose that $\boldsymbol{u}(\theta^1, \theta^2)$ is given by its components with respect to the varying base vectors $\boldsymbol{a}_\alpha(\theta^1, \theta^2)$, where θ^1, θ^2 are *global curvilinear* coordinates, and if we take into account the decomposition (2.39), i.e. $\boldsymbol{u} = v^\beta \boldsymbol{a}_\beta + w\boldsymbol{n}$, we can write, by analogy to (2.56),

$$\nabla_{\mathrm{n}} \otimes \boldsymbol{u} = \left(\frac{\Delta v^{\mu}}{\Delta \theta^{\alpha}} + \Gamma^{\mu}_{\alpha\beta} v^{\beta} - B_{\alpha}{}^{\mu} w\right) \boldsymbol{a}^{\alpha} \otimes \boldsymbol{a}_{\mu} + \left(B_{\alpha\beta} v^{\beta} + \frac{\Delta w}{\Delta \theta^{\alpha}}\right) \boldsymbol{a}^{\alpha} \otimes \boldsymbol{n}. \tag{4.21}$$

Here the change of the basis appears in the Christoffel symbols $\Gamma^{\mu}_{\alpha\beta}$ and in the curvature factors $B_{\alpha}{}^{\mu}$, $B_{\alpha\beta}$.

In order to avoid the present cumbersome relations, as well as the explicit computation of the displacement, it is often preferable to measure increments ΔD of the optical-path difference and then to express the derivative of the displacement directly as a function of this quantity. This procedure will be treated in Sect. 4.2.5, devoted to the determination of strain with the derivatives of D.

4.2 Derivatives of the Optical-Path Difference in Standard Holographic Interferometry

In this section, we shall deal with the aspects of fringe formation relating to a derivative of the optical-path difference; this means that the derivative of the displacement vector will also be present. First, we proceed with the simple description of the interference we used so far and calculate the fringe spacing and direction. Next, we shall take up another, more complete theory, which considers all of the light rays reflected by a small region on the object surface. Depending on how the optical-path difference of these rays varies, the fringes become more or less visible. Therefore, we must also speak of their contrast and their localization. Having thus explained the optical phenomenon, we can then see how the deformation may be measured with it.

4.2.1 Fringe Spacing and Direction

Determining the fringe spacing and direction means calculating the variation of the fringe-order n. To express this fringe-order, we use again the linearized relation (4.6) of the optical-path difference $D = n\lambda$. As we pointed out at the end of Sect. 4.1.1, D is a function of two variables, namely of the position vector $\boldsymbol{r}$ of P on the object surface and of the observation direction $\boldsymbol{k}$. D can therefore change in several ways; so in each situation examined, we must indicate how these two variables change; especially, their interdependence must be expressed.

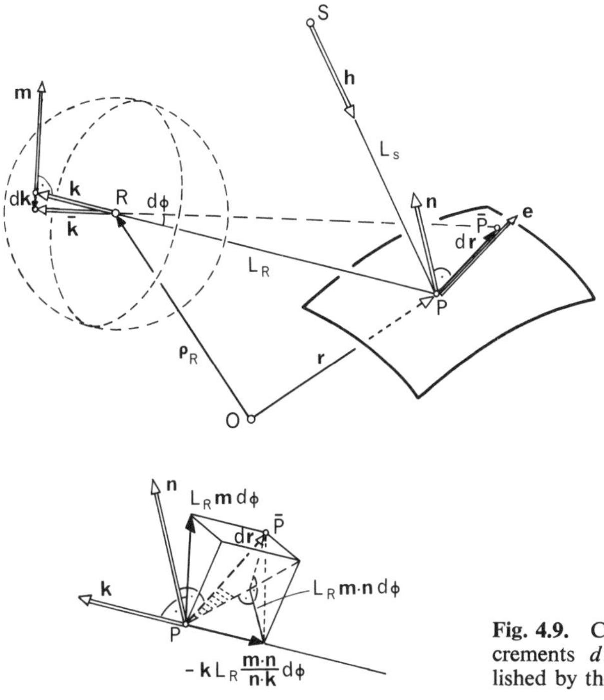

Fig. 4.9. Connection between the increments $d\boldsymbol{r}$ and $d\boldsymbol{k} = -\boldsymbol{m}d\phi$ established by the collineation with center R

First suppose that the fringes are observed *from a fixed point* R (Fig. 4.9); this is the case, for instance, when a telescope or a camera is "placed at R", which means that the rays that enter the instrument go through R. Then, $\boldsymbol{r}$ and $\boldsymbol{k}$ are related by

$$\boldsymbol{r} = \boldsymbol{\rho}_{\mathrm{R}} - L_{\mathrm{R}}\boldsymbol{k}, \tag{4.22}$$

where $\boldsymbol{\rho}_{\mathrm{R}}$ is the position vector of R and L_{R}, the distance RP. This denotes the connection established by the *collineation center* R between the unit sphere centered at R, on which $\boldsymbol{k}$ varies, and the object surface on which $\boldsymbol{r}$ varies (see Sect. 2.1.5). This definition of D on both surfaces is justified because, as shown by (4.6), D does not change along a line RP; this is what *Stetson* has called the "observer-projection" theorem [4.164]. Further, it might be added that, as in the preceding sections, we do not take account of the visibility of the fringes so far but only of their order. We can thus write for the optical-path difference (4.6)

$$D_R(\boldsymbol{r}) = D[\boldsymbol{r}, \boldsymbol{k}(\boldsymbol{r})] = \boldsymbol{u}(\boldsymbol{r})\cdot[\boldsymbol{k}(\boldsymbol{r}) - \boldsymbol{h}(\boldsymbol{r})], \tag{4.23}$$

or

$$D_R(\boldsymbol{k}) = D[\boldsymbol{r}(\boldsymbol{k}), \boldsymbol{k}] = \boldsymbol{u}(\boldsymbol{k})\cdot[\boldsymbol{k} - \boldsymbol{h}(\boldsymbol{k})]; \tag{4.24}$$

the index R draws attention to the fixed point. Consequently, the increment can assume the forms

$$dD_R = d\boldsymbol{r}\cdot\nabla_n D_R = d\boldsymbol{k}\cdot\nabla_k D_R, \tag{4.25}$$

where ∇_n denotes the two-dimensional derivative operator on the surface of the object, the unit normal of which is called $\boldsymbol{n}$, and where ∇_k denotes the similar operator on the unit sphere. $d\boldsymbol{r}$ and $d\boldsymbol{k}$ are increments situated in the respective planes of these operators, i.e. the tangent planes. Let us also introduce a unit vector $\boldsymbol{m}$ parallel to $d\boldsymbol{k}$ and a unit vector $\boldsymbol{e} = d\boldsymbol{r}/ds$ parallel to $d\boldsymbol{r}$. We have then

$$d\boldsymbol{r} = \boldsymbol{e}ds, \qquad d\boldsymbol{k} = -\boldsymbol{m}d\phi.$$

These increments are not independent of each other. Indeed, similar to (2.74), we can write (see also Fig. 4.9)

$$d\boldsymbol{r} = -L_R \boldsymbol{M}^T d\boldsymbol{k}. \tag{4.26}$$

Let us recall that $\boldsymbol{M}^T$ is the transpose of the tensor $\boldsymbol{M} = \boldsymbol{I} - \boldsymbol{n}\otimes\boldsymbol{k}/\boldsymbol{n}\cdot\boldsymbol{k}$. This last tensor projects obliquely along $\boldsymbol{n}$ onto the plane normal to $\boldsymbol{k}$, whereas $\boldsymbol{M}^T$ marks an oblique projection along $\boldsymbol{k}$ onto the plane tangent to the object surface (Fig. 4.10). By use of these relations, we are now able to express dD_R as a function of any increment and also of any derivative operator. For our purposes, the most convenient forms are

$$\frac{dD_R}{d\phi} = L_R \boldsymbol{m}\cdot\boldsymbol{M}\nabla_n D_R, \qquad \frac{dD_R}{ds} = \boldsymbol{e}\cdot\nabla_n D_R, \tag{4.27}$$

where the gradient of the optical-path difference, as given by (4.23), is, in accordance with the rules (2.30),

$$\nabla_n D_R = (\nabla_n \otimes \boldsymbol{u})\boldsymbol{g} - (\nabla_n \otimes \boldsymbol{h})\boldsymbol{u} + (\nabla_n \otimes \boldsymbol{k})|_R \boldsymbol{u}. \tag{4.28}$$

In the first term, we recognize the derivative of the displacement vector on the surface, which contains the strains and the rotations [see (2.109)]. Next, for the

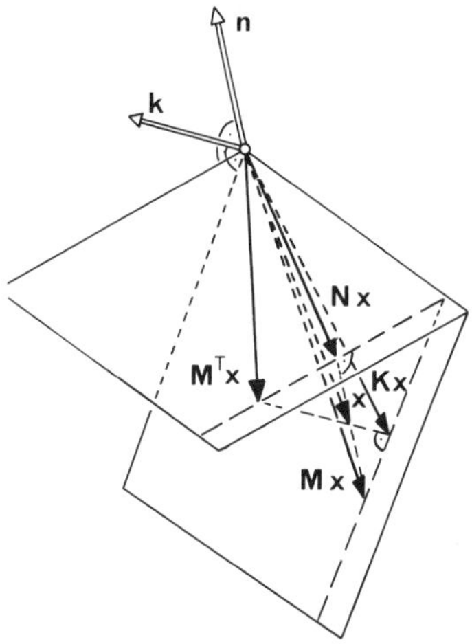

Fig. 4.10. Normal and oblique projections of an arbitrary vector $\boldsymbol{x}$ onto the planes perpendicular to $\boldsymbol{n}$ and $\boldsymbol{k}$

derivatives of the unit vectors $\boldsymbol{h}$ and $\boldsymbol{k}$, which give the directions between the fixed points S and R and the point P which varies on the object surface, we can write, similar to (2.73),

$$\nabla_{\mathrm{n}} \otimes \boldsymbol{h} = \boldsymbol{N}\nabla \otimes \boldsymbol{h} = \frac{1}{L_{\mathrm{S}}}\boldsymbol{NH}, \tag{4.29}$$

$$\nabla_{\mathrm{n}} \otimes \boldsymbol{k}|_{\mathrm{R}} = \boldsymbol{N}\nabla \otimes \boldsymbol{k}|_{\mathrm{R}} = -\frac{1}{L_{\mathrm{R}}}\boldsymbol{NK}, \tag{4.30}$$

where the normal projections $\boldsymbol{H} = \boldsymbol{I} - \boldsymbol{h} \otimes \boldsymbol{h}$, $\boldsymbol{K} = \boldsymbol{I} - \boldsymbol{k} \otimes \boldsymbol{k}$ and $\boldsymbol{N} = \boldsymbol{I} - \boldsymbol{n} \otimes \boldsymbol{n}$ on the planes perpendicular respectively to $\boldsymbol{h}$, $\boldsymbol{k}$ and $\boldsymbol{n}$ appear (Fig. 4.10). Thus, the gradient of the optical-path difference, when the point R is fixed, is

$$\nabla_{\mathrm{n}} D_{\mathrm{R}} = (\nabla_{\mathrm{n}} \otimes \boldsymbol{u})\boldsymbol{g} - \frac{1}{L_{\mathrm{S}}}\boldsymbol{NHu} - \frac{1}{L_{\mathrm{R}}}\boldsymbol{NKu}; \tag{4.31}$$

with the abbreviation

$$\boldsymbol{w} = (\nabla \otimes \boldsymbol{u})\boldsymbol{g} - \frac{1}{L_{\mathrm{S}}}\boldsymbol{Hu}, \tag{4.32}$$

(4.31) becomes

$$\nabla_{\mathrm{n}} D_{\mathrm{R}} = \boldsymbol{N}\left(\boldsymbol{w} - \frac{1}{L_{\mathrm{R}}}\boldsymbol{Ku}\right). \tag{4.33}$$

Because of the simplification $\boldsymbol{MN} = \boldsymbol{M}$, $\boldsymbol{MK} = \boldsymbol{K}$ of the successive projections (see Fig. 4.10), we finally obtain the two forms

$$\frac{dD_R}{d\phi} = \lambda \frac{dn_R}{d\phi} = \boldsymbol{m} \cdot (L_R \boldsymbol{M} \boldsymbol{w} - \boldsymbol{K} \boldsymbol{u}), \tag{4.34}$$

$$\frac{dD_R}{ds} = \lambda \frac{dn_R}{ds} = \boldsymbol{e} \cdot \left(\boldsymbol{N} \boldsymbol{w} - \frac{1}{L_R} \boldsymbol{N} \boldsymbol{K} \boldsymbol{u} \right), \tag{4.35}$$

(see e.g. [Ref. 4.165, p. 557; Ref. 4.166, Eq. (9); Ref. 4.193, Eq. (36)]). Equation (4.34) expresses the derivative of the fringe-order with respect to the parameter ϕ in a direction $\boldsymbol{m}$ situated in the plane perpendicular to $\boldsymbol{k}$ at the distance 1 from R. On the other hand, (4.35) gives the derivative of n_R with respect to the parameter s, in the direction $\boldsymbol{e}$ situated in the plane tangent to the object surface. If we set $\Delta n_R = 1$, we obtain the *fringe spacing*, which is $L_R \Delta\phi$ in the plane perpendicular to $\boldsymbol{k}$ passing through P (in the plane where the fringes are localized, at the distance L behind P, we would have $(L_R + L)\Delta\phi$). On the object surface, the fringe spacing is simply Δs. Moreover, the *distance* between the fringes, i.e. the spacing measured perpendicularly to the fringes, is, for example in the plane normal to $\boldsymbol{k}$ passing through P,

$$\min. |L_R \Delta\phi| \simeq \frac{L_R \lambda}{|L_R \boldsymbol{M} \boldsymbol{w} - \boldsymbol{K} \boldsymbol{u}|}, \tag{4.36}$$

because $\boldsymbol{m}$ is now parallel to $L_R \boldsymbol{M} \boldsymbol{w} - \boldsymbol{K} \boldsymbol{u}$. Finally, the *direction* of a fringe is found from the condition

$$dD_R = 0.$$

The vector $\boldsymbol{m}$ that corresponds to this direction is then perpendicular to the vector $L_R \boldsymbol{M} \boldsymbol{w} - \boldsymbol{K} \boldsymbol{u}$ and, similarly, $\boldsymbol{e}$ appears perpendicular to $\boldsymbol{N} \boldsymbol{w} - \boldsymbol{N} \boldsymbol{K} \boldsymbol{u} / L_R$. As proposed by *Stetson* [4.167, 168], these two vectors perpendicular to the fringe can be called *fringe-vectors*.

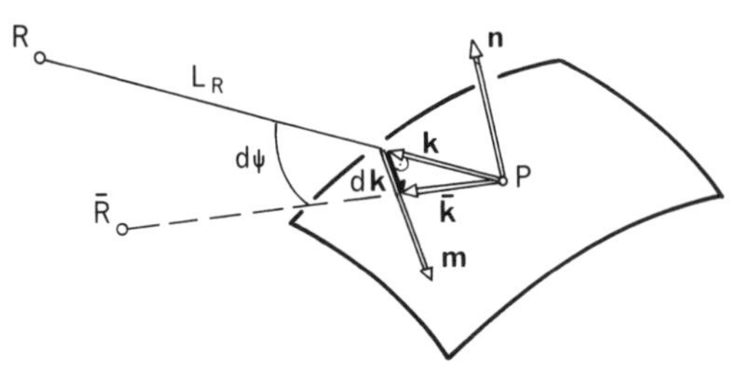

Fig. 4.11. Variation of the observer's position from R to R̄, with a fixed object point P

As a second variation of the fringe-order, let us consider the case in which the observer changes his position but still looks at the *same point* P (Fig. 4.11);

this is in fact the situation we dealt with in Sect. 4.1. Now, because $\boldsymbol{r}$, and thus also $\boldsymbol{u}$ and $\boldsymbol{h}$, are constant, we can write the increment of D_P only in the form

$$dD_P = d\boldsymbol{k} \cdot \boldsymbol{\nabla}_k D_P; \tag{4.37}$$

the index P indicates that here P is the fixed point. With

$$d\boldsymbol{k} = \boldsymbol{m} d\psi,$$

and with the derivative of $\boldsymbol{k}$ on the unit sphere centered in P

$$\boldsymbol{\nabla}_k \otimes \boldsymbol{k}|_P = \boldsymbol{K}, \tag{4.38}$$

we obtain (see e.g. [Ref. 4.9, Eq. (28)] or [Ref. 4.120, Eq. (28)])

$$\frac{dD_P}{d\psi} = \lambda \frac{dn_P}{d\psi} = \boldsymbol{m} \cdot \boldsymbol{K}\boldsymbol{u} = \boldsymbol{m} \cdot \boldsymbol{u}. \tag{4.39}$$

$\boldsymbol{K}$ can be omitted here because $\boldsymbol{m}$ lies in the plane perpendicular to $\boldsymbol{k}$. In the fringe pattern that corresponds to the present situation, i.e. that observed when a small aperture is placed on the real image of P (see Fig. 4.4), the *direction* of a fringe is given by

$$dD_P = 0,$$

i.e. the vector $\boldsymbol{m}$ that corresponds to this direction is perpendicular to $\boldsymbol{K}\boldsymbol{u}$. Finally, the angular distance between the fringes is

$$\min |\Delta\psi| \simeq \frac{\lambda}{|\boldsymbol{K}\boldsymbol{u}|}. \tag{4.40}$$

We have thus established two kinds of derivatives of the optical-path difference. In Sect. 4.2.5, we shall see how they can be used to determine the deformation.

4.2.2 Fringe Contrast or Visibility, Concepts of Localization

So far, we have regarded the interference fringes as being formed, in each observation direction, by only two light rays, reflected by one and the same material point on the undeformed and the deformed object surface. In the present section, we shall investigate how the neighboring rays add to the preceding ones. As we shall see, depending on the examination locus of the phenomenon, the contribution of the rays produces fringes more or less visible or, in other words, fringes with higher or lower contrast, hence, the fringes are said to be localized at certain places, yet to be defined. This point has been studied by several authors [Ref. 3.14, Sect. 15.3; 4.9, 26, 120, 121, 157, 164–204], who used the concepts of diffraction and coherence theories or also purely geometrical conditions. Here, we follow the first path which also includes the second.

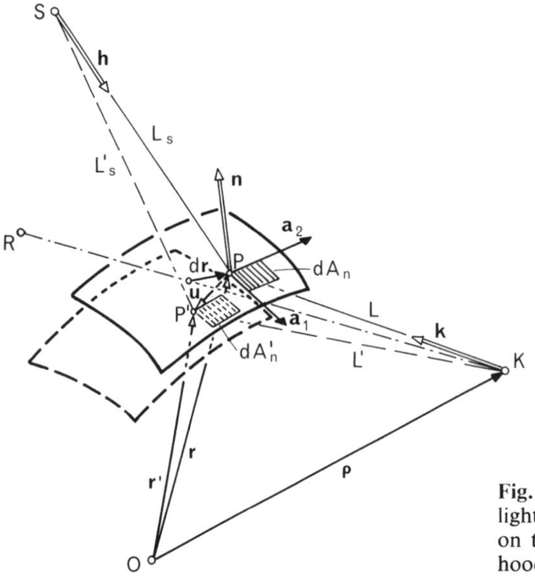

Fig. 4.12. Interference at K of two light rays reflected by points P and P′ on the object surface, in the neighborhood of the central ray KR

As in Sect. 4.1.1, we consider the interference that occurs at some point K, with position vector $\boldsymbol{\rho}$, here being supposed to lie, for instance, behind the object surface (Fig. 4.12). This point is reached by light rays emitted by the point source S and diffusely reflected by the rough surfaces of the object in its undeformed and deformed configurations (as in the preceding, the holographic reconstruction is supposed to be ideal, so that we need not distinguish between the object surfaces and their images).

To describe this diffuse reflection, we will adopt the following model (see e.g. [Ref. 4.184, Sect. II.3.1; Ref. 4.192, Chap. 3; Ref. 4.201, Chap. IV]). First, we assume the existence of two *smooth mean surfaces* (Fig. 4.13), for instance the mean undeformed surface A, on which the points P, with their position vectors $\boldsymbol{r}$, are defined. The actual surface A^*, with the points P*, has then a microprofile that varies in a random way from the mean surface. We envisage an ensemble of undeformed surfaces that have the same mean position but different detailed structures, and similarly an ensemble of deformed surfaces, the detailed structure at corresponding points P and P′ on a pair of such surfaces being identical. For this situation, we describe the reflection by means of a *random complex reflection function* $G(\boldsymbol{r})$ [Ref. 3.5, p. 401]. $G(\boldsymbol{r})$ is defined as the

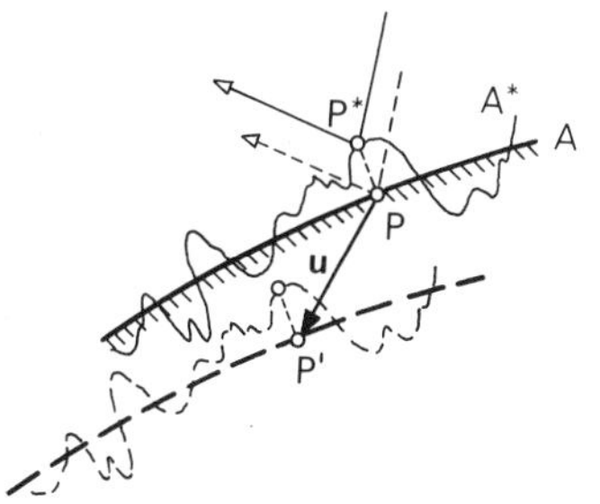

Fig. 4.13. Model of the roughness of the undeformed and deformed surfaces

ratio of the reflected amplitude to the incident amplitude at a point P. In our simplified model, we disregard the dependence on the directions of illumination and reflection, and also any nonuniform reflectance of the surface. The influences of direction will be considered only by multiplying $G(\boldsymbol{r})$ by another, deterministic function. For a more complete treatment of the properties of such surfaces, the reader may refer to the previously quoted works and also to [4.205–209]; as for us, we shall only mention two assumptions. First, the real and imaginary parts of $G(\boldsymbol{r})$ are independent and have zero mean values, which we express by the expected value or ensemble average

$$E[G(\boldsymbol{r})] = 0, \tag{4.41}$$

this being fulfilled at any point of the surface (stationarity with respect to $\boldsymbol{r}$). Second, the autocorrelation of $G(\boldsymbol{r})$ can be approximated by the Dirac function δ (see the experimental verifications of *Monneret* [Ref. 4.184, p. 34] and of *Goldfischer* [4.206]), i.e. we have

$$E[G(\boldsymbol{r})G^*(\bar{\boldsymbol{r}})] = C\delta(\boldsymbol{r} - \bar{\boldsymbol{r}}), \tag{4.42}$$

where C is a constant. This property shows the important fact that each point of the surface can be regarded as an independent scatterer, which thus cannot be confounded with its neighbors; unlike the case of a mirror surface, a movement of the rough surface without any change of its mean surface can be recognized.

Now, the light emitted by S produces in P a disturbance of the form

$$\frac{S}{L_S}\exp\left(\frac{2\pi i}{\lambda}L_S\right),$$

where S is a complex constant that takes into account the characteristics of S, and where L_S is the distance SP. By analogy to the *Huygens-Fresnel* principle [Ref. 3.5, p. 370], we consider each point of the surface as an elemental source that emits a *spherical wave*, so that the total amplitude at K, due to the part $A_n \subset A$ that sends light into the aperture $A_{\mathring{n}}$ of the observing system to form the image of K (Fig. 4.14), can be written as

$$U(\rho) = \iint\limits_{A_n} \frac{S}{L_S|L|} K(\boldsymbol{n}\cdot\boldsymbol{h}, \boldsymbol{n}\cdot\boldsymbol{k}) G(\boldsymbol{r}) \exp\left[\frac{2\pi i}{\lambda}(L_S - L)\right] dA_n. \tag{4.43}$$

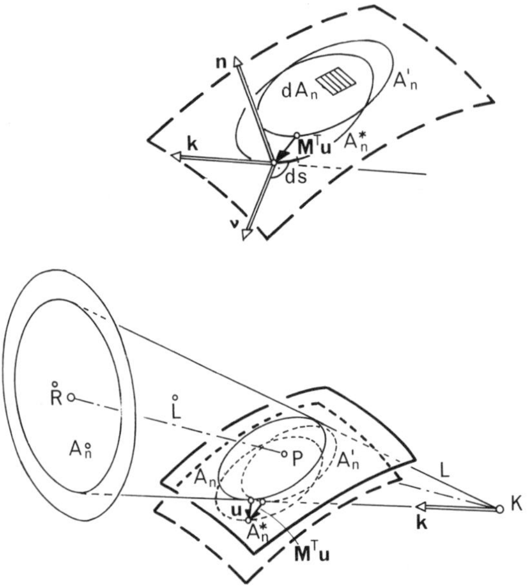

Fig. 4.14. Formation of a fringe at K by all light rays reflected by the parts A_n of the undeformed surface and A_n' of the deformed surface and entering the observation system

K is a deterministic inclination factor and $L = \boldsymbol{k}\cdot(\boldsymbol{r} - \boldsymbol{\rho})$ is the distance PK ($|L| \gg \lambda$, Fig. 4.12), as previously, positive when K is behind the object surface and negative in the opposite case. The surface element dA_n, with the unit normal $\boldsymbol{n}$, can be expressed with respect to the surface coordinates θ^1, θ^2 and the tangential base vectors $\boldsymbol{a}_1$, $\boldsymbol{a}_2$ by [see (2.112)] $dA_n = |\boldsymbol{a}_1 \times \boldsymbol{a}_2| d\theta^1 d\theta^2$. The expression (4.43) is similar to those encountered for instance in [Ref. 4.9, Eq. (14); Ref. 4.157, Eq. (1); Ref. 4.184, Eq. (II.2); Ref. 4.201, Eq. (IX.1)].

Apart from the foregoing, another representation of the diffusely reflected wave field is used, which consists of a superposition of *plane waves* [4.169, 170; Ref. 4.184, Eq. (II.31); Ref. 4.192, Eq. (3.15); 4.193, 200; Ref. 4.201, Eq. (II.2)]. The amplitude produced at K by the whole illuminated part of the undeformed object surface is then

$$\hat{U}(\rho) = \hat{U}(\boldsymbol{r} - L\boldsymbol{k}) = \iint\limits_{F_n} \hat{u}(\boldsymbol{\kappa}) \exp\left(-\frac{2\pi i}{\lambda}\boldsymbol{\kappa}\cdot\boldsymbol{k}L\right) dF_n; \tag{4.44}$$

dF_n, which is often written as $d\kappa_x d\kappa_y$ (or even as $d\boldsymbol{\kappa}$), represents a surface element in the infinitely extended spatial frequency plane F_n (Fig. 4.15). This plane is

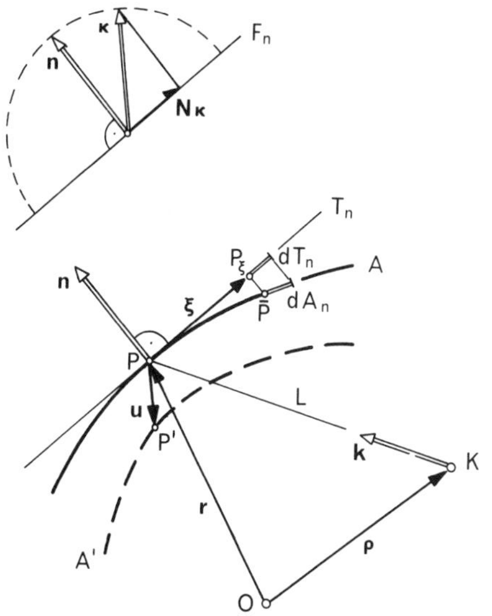

Fig. 4.15. Planes T_n and F_n to which the Fourier representation of the wave fields refers (two-dimensional sketch)

chosen to be paralled to the plane T_{n} tangent to the object surface at some fixed point P and has thus a fixed unit normal $\boldsymbol{n}$. Each plane-wave component is distinguished, first, by its propagation direction $\boldsymbol{\kappa}$, the unit vector $\boldsymbol{\kappa}$ having only two independent components, namely those of the spatial frequency $N\boldsymbol{\kappa}$ (e.g. κ_x, κ_y), which is the normal projection of $\boldsymbol{\kappa}$ onto F_{n}. Second, the phase of a plane wave in K is given by $-\boldsymbol{k}L = \boldsymbol{\rho} - \boldsymbol{r}$, which indicates the position of K with respect to P. Third, the specific complex amplitude is $\hat{u}(\boldsymbol{\kappa}) \equiv \hat{u}(N\boldsymbol{\kappa})$. The latter function is determined by the boundary conditions, i.e. by the wave field on the object surface. Indeed, if instead of K a point $P_\xi \in T_{\mathrm{n}}$, with position vector $\boldsymbol{\xi} \equiv N\boldsymbol{\xi}$ relative to P, is chosen, the amplitude

$$\hat{U}(\boldsymbol{\xi}) = \iint_{F_{\mathrm{n}}} \hat{u}(\boldsymbol{\kappa}) \exp\left(\frac{2\pi\mathrm{i}}{\lambda}\boldsymbol{\kappa}\cdot\boldsymbol{\xi}\right) dF_{\mathrm{n}} = \iint_{F_{\mathrm{n}}} \hat{u}(N\boldsymbol{\kappa}) \exp\left[\frac{2\pi\mathrm{i}}{\lambda}(N\boldsymbol{\kappa})\cdot\boldsymbol{\xi}\right] dF_{\mathrm{n}} \quad (4.45)$$

is supposed to be a given, stochastic function on the (plane) part of T_{n} that corresponds to the near, illuminated (curved) part of the object. Elsewhere, the amplitude is supposed to be zero. Equation (4.45) represents the inverse of a *Fourier transform*; the amplitude spectrum is thus given by

$$\hat{u}(\boldsymbol{\kappa}) = \frac{1}{\lambda^2} \iint_{T_{\mathrm{n}}} \hat{U}(\boldsymbol{\xi}) \exp\left(-\frac{2\pi\mathrm{i}}{\lambda}\boldsymbol{\kappa}\cdot\boldsymbol{\xi}\right) dT_{\mathrm{n}}. \quad (4.46)$$

Both planes F_{n} and T_{n} are infinitely extended (thus, usually integration limits $\pm\infty$ are written) and (4.45, 46) are direct and inverse *integral transformations* between these two-dimensional spaces, in contrast with the geometric linear transformations between the tangent planes that we have encountered previously. $\hat{u}(\boldsymbol{\kappa})$ contains the amplitudes of the travelling part of the wave field, i.e. with $|N\boldsymbol{\kappa}| \leqslant 1$, and of the evanescent part, which does not contribute to the fringe formation and can therefore be omitted in (4.44). To take into account the filtering effect of the optical system used to observe the wave field, an aperture function $\mathring{\Omega}(\boldsymbol{\kappa}, \boldsymbol{\rho})$ must be added, which is defined as

$$\mathring{\Omega}(\boldsymbol{\kappa}, \boldsymbol{\rho}) = \begin{cases} 1 \text{ for the transmitted spatial frequencies,} \\ 0 \text{ for the stopped spatial frequencies.} \end{cases}$$

The resulting complex amplitude at K is then

$$U(\boldsymbol{\rho}) = \iint_{F_n} \mathring{\Omega}(\boldsymbol{\kappa}, \boldsymbol{\rho})\hat{u}(\boldsymbol{\kappa}) \exp\left(-\frac{2\pi i}{\lambda}\boldsymbol{\kappa}\cdot\boldsymbol{k}L\right) dF_n. \tag{4.47}$$

Returning to (4.43), let us now express the amplitude due to the deformed object surface in a similar form

$$U'(\boldsymbol{\rho}) = \iint_{A_n'} \frac{S}{L_S'|L'|} K(\boldsymbol{n}'\cdot\boldsymbol{h}', \boldsymbol{n}'\cdot\boldsymbol{k}') G'(\boldsymbol{r}') \exp\left[\frac{2\pi i}{\lambda}(L_S' - L')\right] dA_n'. \tag{4.48}$$

Because $U'(\boldsymbol{\rho})$ and $U(\boldsymbol{\rho})$ are produced by the same object in two very close configurations, these two amplitudes can be related to each other. Therefore, we look at the functions under the integral symbol at the same material point on the deformed surface as on the undeformed one, i.e. with the same convected coordinates θ^1, θ^2. Then, the microprofiles being the same, we have first

$$G'(\boldsymbol{r}') = G(\boldsymbol{r}) = G(\theta^1, \theta^2).$$

For the quantities that depend on the deformation, we can write [see (4.5), (3.42), (2.110, 115)]

$$L_S' \simeq L_S + \boldsymbol{u}\cdot\boldsymbol{h}, \qquad L' \simeq L + \boldsymbol{u}\cdot\boldsymbol{k},$$

$$\boldsymbol{h}' \simeq \boldsymbol{h} + \frac{1}{L_S}\boldsymbol{H}\boldsymbol{u}, \qquad \boldsymbol{k}' \simeq \boldsymbol{k} + \frac{1}{L}\boldsymbol{K}\boldsymbol{u},$$

$$\boldsymbol{n}' \simeq \boldsymbol{n} - \boldsymbol{\omega},$$

$$dA_n' \simeq (1 + \operatorname{tr}\boldsymbol{\gamma})dA_n,$$

where $\boldsymbol{u}$ is the displacement vector, $\boldsymbol{H}$ (or $\boldsymbol{K}$) the normal projection along $\boldsymbol{h}$ (or $\boldsymbol{k}$), $\boldsymbol{\omega}$ the inclination vector and $\boldsymbol{\gamma}$ the strain tensor. Because the deformation is small, we shall assume that the factor $K(\boldsymbol{n}' \cdot \boldsymbol{h}', \boldsymbol{n}' \cdot \boldsymbol{k}')dA'_n/L'_S|L'|$ is equal to its counterpart on the undeformed surface; what is thus neglected affects only the amplitude of the wave in K. For its phase, we shall use the linear approximation established in Sect. 4.1.1,

$$L'_S - L' \simeq L_S - L + \boldsymbol{u} \cdot (\boldsymbol{h} - \boldsymbol{k}) = L_S - L - D.$$

For the area of integration (Fig. 4.14), we write, with a unit vector $\boldsymbol{v}$ on the contour ∂A_n of A_n and the length s on this contour,

$$\iint_{A_n'} (\cdots) dA'_n \simeq \iint_{A_n} (\cdots) dA_n - \oint_{\partial A_n} (\cdots) \boldsymbol{v} \cdot \boldsymbol{M}^T \boldsymbol{u} ds.$$

$\boldsymbol{M}^T\boldsymbol{u}$, which is the oblique projection of $\boldsymbol{u}$ along $\boldsymbol{k}$ onto the object surface, expresses the displacement of each point of the contour. The approximation of the integration area of A'_n by A_n is thus warrantable, provided that

$$|\boldsymbol{K}\boldsymbol{u}| \ll \frac{L\mathring{r}}{L + \mathring{L}}, \tag{4.49}$$

i.e. provided that the lateral displacement of ∂A_n is much smaller than a characteristic dimension of A_n (overlapping of the two areas). $\mathring{r}$ may be, for instance, the radius of a circular aperture that is situated at the distance $\mathring{L}$ from the object. Thus, according to the foregoing assumptions, light rays reflected by the same material point on the undeformed and deformed surfaces differ from each other by only a *phase factor* [Ref. 4.9, Eq. (17); Ref. 4.157, Eq. (2); Ref. 4.165, Eq. (7); Ref. 4.169, Eq. (13); Ref. 4.170, Eq. (3); Ref. 4.184, Eq. (II.18); Ref. 4.193, Eq. (11)], i.e. we get

$$U'(\rho) = \iint_{A_n} \frac{S}{L_S|L|} K(\boldsymbol{n} \cdot \boldsymbol{h}, \boldsymbol{n} \cdot \boldsymbol{k}) G(\boldsymbol{r}) \exp\left[\frac{2\pi i}{\lambda}(L_S - L - D)\right] dA_n. \tag{4.50}$$

Thus, similar to what we saw for the deformation in Sect. 2.2, the quantities that relate to the deformed surface have been expressed as functions of those that relate to the undeformed surface, which is the reference configuration.

Furthermore, because the aperture $A_{\mathring{n}}$ of the observing system and consequently also A_n are small, we can replace the optical-path difference D by the approximation

$$\bar{D}_K = D + dD_K.$$

From now on, we call P the (fixed) point of the undeformed object surface situated on the *central ray* K$\mathring{\text{R}}$ (Fig. 4.16) and $\bar{\text{P}}$ the variable neighboring point

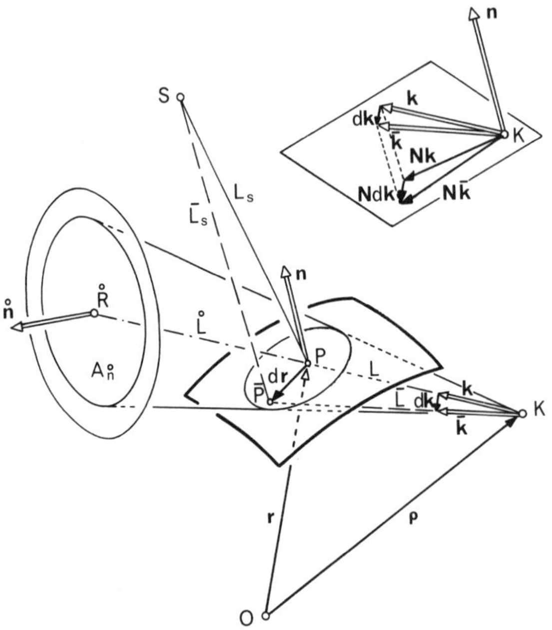

Fig. 4.16. Formation of a fringe at K: expressing the quantities relating to $\bar{\mathrm{P}}$ as functions of those relating to P. Connections between $\boldsymbol{dr}$, $\boldsymbol{dk}$ and $\boldsymbol{Ndk}$

separated from P by the increment $\boldsymbol{dr}$. Quantities at $\bar{\mathrm{P}}$ are marked by a bar as well. To calculate the increment dD_{K}, we proceed in the same way as in Sect. 4.2.1, observing however that K is the point common to all considered light rays, of which the index K reminds us. $\bar{D}_{\mathrm{K}}$ is a function of $\bar{\boldsymbol{r}}$ and $\bar{\boldsymbol{k}}$ (see Sect. 4.1.1), which are now connected by

$$\bar{\boldsymbol{r}} = \boldsymbol{\rho} + \bar{L}\bar{\boldsymbol{k}}. \tag{4.51}$$

Similar to (4.25), we can thus write the two forms

$$dD_{\mathrm{K}} = \boldsymbol{dr} \cdot \boldsymbol{\nabla}_{\mathrm{n}} D_{\mathrm{K}} = \boldsymbol{dk} \cdot \boldsymbol{\nabla}_{\mathrm{k}} D_{\mathrm{K}}. \tag{4.52}$$

Instead of the increments $\boldsymbol{dr}$ on the object surface and $\boldsymbol{dk}$ on the unit sphere around K, some authors use the increment $\boldsymbol{Ndk} = d(\boldsymbol{Nk})$ situated in a plane through K parallel to the plane tangent to the object surface in P (Fig. 4.16). In a cartesian coordinate system with axes x and y in this plane, they thus have the components $\boldsymbol{dr} \triangleq (dx, dy, 0)$ and $d(\boldsymbol{Nk}) \triangleq (dk_x, dk_y, 0)$. Between these three increments, there exists, first, the relation similar to (2.74)

$$\boldsymbol{dr} = L\boldsymbol{M}^{\mathrm{T}}\boldsymbol{dk}, \tag{4.53}$$

because of the collineation with center K, and second, with the oblique projection $\boldsymbol{M}$ that acts as the "inverse" of the normal projection $\boldsymbol{N}$ (see Fig. 4.10), $d\boldsymbol{k} = \boldsymbol{M}d(\boldsymbol{N}\boldsymbol{k})$; thus we obtain

$$d\boldsymbol{r} = L\boldsymbol{M}^{\mathrm{T}}\boldsymbol{M}d(\boldsymbol{N}\boldsymbol{k}). \tag{4.54}$$

Expressing now dD_{K} by means of the derivative on the object surface, which is the most suitable derivative, we have

$$dD_{\mathrm{K}} = Ld\boldsymbol{k}\cdot\boldsymbol{M}\boldsymbol{\nabla}_{\mathrm{n}}D_{\mathrm{K}} = Ld(\boldsymbol{N}\boldsymbol{k})\cdot\boldsymbol{M}^{\mathrm{T}}\boldsymbol{M}\boldsymbol{\nabla}_{\mathrm{n}}D_{\mathrm{K}}. \tag{4.55}$$

The oblique projections $\boldsymbol{M}$ and $\boldsymbol{M}^{\mathrm{T}}$ appear, for instance, in the work of *Stetson* [Ref. 4.164, Eqs. (7, 35); 4.173], where $\boldsymbol{M}^{\mathrm{T}}$ is called "shadow", and, in forms similar to (2.61), in those of *Walles* [Ref. 4.193, Eq. (16)] and of *Přikryl* [Ref. 4.198, Eq. (12)]. The derivative $\boldsymbol{\nabla}_{\mathrm{n}}D_{\mathrm{K}}$ is calculated in the same way as in Sect. 4.2.1; with

$$\boldsymbol{\nabla}_{\mathrm{n}} \otimes \boldsymbol{k}|_{\mathrm{K}} = \boldsymbol{N}\boldsymbol{\nabla} \otimes \boldsymbol{k}|_{\mathrm{K}} = \frac{1}{L}\boldsymbol{N}\boldsymbol{K}, \tag{4.56}$$

we obtain in the present case

$$\boldsymbol{\nabla}_{\mathrm{n}}D_{\mathrm{K}} = (\boldsymbol{\nabla}_{\mathrm{n}} \otimes \boldsymbol{u})\boldsymbol{g} - \frac{1}{L_{\mathrm{S}}}\boldsymbol{N}\boldsymbol{H}\boldsymbol{u} + \frac{1}{L}\boldsymbol{N}\boldsymbol{K}\boldsymbol{u}, \tag{4.57}$$

or with (4.32)

$$\boldsymbol{\nabla}_{\mathrm{n}}D_{\mathrm{K}} = \boldsymbol{N}\left(\boldsymbol{w} + \frac{1}{L}\boldsymbol{K}\boldsymbol{u}\right). \tag{4.58}$$

For the complex amplitude in K due to the deformed surface, we have then (see also [Ref. 4.9, Eq. (24); Ref. 4.192, Eq. (4.34)])

$$U'(\boldsymbol{\rho}) = S\exp\left(\frac{-2\pi\mathrm{i}}{\lambda}D\right)\iint_{A_{\mathrm{n}}}\frac{1}{\bar{L}_{\mathrm{S}}|\bar{L}|}K(\bar{\boldsymbol{n}}\cdot\bar{\boldsymbol{h}}, \bar{\boldsymbol{n}}\cdot\bar{\boldsymbol{k}})G(\bar{\boldsymbol{r}}) \times \exp\left[\frac{2\pi\mathrm{i}}{\lambda}(\bar{L}_{\mathrm{S}} - \bar{L} - dD_{\mathrm{K}})\right]dA_{\mathrm{n}}. \tag{4.59}$$

Following *Walles* [Ref. 4.192, Sect. 10.3; Ref. 4.193, Sect. 4], we can now express $U'(\boldsymbol{\rho})$ in a third manner, which develops the concept of the *homologous rays*, introduced by *Viénot* and his associates [4.180–185]. Remembering that $\boldsymbol{M}\boldsymbol{N} = \boldsymbol{M}$ and $\boldsymbol{M}\boldsymbol{K} = \boldsymbol{K}$, we write first with (4.55), (4.58) and the relation

$$dk \cdot k = 0$$

$$\bar{D}_K = D + dk \cdot (LMw + Ku) = D + dk \cdot (LMw + u).$$

Then, replacing dk by $\bar{k} - k$ and observing that Mw is perpendicular to k, we obtain

$$\bar{D}_K = u \cdot g + (\bar{k} - k) \cdot (LMw + u) = -u \cdot h + \bar{k} \cdot (u + LMw), \tag{4.60}$$

so that, with $\bar{L} = \bar{k} \cdot (\bar{r} - \rho)$, (4.50) becomes

$$U'(\rho) = \exp\left(\frac{2\pi i}{\lambda} u \cdot h\right) \iint_{A_n} \frac{S}{\bar{L}_S |\bar{L}|} K(\bar{n} \cdot \bar{h}, \bar{n} \cdot \bar{k}) G(\bar{r}) \times \exp\left[\frac{2\pi i}{\lambda}(\bar{L}_S - \bar{k} \cdot \bar{r})\right] \exp\left[\frac{2\pi i}{\lambda} \bar{k} \cdot (\rho - u - LMw)\right] dA_n. \tag{4.61}$$

If we compare this expression of $U'(\rho)$ with that of $U(\rho)$ (4.43), we can establish that

$$U'(\rho) = \exp\left(\frac{2\pi i}{\lambda} u \cdot h\right) U(\rho - u - LMw),$$

or also that

$$U'(\rho + u + LMw) = \exp\left(\frac{2\pi i}{\lambda} u \cdot h\right) U(\rho). \tag{4.62}$$

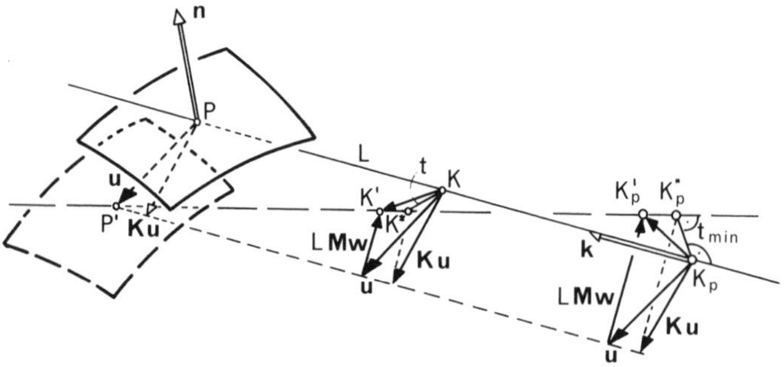

Fig. 4.17. Homologous rays PK and P′K′ relating to the pair of points P and P′ observed in the direction k

This result may be interpreted as follows: the deformation of the object surface produces a modification of the reflected wave front such that the vibration produced at K, with position vector ρ, by the undeformed surface is the same, except for the phase factor, as that produced at K′, with position vector $\boldsymbol{\rho} + \boldsymbol{u} + L\boldsymbol{M}\boldsymbol{w}$, by the deformed surface. Rays such as PK and P′K′ are said to be homologous (Fig. 4.17). The distance t between them measured in K perpendicular to $\boldsymbol{k}$ is

$$t = |\boldsymbol{K}\boldsymbol{u} + L\boldsymbol{M}\boldsymbol{w}|. \tag{4.63}$$

Generally, for a given direction $\boldsymbol{k}$, the homologous rays are skew; thus, no point K exists where the distance t between them is zero. Let us recall also the assumptions that have made this interpretation possible, namely that the object deformation as well as the aperture of the observing system are small.

After having calculated the amplitudes produced at the point K by the undeformed and the deformed object, we can now write the *total complex amplitude* that results from the superposition of these two coherent fields at K,

$$U_{\rm t}(\boldsymbol{\rho}) = U(\boldsymbol{\rho}) + U'(\boldsymbol{\rho}). \tag{4.64}$$

Because of surface roughness, these amplitudes are random functions, which, because of (4.41), have the mean value

$$E[U_{\rm t}(\boldsymbol{\rho})] = E[U(\boldsymbol{\rho})] = E[U'(\boldsymbol{\rho})] = 0. \tag{4.65}$$

In order to eliminate, in the following, statistical variations of the quantities, we work preferably with expected values (for a treatment of the statistical properties of the interference phenomenon, the reader may refer to the works we mentioned in connection with the properties of the rough surface). Thus, we calculate the *intensity* of the total field as [Ref. 4.192, Sect. 2.3 and Chap. 5; Ref. 4.204, Sects. 2, 3]

$$\begin{aligned} J(\boldsymbol{\rho}) &= \frac{1}{2}(E[UU^*] + E[U'U'^*] + E[U^*U'] + E[UU'^*]) \\ &= I + I' + \Gamma + \Gamma^*, \end{aligned} \tag{4.66}$$

where $I = \frac{1}{2}E[UU^*]$ represents the mean intensity produced by the undeformed object alone, which can be expressed by the *autocorrelation* of the "image" amplitude $E[U(\boldsymbol{\rho})U^*(\bar{\boldsymbol{\rho}})]$ when $\bar{\boldsymbol{\rho}} = \boldsymbol{\rho}$. Similarly, $E[U^*U']$ is equal to the *cross-correlation* of the two amplitudes U and U' when $\bar{\boldsymbol{\rho}} = \boldsymbol{\rho}$. Explicitly, we have for example with (4.43), P and $\bar{\rm P}$ now being variable points of $A_{\rm n}$,

$$\begin{aligned} E[UU^*] = E\Bigg[&\iint_{A_{\rm n}} \iint_{A_{\rm n}} \frac{S}{L_{\rm S}|L|} K(\boldsymbol{n}\cdot\boldsymbol{h}, \boldsymbol{n}\cdot\boldsymbol{k}) G(\boldsymbol{r}) \exp\left[\frac{2\pi {\rm i}}{\lambda}(L_{\rm S} - L)\right] \\ &\times \frac{S^*}{\bar{L}_{\rm S}|\bar{L}|} K^*(\bar{\boldsymbol{n}}\cdot\bar{\boldsymbol{h}}, \bar{\boldsymbol{n}}\cdot\bar{\boldsymbol{k}}) G^*(\bar{\boldsymbol{r}}) \exp\left[\frac{-2\pi {\rm i}}{\lambda}(\bar{L}_{\rm S} - \bar{L})\right] dA_{\rm n} d\bar{A}_{\rm n} \Bigg]. \end{aligned}$$

Because of the property (4.42), this becomes (similarly to [Ref. 4.209, Eq. (12)])

$$E[UU^*] = \iint\limits_{A_n} \frac{C|S|^2}{L_S^2 L^2} |K(\boldsymbol{n}\cdot\boldsymbol{h}, \boldsymbol{n}\cdot\boldsymbol{k})|^2 dA_n. \tag{4.67}$$

In the same way, we obtain with (4.50)

$$E[U'U'^*] = E[UU^*] \quad \text{or} \quad I' = I, \tag{4.68}$$

and with (4.43, 59)

$$E[U^*U'] = \exp\left(\frac{-2\pi i}{\lambda} D\right) \iint\limits_{A_n} \frac{C|S|^2}{L_S^2 L^2} |K(\boldsymbol{n}\cdot\boldsymbol{h}, \boldsymbol{n}\cdot\boldsymbol{k})|^2 \exp\left(\frac{-2\pi i}{\lambda} dD_K\right) dA_n. \tag{4.69}$$

If we now put

$$\frac{1}{2} E[U^*U'] = \Gamma = |\Gamma| \exp\left[-\frac{2\pi i}{\lambda}(D + \delta)\right],$$

where δ summarizes the influence of dD_K, we can write

$$J(\rho) = 2\left\{I + |\Gamma| \cos\left[\frac{2\pi}{\lambda}(D + \delta)\right]\right\}. \tag{4.70}$$

Finally, the *visibility* or *contrast* of the interference fringes at K is defined by

$$V(\rho) = \frac{J_{max} - J_{min}}{J_{max} + J_{min}},$$

where J_{max} and J_{min} indicate, respectively, the maximum and the minimum of the intensity in the immediate vicinity of K [Ref. 3.5, p. 267]. In the present case, we have

$$J_{max} = 2(I + |\Gamma|), \qquad J_{min} = 2(I - |\Gamma|), \tag{4.71}$$

so that the visibility takes the simple form (see also e.g. [Ref. 4.184, Eq. (II.25); Ref. 4.189, Eq. (7); Ref. 4.192, Eq. (2.12); Ref. 4.203, Eq. (8); Ref. 4.204, Eq. (20)])

$$V(\rho) = \frac{|\Gamma|}{I}. \tag{4.72}$$

This quantity depends thus on the one hand on the characteristics of the two object configurations as well as on the aperture of the observing system; on the other hand it depends on the position $\boldsymbol{\rho}$ of the point K where the fringes are observed. The fringes are therefore said to be *localized* at those points K where their visibility reaches a maximum, two definitions of fringe localization being however possible (see for example [4.165, 189, 191, 193, 194]):

a) Each pair of rays that come from the same material point on the undeformed and the deformed surface may be required to have the same optical-path difference. This demands that $\bar{D}_K$ remain constant all over A_n, i.e., to a first approximation, that D_K be stationary at the central point P: $dD_K = 0 \ \forall \ d\boldsymbol{r}$, or also

$$\nabla_n D_K = 0, \quad \text{or} \quad \nabla_k D_K = 0. \tag{4.73}$$

This purely geometrical condition, which applies here to interferometry by means of holography (see e.g. [4.9, 120, 164, 178, 179, 195, 198, 199]), is identical to that utilized long since in classical interferometry [4.210, 211]. When it is fulfilled, (4.69) can be written as

$$\begin{aligned} E[U^*U'] &= \exp\left(-\frac{2\pi i}{\lambda}D\right)E[UU^*], \\ \Gamma &= I\exp\left(-\frac{2\pi i}{\lambda}D\right). \end{aligned} \tag{4.74}$$

Hence the intensity becomes

$$J(\boldsymbol{\rho}) = 2I\left[1 + \cos\left(\frac{2\pi}{\lambda}D\right)\right], \tag{4.75}$$

and the visibility attains the value $V(\boldsymbol{\rho}) = 1$. Thus, in this particular case, by virtue of the assumptions made above, the contrast is the maximum possible; therefore, we shall speak of *complete localization* of the fringes. We can also characterize this situation in other words by stating that the *homologous rays intersect there*, because $LM\nabla_n D_K = 0$ implies that the distance between them is zero [see (4.58, 63)].

b) In a given observation direction $\boldsymbol{k}$, the point K_p may be sought where the visibility attains its absolute maximum with respect to this direction [4.192]. A necessary but not sufficient condition is

$$\left.\frac{dV}{dL}\right|_{\boldsymbol{k}=\text{const}} = 0. \tag{4.76}$$

As we shall see in the next section, this approximately amounts to stating that at K_p the *homologous rays have the shortest distance* (Fig. 4.17) [Ref. 4.184, p. 42]. Generally, this distance is not zero; furthermore, $\nabla_n D_K \neq 0$ and $V(\rho) < 1$; therefore, we call this case *partial localization.* If the same object point P is observed from different directions $\boldsymbol{k}$, a set $\{K_p\}$, which forms a surface of partial localization related to P, is obtained. As we shall see in the following, complete localization is found on a line situated on this surface, *a* being a particular case of *b*.

Before investigating further the two types of localization, we must add some explanation about the simplified description of the interference phenomenon utilized in Sects. 4.1 and 4.2.1. There, we considered only two rays among the whole pencil of rays that form a fringe. These two rays cannot be isolated from the remainder of the beam. The implicit hypothesis was, therefore, that the whole pencil contributed in the same way to the fringe formation. Strictly speaking, this is possible only in case of complete localization; indeed, (4.75) shows that then the fringe-order depends on only D. However, this approximation is also suited to the general case, provided that the aperture of the observing system is very small, so that dD_K, i.e. δ in (4.70), can be neglected. Then, fringes can be seen everywhere, however with only low brightness and sometimes with a middling contrast. In this sense, we could define the fringe-order at each point K along an observation direction. If, on the contrary, dD_K and δ are not negligible because the aperture is not small, their presence can produce a so-called contrast inversion (see e.g. [4.190, 204]), i.e. a change of the fringe-order from the expected order, which is given by D alone.

4.2.3 Fringe Visibility for Different Apertures, Partial Localization

In this section, we will calculate, for the cases of small circular and rectangular entrance pupils of the observing system, the visibility of the fringes and also the distance from the object surface at which the visibility is maximum, in other words, at which the fringes are partially localized. The starting point is (4.72), which, with (4.67, 69), is

$$V(\rho) = \frac{\left| \iint\limits_{A_n} \frac{C|S|^2}{\bar{L}_S^2 \bar{L}^2} |K(\bar{\boldsymbol{n}} \cdot \bar{\boldsymbol{h}}, \bar{\boldsymbol{n}} \cdot \bar{\boldsymbol{k}})|^2 \exp\left(\frac{-2\pi i}{\lambda} dD_K\right) dA_n \right|}{\iint\limits_{A_n} \frac{C|S|^2}{\bar{L}_S^2 \bar{L}^2} |K(\bar{\boldsymbol{n}} \cdot \bar{\boldsymbol{h}}, \bar{\boldsymbol{n}} \cdot \bar{\boldsymbol{k}})|^2 dA_n}, \tag{4.77}$$

where the quantities with a bar relate again to the variable point $\bar{P}$ in the neighborhood of the fixed central point P. It must be noted that the area A_n on the object surface depends on not only the position of the entrance pupil but also on the position of the point K on which the observing system is focused. To eliminate these influences on the integration limits, we transform the integration over A_n into one over the area A_n of the entrance pupil (Fig. 4.18)

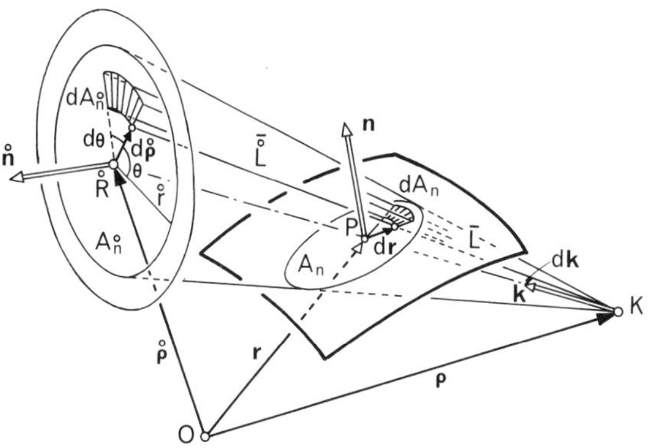

Fig. 4.18. Formation of a fringe at K by all light rays that are reflected by the part A_n of the object surface and that pass through the circular entrance pupil $A_{\mathring{n}}$ of the observing system

[Ref. 4.192, Sect. 7.2]. The surface element changes according to

$$dA_n = \left(\frac{\bar{L}}{\bar{L} + \bar{\mathring{L}}}\right)^2 \frac{\mathring{\boldsymbol{n}}\cdot\bar{\boldsymbol{k}}}{\bar{\boldsymbol{n}}\cdot\bar{\boldsymbol{k}}} dA_{\mathring{n}}, \tag{4.78}$$

and, between the increments, there exists the relation similar to (2.70)

$$d\boldsymbol{r} = \frac{L}{L + \mathring{L}} \mathbf{M}^{\mathrm{T}} \mathring{\mathbf{N}} d\mathring{\boldsymbol{\rho}}, \tag{4.79}$$

so that, in addition to the formulae (4.52), we can use

$$dD_{\mathrm{K}} = d\boldsymbol{r}\cdot\boldsymbol{\nabla}_{\mathrm{n}} D_{\mathrm{K}} = \frac{L}{L + \mathring{L}} d\mathring{\boldsymbol{\rho}}\cdot\mathring{\mathbf{N}}\mathbf{M}\boldsymbol{\nabla}_{\mathrm{n}} D_{\mathrm{K}}. \tag{4.80}$$

Moreover, because A_n and $A_{\mathring{n}}$ are supposed to be small, we assume that the amplitude factors are constant, i.e. $\bar{\boldsymbol{n}} \simeq \boldsymbol{n}$, $\bar{L} \simeq L$, etc. We obtain, then, the general expression

$$V(\boldsymbol{\rho}) = \frac{1}{A_{\mathring{n}}}\left|\iint\limits_{A_{\mathring{n}}} \exp\left[\frac{-2\pi \mathrm{i} L}{\lambda(L + \mathring{L})} d\mathring{\boldsymbol{\rho}}\cdot\mathring{\mathbf{N}}\mathbf{M}\boldsymbol{\nabla}_{\mathrm{n}} D_{\mathrm{K}}\right] dA_{\mathring{n}}\right|. \tag{4.81}$$

Let us now first envisage a *circular entrance pupil* of radius $\mathring{r}$. With the polar coordinates $r = |d\mathring{\boldsymbol{\rho}}|$ and $\theta = \arccos(d\mathring{\boldsymbol{\rho}}\cdot\mathring{\mathbf{N}}\mathbf{M}\boldsymbol{\nabla}_{\mathrm{n}} D_{\mathrm{K}}/r|\mathring{\mathbf{N}}\mathbf{M}\boldsymbol{\nabla}_{\mathrm{n}} D_{\mathrm{K}}|)$ in $A_{\mathring{n}}$, (4.81) becomes

$$V(\boldsymbol{\rho}) = \frac{1}{\pi\mathring{r}^2}\left|\int_0^{2\pi}\int_0^{\mathring{r}} \exp\left[\frac{-2\pi \mathrm{i} L}{\lambda(L + \mathring{L})} |\mathring{\mathbf{N}}\mathbf{M}\boldsymbol{\nabla}_{\mathrm{n}} D_{\mathrm{K}}| r\cos\theta\right] r dr d\theta\right|.$$

This integral is similar to that which governs the Fraunhofer diffraction at a circular aperture [Ref. 3.5, p. 395] so that, by the same reasoning, we can write

$$V(\rho) = \frac{2\left|\mathrm{J}_1\left[\dfrac{-2\pi L\mathring{r}}{\lambda(L+\mathring{L})}|\mathring{N}MV_{\mathrm{n}}D_{\mathrm{K}}|\right]\right|}{\dfrac{2\pi L\mathring{r}}{\lambda(L+\mathring{L})}|\mathring{N}MV_{\mathrm{n}}D_{\mathrm{K}}|}, \tag{4.82}$$

where J_1 is the Bessel function of the first order and of the first kind [Ref. 4.178, Eq. (7); Ref. 4.179, Eqs. (20, 21); Ref. 4.190, Eq. (3); Ref. 4.192, Eq. (7.8); Ref. 4.201, Eq. (XIV.6); Ref. 4.204, Eq. (29)]. With this result, let us calculate the distance of localization that fulfills (4.76). For that purpose, we derive (4.82) with respect to L. However, as shown by Fig. 4.19, an indeterminacy arises. To

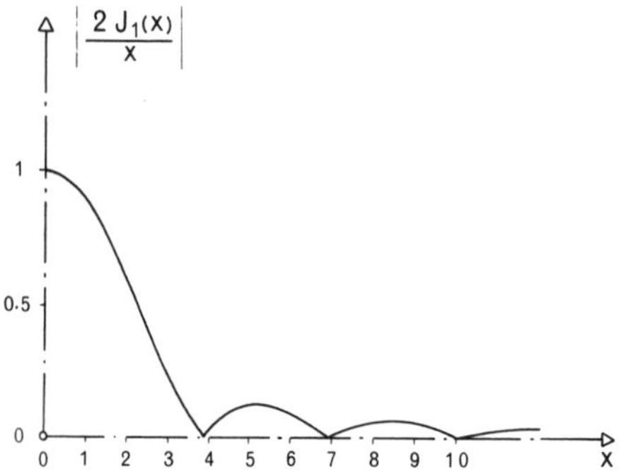

Fig. 4.19. Graph of the function $2\mathrm{J}_1(x)/x$ that appears in the visibility formula in the case of a circular entrance pupil

remove this, we suppose that, at least in some interval, the argument of the function J_1 is much smaller than the first zero so that, in order to maximize V, we must minimize the argument. This procedure is warrantable if $\mathring{r}$ is sufficiently small. Moreover, two cases must be distinguished:

a) $L + \mathring{L} = \text{const}$, i.e. the optical system is moved along $\boldsymbol{k}$ while it remains focused on the same distance $L + \mathring{L}$. With (4.58), we obtain then the condition

$$\frac{d}{dL}|L\mathring{N}M\boldsymbol{w} + \mathring{N}K\boldsymbol{u}|_{L=L_0} = 0, \tag{4.83}$$

which is fulfilled when [Ref. 4.192, Eq. (7.12); Ref. 4.193, Eq. (31)]

$$L_0 = \frac{-(\mathring{N}M\boldsymbol{w})\cdot(\mathring{N}K\boldsymbol{u})}{(\mathring{N}M\boldsymbol{w})^2}. \tag{4.84}$$

In case the fringes are observed along the axis of the optical system, i.e. in case

$\mathring{\boldsymbol{n}}$ is parallel to $\boldsymbol{k}$, we have $\mathring{N}\boldsymbol{M} = \boldsymbol{M}$, $\mathring{N}\boldsymbol{K} = \boldsymbol{K}$, so that then $L_0 = L_p$, where

$$L_p = \frac{-\boldsymbol{Mw}\cdot\boldsymbol{Ku}}{(\boldsymbol{Mw})^2}. \quad (4.85)$$

Looking back at (4.63), we see that L_p defines the pair of points K_p, K_p^*, which is the locus of the shortest distance of the homologous rays. This situation, in which the location of the entrance pupil in no way affects the distance of localization, will be called the *standard case of partial localization.*

b) $\mathring{L} = \text{const}$, i.e. the entrance pupil does not move but the optical system is focused on several points along $\boldsymbol{k}$ [Ref. 4.192, Chap. 11]. Again with $\mathring{\boldsymbol{n}}$ parallel to $\boldsymbol{k}$, we get another localization distance L_q, which is determined by

$$\frac{d}{dL}\left|\frac{L\boldsymbol{Mw} + \boldsymbol{Ku}}{L + \mathring{L}}\right|_{L=L_q} = 0, \quad (4.86)$$

that is to say,

$$L_q = -\frac{(\mathring{L}\boldsymbol{Mw} - \boldsymbol{Ku})\cdot\boldsymbol{Ku}}{(\mathring{L}\boldsymbol{Mw} - \boldsymbol{Ku})\cdot\boldsymbol{Mw}}. \quad (4.87)$$

For large $\mathring{L}$, we have $L_q \simeq L_p$.

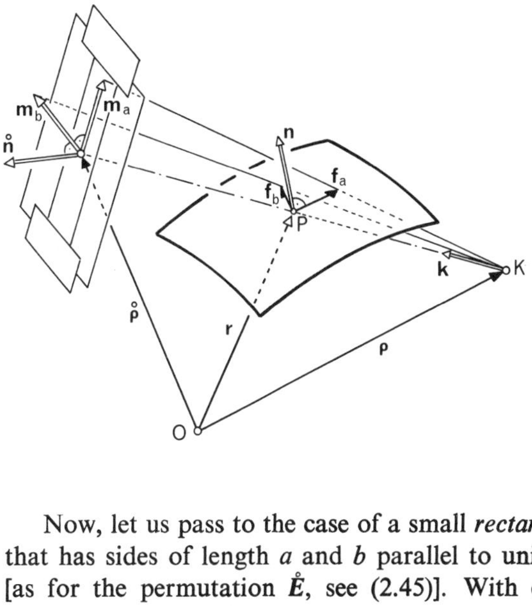

Fig. 4.20. Formation of a fringe at K in the case of a variable rectangular entrance pupil

Now, let us pass to the case of a small *rectangular entrance pupil* (Fig. 4.20) that has sides of length a and b parallel to unit vectors $\boldsymbol{m}_a$ and $\boldsymbol{m}_b = -\mathring{\boldsymbol{E}}\boldsymbol{m}_a$ [as for the permutation $\mathring{\boldsymbol{E}}$, see (2.45)]. With cartesian coordinates x and y

parallel to the edges, we have $d\mathring{\rho} = x\boldsymbol{m}_a + y\boldsymbol{m}_b$ so that (4.81) becomes

$$V(\rho) = \frac{1}{ab}\left| \int_{-a/2}^{a/2} \int_{-b/2}^{b/2} \exp\left[\frac{-2\pi \mathrm{i} L}{\lambda(L+\mathring{L})}(x\boldsymbol{m}_a \cdot \mathring{\boldsymbol{N}}\boldsymbol{M}\boldsymbol{V}_{\mathrm{n}}D_{\mathrm{K}} + y\boldsymbol{m}_b \cdot \mathring{\boldsymbol{N}}\boldsymbol{M}\boldsymbol{V}_{\mathrm{n}}D_{\mathrm{K}})\right] dx dy \right|.$$

To simplify the formulae, we introduce the vector

$$\boldsymbol{f}_a = \frac{L}{L+\mathring{L}}\boldsymbol{M}^{\mathrm{T}}\mathring{\boldsymbol{N}}\boldsymbol{m}_a = \frac{L}{L+\mathring{L}}\boldsymbol{M}^{\mathrm{T}}\boldsymbol{m}_a, \tag{4.88}$$

which is a nonunit vector situated in the plane tangent to the object surface and which corresponds to the unit vector $\boldsymbol{m}_a$ by virtue of the collineation with the center K. Similarly, to $\boldsymbol{m}_b$ corresponds a vector $\boldsymbol{f}_b$. The result of the integration is then [Ref. 4.192, Eq. (8.7); Ref. 4.201, Eq. (IX, 18); Ref. 4.203, Eq. (9)]

$$V(\rho) = \left| \frac{\sin\left[\frac{\pi a}{\lambda}\boldsymbol{f}_a \cdot \boldsymbol{V}_{\mathrm{n}}D_{\mathrm{K}}\right]}{\frac{\pi a}{\lambda}\boldsymbol{f}_a \cdot \boldsymbol{V}_{\mathrm{n}}D_{\mathrm{K}}} \; \frac{\sin\left[\frac{\pi b}{\lambda}\boldsymbol{f}_b \cdot \boldsymbol{V}_{\mathrm{n}}D_{\mathrm{K}}\right]}{\frac{\pi b}{\lambda}\boldsymbol{f}_b \cdot \boldsymbol{V}_{\mathrm{n}}D_{\mathrm{K}}} \right|. \tag{4.89}$$

Here, we do not further investigate this general case (for this subject refer, e.g., to [Ref. 4.192, Sect. 8.3]) but look at only the case in which the rectangular aperture becomes a *slit*, i.e. when $b \to 0$. The formula (4.89) then becomes

$$V(\rho) = \left| \frac{\sin\left[\frac{\pi a}{\lambda}\boldsymbol{f}_a \cdot \boldsymbol{V}_{\mathrm{n}}D_{\mathrm{K}}\right]}{\frac{\pi a}{\lambda}\boldsymbol{f}_a \cdot \boldsymbol{V}_{\mathrm{n}}D_{\mathrm{K}}} \right|. \tag{4.90}$$

Of course, if the argument vanishes, i.e. with a unit vector $\boldsymbol{e}_a$ parallel to $\boldsymbol{f}_a$, if

$$\boldsymbol{e}_a \cdot \boldsymbol{V}_{\mathrm{n}}D_{\mathrm{K}} = 0, \tag{4.91}$$

the visibility is equal to 1. The distance of localization is then with (4.58) [Ref. 4.164, Eq. (38)]

$$L_a = -\frac{\boldsymbol{e}_a \cdot \boldsymbol{N}\boldsymbol{K}\boldsymbol{u}}{\boldsymbol{e}_a \cdot \boldsymbol{N}\boldsymbol{w}} = -\frac{\boldsymbol{m}_a \cdot \boldsymbol{K}\boldsymbol{u}}{\boldsymbol{m}_a \cdot \boldsymbol{M}\boldsymbol{w}}. \tag{4.92}$$

This result is the same as that given by (4.84) when the circular aperture is rotated around its diameter of direction $\boldsymbol{m}_a$ until $\mathring{\boldsymbol{n}}$ becomes perpendicular to $\boldsymbol{k}$, because then the vector $\mathring{\boldsymbol{N}}\boldsymbol{M}\boldsymbol{w}$ is parallel to $\boldsymbol{m}_a$ (the circular edge appears here as a very narrow ellipse). It must also be remarked that, generally, in order for the condition (4.91) to be fulfilled, it is sufficient that the increment of the optical-path difference vanish in the only direction e_a on the object surface, which direction corresponds to that of the slit.

In conclusion, we may add that the different distances of partial localization that we have calculated so far, depend on the shape and, with one exception, on the position of the entrance pupil. On the contrary, in case of complete localization $\nabla_n D_K = 0$, all of these distances become equal and thus independent of the aperture. This case will be studied in more detail in the next section.

4.2.4 Complete Localization, Normality Theorem, Shape of the Line of Complete Localization

As explained in Sect. 4.2.2, the fringes are said to be completely localized when condition (4.73) is fulfilled, i.e., with (4.58), if (Fig. 4.21)

$$\left[\boldsymbol{N}\boldsymbol{w} + \frac{1}{L}\boldsymbol{N}\boldsymbol{K}\boldsymbol{u}\right]_{\substack{L=L_c\\ \boldsymbol{k}=\boldsymbol{k}_c}} = 0, \quad \text{or} \quad [L\boldsymbol{M}\boldsymbol{w} + \boldsymbol{K}\boldsymbol{u}]_{\substack{L=L_c\\ \boldsymbol{k}=\boldsymbol{k}_c}} = 0 \tag{4.93}$$

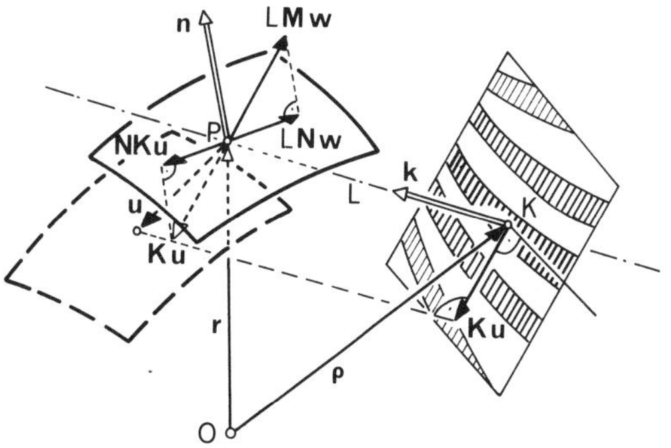

Fig. 4.21. Directions of the vectors that appear in the condition of complete localization

(see e.g. [Ref. 4.9, Eq. (25); Ref. 4.164, Eq. (29); Ref. 4.165, Eq. (12); Ref. 4.178, Eqs. (2, 3); Ref. 4.186, Eq. (5); Ref. 4.195, Eq. (6); Ref. 4.198, Eq. (9); Ref. 4.199, Eqs. (5, 6)]). Owing to the definition (4.32), we know that strain and rotation are contained in the term

$$\boldsymbol{N}\boldsymbol{w} = (\nabla_n \otimes \boldsymbol{u})\boldsymbol{g} - \frac{1}{L_S}\boldsymbol{N}\boldsymbol{H}\boldsymbol{u}.$$

In the following, the properties of this particular case of localization are to be investigated; to simplify the formulae, we do not repeat the index c any more.

First let us recall that completely localized fringes have the *best possible contrast*, i.e. $V(\rho) = 1$, and that they are bright, because the whole neighborhood of the considered point P on the object surface can contribute to the interference phenomenon. These qualities are present independently of the shape and position of the aperture of the observing system.

Second, when inserting (4.93) into the relation (4.34) that specifies the fringe direction, we obtain

$$\frac{dD_{\mathrm{R}}}{d\phi} = -\frac{L_{\mathrm{R}} + L}{L}\boldsymbol{m}\cdot\boldsymbol{K}\boldsymbol{u}. \tag{4.94}$$

The same reasoning as in Sect. 4.2.1 ends with the statement that the fringes are perpendicular to $\boldsymbol{Ku}$, which leads us to the *normality theorem* (see e.g. [Ref. 4.165, p. 557; Ref. 4.167, p. 6; 4.175; Ref. 4.180, p. 140; Ref. 4.184, p. 109; Ref. 4.193, p. 911; Ref. 4.195, p. 229]) comparable to that in classical interferometry, where the fringes are perpendicular to the plane formed by the two interfering rays [Ref. 3.5, p. 261 et seq.; Ref. 4.212, p. 107]: *if and only if the fringes are observed in a direction of complete localization, they are normal to the projection of the displacement vector onto the plane perpendicular to the observation direction* (Fig. 4.21). We can notice also that, contrary to the general case [see (4.34)], the fringe direction does not depend on the observer's position. Moreover, although the displacement gradient $\boldsymbol{\nabla}_{\mathrm{n}} \otimes \boldsymbol{u}$ is present in (4.34) as well as in (4.93), the fringe direction is given by the displacement vector alone.

Third, let us add a remark about the presence of the projection $\boldsymbol{N}$ in the condition $\boldsymbol{\nabla}_{\mathrm{n}} D_{\mathrm{K}} = 0$. Supposing that $\bar{D}_{\mathrm{K}}$ is defined in a three-dimensional vicinity of P, we consider its series development along an arbitrary curve $\boldsymbol{r}(s)$ on the object surface [see (2.31)]

$$\bar{D}_{\mathrm{K}} \simeq D + d\boldsymbol{r}\cdot\boldsymbol{\nabla} D_{\mathrm{K}} + \frac{1}{2}[d\boldsymbol{r}\cdot(\boldsymbol{\nabla}\otimes\boldsymbol{\nabla} D_{\mathrm{K}})d\boldsymbol{r} + d^2\boldsymbol{r}\cdot\boldsymbol{\nabla} D_{\mathrm{K}}].$$

With $d^2\boldsymbol{r} = \boldsymbol{b}ds^2 + (d\boldsymbol{r}\cdot\boldsymbol{B}d\boldsymbol{r})\boldsymbol{n}$ [see (2.59, 52)], where $\boldsymbol{b}$ is the geodesic-curvature vector and $\boldsymbol{B}$, the normal-curvature tensor, we have with $d\boldsymbol{r} \equiv d\boldsymbol{r}\boldsymbol{N}$, $\boldsymbol{b} \equiv \boldsymbol{b}\boldsymbol{N}$

$$\bar{D}_{\mathrm{K}} \simeq D + d\boldsymbol{r}\cdot\boldsymbol{\nabla}_{\mathrm{n}} D_{\mathrm{K}} + \frac{1}{2}d\boldsymbol{r}\cdot[\boldsymbol{\nabla}\otimes\boldsymbol{\nabla} D_{\mathrm{K}} + \boldsymbol{B}(\boldsymbol{n}\cdot\boldsymbol{\nabla} D_{\mathrm{K}})]d\boldsymbol{r} + \frac{1}{2}\boldsymbol{b}\cdot\boldsymbol{\nabla}_{\mathrm{n}} D_{\mathrm{K}} ds^2,$$

and, in particular, when the condition of complete localization is fulfilled,

$$\bar{D}_{\mathrm{K}} \simeq D + \frac{1}{2}d\boldsymbol{r}\cdot[\boldsymbol{\nabla}\otimes\boldsymbol{\nabla} D_{\mathrm{K}} + \boldsymbol{B}(\boldsymbol{n}\cdot\boldsymbol{\nabla} D_{\mathrm{K}})]d\boldsymbol{r}.$$

If the curvature $\boldsymbol{B}$ is large, as on the edge or in a corner of an object, the second term in the brackets prevails, so that the condition $\boldsymbol{n}\cdot\boldsymbol{\nabla} D_{\mathrm{K}} = 0$ may be added to

$N\nabla D_K = 0$ to form a three-dimensional localization condition

$$\nabla D_K = \boldsymbol{w} + \frac{1}{L}\boldsymbol{K}\boldsymbol{u} = 0. \tag{4.95}$$

Because $\bar{D}_K$ would then be stationary in P, independently of this strong curvature, we could call (4.95) a condition of *absolute localization* [Ref. 4.165, Eq. (19)].

Fourth, let us consider the whole set $\{K_c\}$ of the points of complete localization associated to P. One vectorial relation (4.93) constitutes two nonlinear equations for the vector-valued function $\boldsymbol{k}(L)$, which defines a curve on the unit sphere $|\boldsymbol{k}| = 1$. Therefore, the position vector of K

$$\boldsymbol{\rho}(L) = \boldsymbol{r} - L\boldsymbol{k}(L), \tag{4.96}$$

where $\boldsymbol{r}$ is constant, defines in space a *line of complete localization* relative to P. This curve generally cannot be found easily; nevertheless, useful general indications as well as particular examples can be given. Before doing this, let us point out that the line of complete localization as well as the surface of partial localization of which we speak here are generated by the ensemble of localization points defined by one fixed object point P observed in a variable direction. Similar to what we did for $D(\boldsymbol{r}, \boldsymbol{k})$ in Sect. 4.1.1, we could also define a line or a surface of localization relating to a given position of the observer and to the whole object surface.

For the line concerning one point P, we notice first that, as shown by (4.93), to each direction $\boldsymbol{k}$ of complete localization belongs only one point K, i.e. only one distance L, which is positive when K lies behind the object surface, negative in the opposite case. Furthermore, the condition $L \to \infty$, i.e. $N\boldsymbol{w} = 0$, determines the direction $\boldsymbol{k}_\infty$ of the *asymptote*. Let us examine the behavior of the line in the vicinity of the object surface [Ref. 4.196, p. 809; Ref. 4.197, p. 66 et seq.]. In the preceding developments, we assumed that $|L| \gg u$ and $|L| \gg \lambda$ so that, strictly speaking, the results obtained are not valid near P. However, the following treatment can be interpreted as an extrapolation that is experimentally corroborated. With the notation $d/dL = {}'$, from (4.96) we obtain for the tangent

$$\boldsymbol{\rho}'(L) = -\boldsymbol{k}(L) - L\boldsymbol{k}'(L),$$

i.e. in the particular case of P,

$$\boldsymbol{\rho}'(0) = -\boldsymbol{k}(0). \tag{4.97}$$

On the other hand, (4.93) gives, when $L \to 0$, $N\boldsymbol{K}\boldsymbol{u} = 0$ or $\boldsymbol{K}\boldsymbol{u} = \chi\boldsymbol{n}$. Two possibilities come thus into consideration

$$a)\quad \boldsymbol{k}(0) = \boldsymbol{k}_{\mathrm{u}} = \pm\frac{\boldsymbol{u}}{u}, \qquad \text{since} \quad \boldsymbol{K}_{\mathrm{u}}\boldsymbol{u} = 0,$$

$$b)\quad \boldsymbol{k}(0) = \boldsymbol{k}_{\mathrm{t}} = \pm\frac{\boldsymbol{N}\boldsymbol{u}}{|\boldsymbol{N}\boldsymbol{u}|}, \qquad \text{since} \quad \boldsymbol{N}\boldsymbol{K}_{\mathrm{t}}\boldsymbol{u} = 0 \quad (\boldsymbol{N}\boldsymbol{u} \neq 0).$$

(In the special case $\boldsymbol{N}\boldsymbol{u} = 0$, $\boldsymbol{k}_{\mathrm{t}}$ is also in the tangent plane, but the direction is undetermined). Case *a* means that the line of complete localization intersects the object surface *tangent to the displacement vector* (Fig. 4.22); in other words, the fringes are localized on the object surface when they are observed in the direction of $\boldsymbol{u}$ (see e.g. [4.178, 180, 188–190, 196, 198]). Case *b* states that the line also arrives at P *tangent to the object surface* (Fig. 4.23).

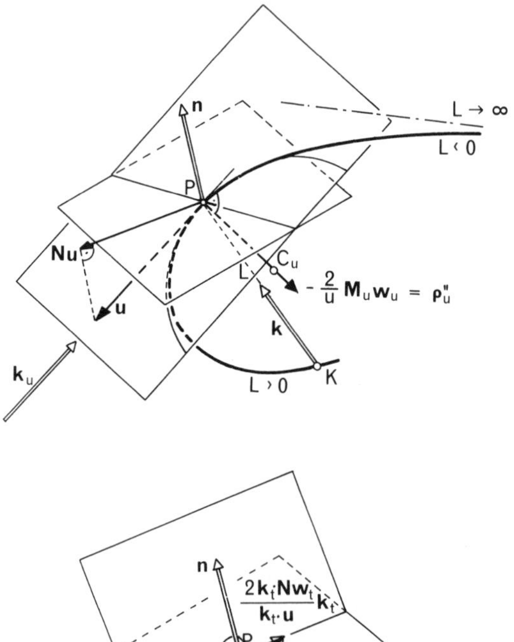

Fig. 4.22. The part of the line of complete localization when $\boldsymbol{k}$ comes near to the direction of the displacement. C_{u}: center of the curvature circle in P

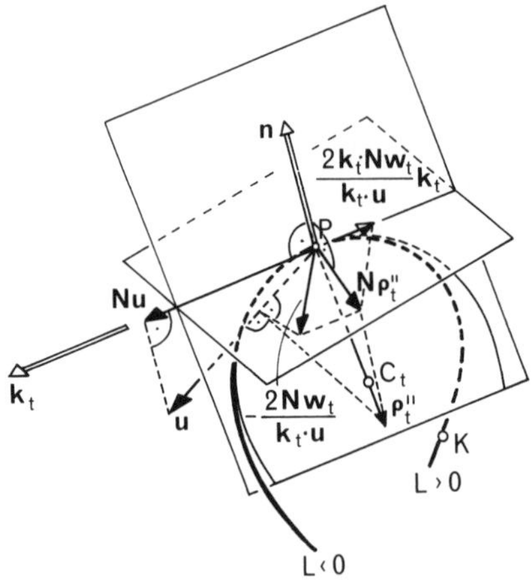

Fig. 4.23. The part of the line of complete localization when $\boldsymbol{k}$ comes near to the tangent plane of the object. C_{t}: center of the curvature circle in P. The left part of the line relates to fictitious positions of the observer located in the interior of the body (extrapolation)

Next, we investigate for these cases the second derivative of $\boldsymbol{\rho}$, which is

$$\boldsymbol{\rho}''(L) = -2\boldsymbol{k}'(L) - L\boldsymbol{k}''(L).$$

At P, we thus get

$$\boldsymbol{\rho}''(0) = -2\boldsymbol{k}'(0). \tag{4.98}$$

Moreover, the derivative of (4.93) gives

$$\boldsymbol{N}\boldsymbol{w} + L\boldsymbol{N}\boldsymbol{w}' + \boldsymbol{N}\boldsymbol{K}'\boldsymbol{u} = 0 \quad \text{or} \quad \boldsymbol{M}\boldsymbol{w} + L(\boldsymbol{M}\boldsymbol{w})' + \boldsymbol{K}'\boldsymbol{u} = 0, \tag{4.99}$$

into which we can introduce the derivative

$$\boldsymbol{K}' = -\boldsymbol{k}' \otimes \boldsymbol{k} - \boldsymbol{k} \otimes \boldsymbol{k}'.$$

So, from the second equation (4.99), we find in case *a*, with $\boldsymbol{k}_u' \cdot \boldsymbol{u} = 0$ and $L \to 0$

$$\boldsymbol{M}_u \boldsymbol{w}_u = \boldsymbol{k}_u' u,$$

so that (4.98) becomes

$$\boldsymbol{\rho}_u'' = -\frac{2}{u}\boldsymbol{M}_u\boldsymbol{w}_u = -\frac{2}{u}\left(\boldsymbol{I} - \frac{\boldsymbol{n} \otimes \boldsymbol{k}_u}{\boldsymbol{n} \cdot \boldsymbol{k}_u}\right)\left[(\boldsymbol{\nabla} \otimes \boldsymbol{u})(\boldsymbol{k}_u - \boldsymbol{h}) - \frac{1}{L_S}\boldsymbol{H}\boldsymbol{u}\right]. \tag{4.100}$$

On the one hand, the vectors $\boldsymbol{\rho}_u''$ and $\boldsymbol{u}$ determine the *osculating plane* of this part of the line at P (Fig. 4.22); on the other hand, because a small L measures, approximately, the arc, Frenet's formula [Ref. 2.8, Eq. (8–3)] shows that the *curvature* at P is $|\boldsymbol{\rho}_u''| = 2|\boldsymbol{M}_u\boldsymbol{w}_u|/u$ [4.196]. In case *b*, because $\boldsymbol{M}$ degenerates when $\boldsymbol{k} \to \boldsymbol{k}_t$, we use the first equation (4.99) and find, with $\boldsymbol{N}\boldsymbol{k}_t = \boldsymbol{k}_t$, and still $\boldsymbol{N}\boldsymbol{u} \neq 0$,

$$\boldsymbol{N}\boldsymbol{w}_t = \boldsymbol{N}\boldsymbol{k}_t'(\boldsymbol{k}_t \cdot \boldsymbol{u}) + \boldsymbol{k}_t(\boldsymbol{k}_t' \cdot \boldsymbol{u}).$$

From here, observing that $\boldsymbol{k}_t'$ and $\boldsymbol{N}\boldsymbol{k}_t'$ are perpendicular to $\boldsymbol{k}_t$, by projection we obtain with (4.98)

$$\boldsymbol{k}_t \cdot \boldsymbol{N}\boldsymbol{w}_t = -\frac{1}{2}\boldsymbol{\rho}_t'' \cdot \boldsymbol{u}, \qquad \boldsymbol{K}_t\boldsymbol{N}\boldsymbol{w}_t = -\frac{1}{2}\boldsymbol{N}\boldsymbol{\rho}_t''(\boldsymbol{k}_t \cdot \boldsymbol{u}).$$

Decomposing $\boldsymbol{\rho}_t'' = \mu \boldsymbol{E}\boldsymbol{k}_t + \nu\boldsymbol{n}$ and $\boldsymbol{u} = \boldsymbol{N}\boldsymbol{u} + \boldsymbol{n}(\boldsymbol{n} \cdot \boldsymbol{u})$, we find from the preceding relations for the coefficients

$$\nu = -\frac{2}{\boldsymbol{u} \cdot \boldsymbol{n}}\boldsymbol{k}_t \cdot \boldsymbol{N}\boldsymbol{w}_t, \qquad \mu\boldsymbol{E}\boldsymbol{k}_t = -\frac{2}{\boldsymbol{u} \cdot \boldsymbol{k}_t}\boldsymbol{K}_t\boldsymbol{N}\boldsymbol{w}_t,$$

so that, finally,

$$\boldsymbol{\rho}_t'' = -2\left(\frac{1}{\boldsymbol{u}\cdot\boldsymbol{k}_t}\boldsymbol{K}_t + \frac{1}{\boldsymbol{u}\cdot\boldsymbol{n}}\boldsymbol{n}\otimes\boldsymbol{k}_t\right)\boldsymbol{N}\boldsymbol{w}_t, \tag{4.101}$$

which again defines an osculating plane and a curvature (Fig. 4.23).

Let us now look at one particular property of the surface of partial localization in the vicinity of the line of complete localization when the latter intersects the object surface. We consider the surface of partial localization given by $|\boldsymbol{k}| = 1$ and by the scalar equation (4.85), namely,

$$(L_p\boldsymbol{M}\boldsymbol{w} + \boldsymbol{K}\boldsymbol{u})\cdot\boldsymbol{M}\boldsymbol{w} = 0.$$

Using curvilinear coordinates θ^1, θ^2 on it, such that for $\theta^2 = 0$, $\theta^1 = L$, we form the partial derivative with respect to θ^2

$$(L_p\boldsymbol{M}\boldsymbol{w} + \boldsymbol{K}\boldsymbol{u})_{,2}\cdot\boldsymbol{M}\boldsymbol{w} + (L_p\boldsymbol{M}\boldsymbol{w} + \boldsymbol{K}\boldsymbol{u})\cdot(\boldsymbol{M}\boldsymbol{w})_{,2} = 0,$$

which becomes, on the line of complete localization,

$$[L_{,2}\boldsymbol{M}\boldsymbol{w} + L(\boldsymbol{M}\boldsymbol{w})_{,2} - (\boldsymbol{k}_{,2}\otimes\boldsymbol{k} + \boldsymbol{k}\otimes\boldsymbol{k}_{,2})\boldsymbol{u}]\cdot\boldsymbol{M}\boldsymbol{w} = 0.$$

With $\boldsymbol{\rho} = \boldsymbol{r} - L_p\boldsymbol{k}$, $\boldsymbol{\rho}_{,2} = -L_{p,2}\boldsymbol{k} - L_p\boldsymbol{k}_{,2}$ and $\boldsymbol{k}\cdot\boldsymbol{M}\boldsymbol{w} = 0$, we can write, still on the line,

$$[LL_{,2}\boldsymbol{M}\boldsymbol{w} + L^2(\boldsymbol{M}\boldsymbol{w})_{,2} + (\boldsymbol{k}\cdot\boldsymbol{u})\boldsymbol{\rho}_{,2}]\cdot\boldsymbol{M}\boldsymbol{w} = 0,$$

and in particular, if $L \to 0$, $\boldsymbol{k} = \boldsymbol{k}_u$,

$$\boldsymbol{\rho}_{,2}\cdot\boldsymbol{M}_u\boldsymbol{w}_u = 0. \tag{4.102}$$

The osculating plane of the line of complete localization thus is at P normal to the surface of partial localization.

To get a better view of the shape of the whole line of complete localization, let us calculate some special cases (see also, e.g., [4.171, 172, 197, 199]). First, we suppose that there exists a *plane of symmetry* S through P in which the vectors $\boldsymbol{h}$, $\boldsymbol{n}$ and $\boldsymbol{u}$ are situated. For the displacement gradient $\nabla_n\otimes\boldsymbol{u} = \gamma + \Omega\boldsymbol{E} + \omega\otimes\boldsymbol{n}$ [see (2.109)], we assume that $\Omega = 0$, $\omega\in S$ and that one principal direction of γ, 1 for instance, lies in S, also. If we call (Fig. 4.24) $\boldsymbol{e}_t$ the unit vector along the intersection of S with the plane tangent to the object, and $\boldsymbol{e}_b$ the unit vector perpendicular to $\boldsymbol{n}$ and $\boldsymbol{e}_t$, we can write, similarly to (2.40)

$$\begin{aligned}(\nabla_n\otimes\boldsymbol{u})\boldsymbol{g} &= [\varepsilon_1(\boldsymbol{e}_t\otimes\boldsymbol{e}_t) + \varepsilon_2(\boldsymbol{e}_b\otimes\boldsymbol{e}_b) + \omega(\boldsymbol{e}_t\otimes\boldsymbol{n})]\boldsymbol{g}\\ &= \varepsilon_1(\boldsymbol{e}_t\cdot\boldsymbol{g})\boldsymbol{e}_t + \varepsilon_2(\boldsymbol{e}_b\cdot\boldsymbol{g})\boldsymbol{e}_b + \omega(\boldsymbol{n}\cdot\boldsymbol{g})\boldsymbol{e}_t.\end{aligned}$$

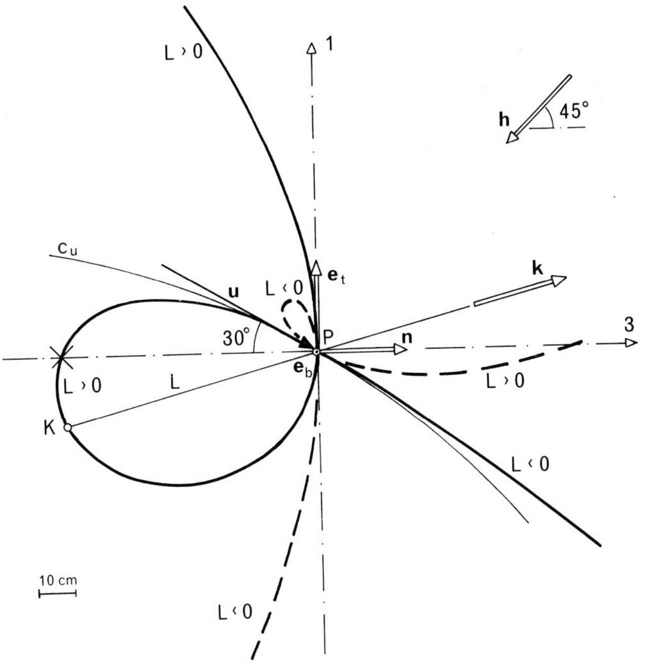

Fig. 4.24. Example of a plane line of complete localization. $u = 100\,\mu$m, $\varepsilon_1 = 10^{-4}$, $\Omega = 0$, $\omega = 0$, $L_s \to \infty$. c_u: curvature circle in P with radius = 2090 mm. x: point of absolute localization. The dashed part relates to positions of the observer behind the object surface

Multiplying the condition of complete localization (4.93) by $\boldsymbol{e}_t$ and $\boldsymbol{e}_b$, we can put it into the form of the two scalar equations

$$\begin{aligned} &\varepsilon_1(\boldsymbol{e}_t \cdot \boldsymbol{g}) + \omega(\boldsymbol{n} \cdot \boldsymbol{g}) - \frac{1}{L_S}\boldsymbol{e}_t \cdot \boldsymbol{H}\boldsymbol{u} + \frac{1}{L}\boldsymbol{e}_t \cdot \boldsymbol{K}\boldsymbol{u} = 0, \\ &\varepsilon_2(\boldsymbol{e}_b \cdot \boldsymbol{k}) + \frac{1}{L}\boldsymbol{e}_b \cdot \boldsymbol{K}\boldsymbol{u} = 0. \end{aligned} \tag{4.103}$$

We observe that, whatever value ε_2 may have, any observation direction in the plane of symmetry is a direction of complete localization: the second relation (4.103) is then identical to zero whereas the first yields

$$L = \frac{-\boldsymbol{e}_t \cdot \boldsymbol{K}\boldsymbol{u}}{\varepsilon_1(\boldsymbol{e}_t \cdot \boldsymbol{g}) + \omega(\boldsymbol{n} \cdot \boldsymbol{g}) - \frac{1}{L_S}\boldsymbol{e}_t \cdot \boldsymbol{H}\boldsymbol{u}}\,. \tag{4.104}$$

If $\varepsilon_2 = 0$, we see immediately that all points of complete localization are in S. Figs. 4.24, 25 present examples of the corresponding line in S [Ref. 4.197, pp.

80, 92]. It has thus the shape of a "*butterfly*" with two wings, the loops, and two antennae. However, the observer, who can be only in front of the object, cannot see it in its entirety.

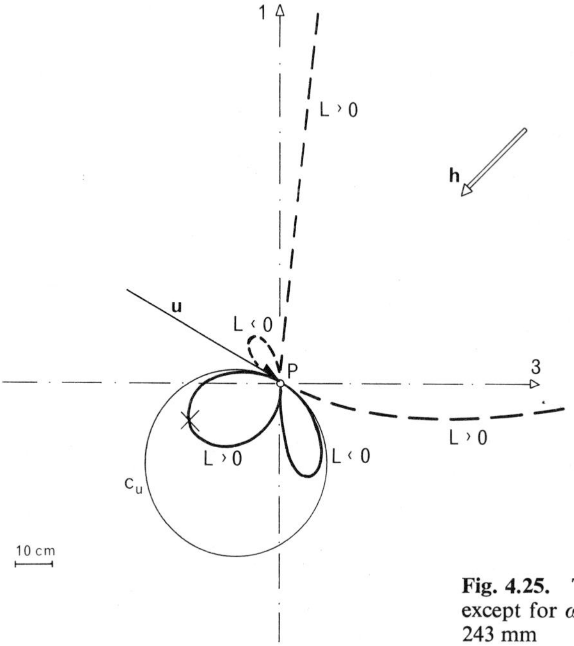

Fig. 4.25. The same case as in Fig. 4.24, except for $\omega = 10^{-4}$ and a radius of c_u = 243 mm

As a second example, let us consider a simple three-dimensional case in which the region considered around P undergoes only a *pivot motion* and a *translation*, i.e. $\Omega \neq 0$, $\boldsymbol{u} \neq 0$, $\gamma = 0$, $\omega = 0$, and where P is illuminated by a collimated beam parallel to $\boldsymbol{n}$, i.e. $\boldsymbol{h} = -\boldsymbol{n}$, $L_S \to \infty$. The condition for complete localization (4.93) becomes

$$\Omega \boldsymbol{E}\boldsymbol{k} + \frac{1}{L}\boldsymbol{N}\boldsymbol{K}\boldsymbol{u} = 0. \tag{4.105}$$

As $\gamma = 0$, the surface element at P behaves like a rigid body that rotates around and glides along an axis Δ parallel to $\boldsymbol{n}$. If we denote by $\boldsymbol{r}$ the position vector of P with respect to the point of intersection O of Δ with the tangent plane (Fig. 4.26), we can write for the displacement

$$\boldsymbol{u} = -\Omega \boldsymbol{E}\boldsymbol{r} + w\boldsymbol{n}.$$

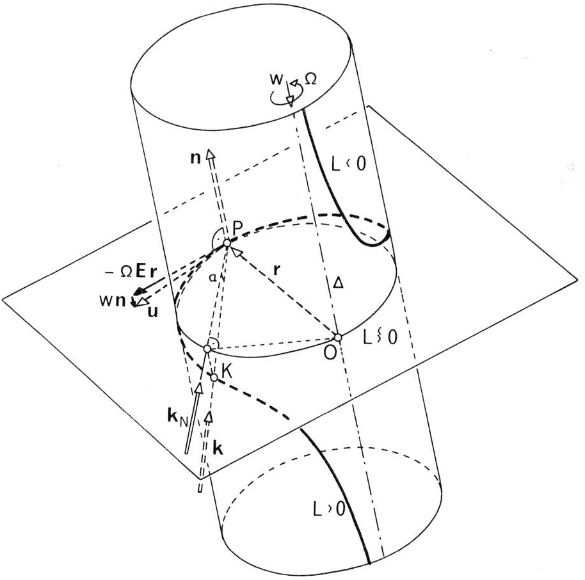

Fig. 4.26. Line of complete localization in the case of a pivot motion and a translation, with $r = 100$ mm, $u = 100.4\,\mu$m, $\Omega = 10^{-3}$, $\boldsymbol{h} = -\boldsymbol{n}$, $L_s \to \infty$

Equation (4.105) is then

$$\Omega \boldsymbol{E}\boldsymbol{k} + \frac{1}{L}[-\Omega \boldsymbol{E}\boldsymbol{r} - \boldsymbol{N}\boldsymbol{k}(-\Omega \boldsymbol{k}\cdot\boldsymbol{E}\boldsymbol{r} + w\boldsymbol{k}\cdot\boldsymbol{n})] = 0.$$

From this result, the line of complete localization may easily be found by the following trick. If, with the angle α of $\boldsymbol{k}$, we define quantities in the tangent plane, namely $\boldsymbol{k}_\mathrm{N} = \boldsymbol{N}\boldsymbol{k}/|\boldsymbol{N}\boldsymbol{k}| = \boldsymbol{N}\boldsymbol{k}/\cos\alpha$, $L_\mathrm{N} = L\cos\alpha$ we find, with $\boldsymbol{E}\boldsymbol{k} = \boldsymbol{E}\boldsymbol{N}\boldsymbol{k} = \boldsymbol{E}\boldsymbol{k}_\mathrm{N}\cos\alpha$,

$$\Omega \boldsymbol{E}\boldsymbol{k}_\mathrm{N} + \frac{1}{L_\mathrm{N}}[-\Omega \boldsymbol{E}\boldsymbol{r} - \cos\alpha\, \boldsymbol{k}_\mathrm{N}(-\Omega\cos\alpha\, \boldsymbol{k}_\mathrm{N}\cdot\boldsymbol{E}\boldsymbol{r} + w\boldsymbol{k}\cdot\boldsymbol{n})] = 0. \tag{4.106}$$

A scalar multiplication with $\boldsymbol{E}\boldsymbol{k}_\mathrm{N}$ gives then, because $\boldsymbol{E}\boldsymbol{k}_\mathrm{N} = -\boldsymbol{k}_\mathrm{N}\boldsymbol{E}$, $\boldsymbol{k}_\mathrm{N}\boldsymbol{E}\cdot\boldsymbol{k}_\mathrm{N} = 0$ and $\boldsymbol{E}^2 = -\boldsymbol{N}$,

$$\Omega - \frac{1}{L_\mathrm{N}}\Omega \boldsymbol{k}_\mathrm{N}\cdot\boldsymbol{r} = 0, \quad \text{i.e.} \quad L_\mathrm{N} = \boldsymbol{k}_\mathrm{N}\cdot\boldsymbol{r}. \tag{4.107}$$

This means that the projection of the line of complete localization is the circle of Thales with diameter OP. On the other hand, multiplying (4.106) with $\boldsymbol{k}_\mathrm{N}$,

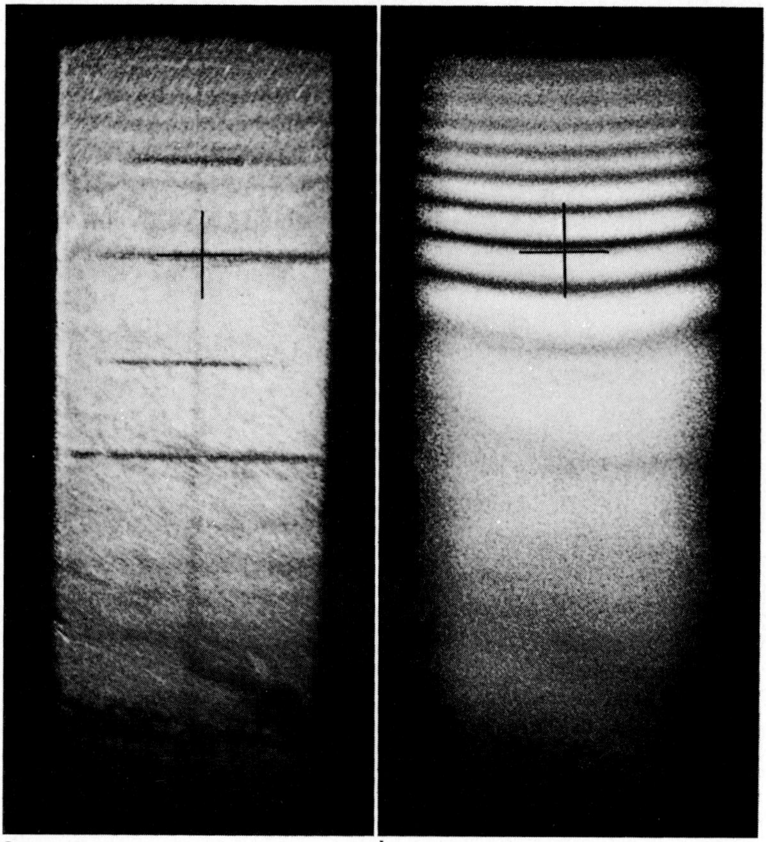

a b

Fig. 4.27a and **b.** Example of fringe pattern. a) Camera focussed on the object surface; the object point marked by a cross is aimed at in a direction of complete localization. b) Camera focussed on the localization point: the fringe is sharp (aperture 1 : 4, $f = 58$ mm)

we get

$$-\Omega \boldsymbol{k}_{\mathrm{N}} \cdot \boldsymbol{E r} - \cos\alpha(-\Omega \cos\alpha\, \boldsymbol{k}_{\mathrm{N}} \cdot \boldsymbol{E r} + w \boldsymbol{k} \cdot \boldsymbol{n}) = 0,$$

or, for the slope $\mathrm{tg}\,\alpha = \boldsymbol{k} \cdot \boldsymbol{n}/\cos\alpha$ of $\boldsymbol{k}$, the two solutions

$$\mathrm{tg}\,\alpha_1 = \frac{-w}{\Omega \boldsymbol{k}_{\mathrm{N}} \cdot \boldsymbol{E r}}, \qquad \mathrm{tg}\,\alpha_2 = 0. \tag{4.108}$$

The line of complete localization forms thus again a butterfly (Fig. 4.26): the two wings are the circle in the tangent plane, whereas the antennae are the observable curves with the rotation axis as an asymptote [Ref. 4.171, Fig. 20; Ref. 4.197, p. 96; Ref. 4.199, p. 335]. Let us add that the right cylinder with the

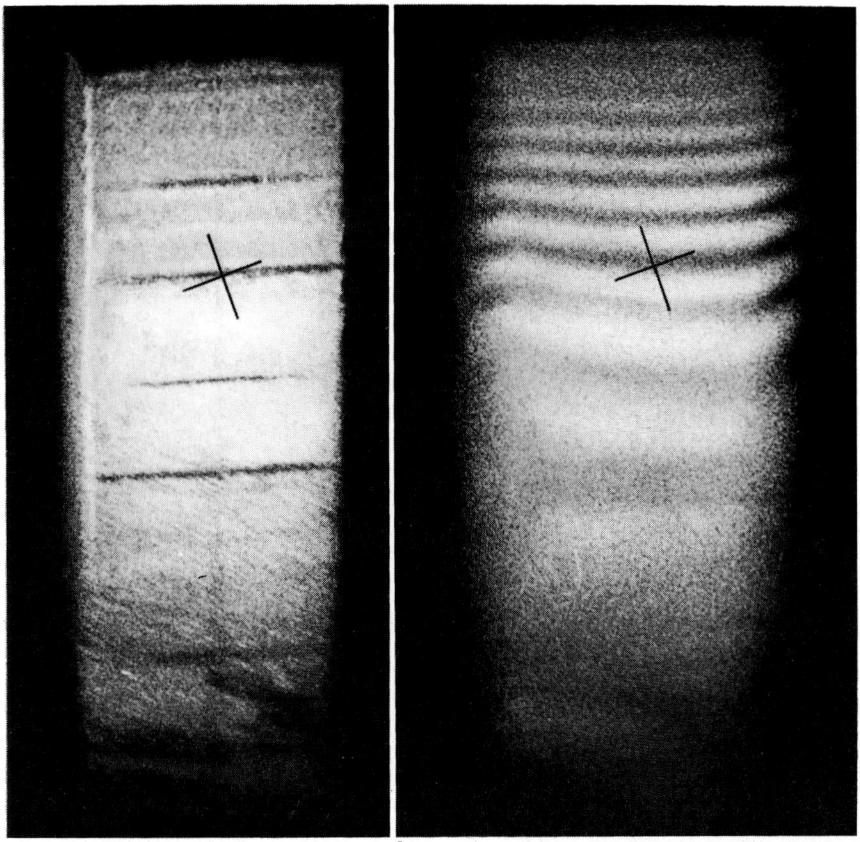

a b

Fig. 4.28a and **b.** Example of fringe pattern. **a)** Camera focussed on the object surface; the object point marked by a cross is aimed at in a direction of partial localization. **b)** Camera focussed on the localization point: the fringe is blurred (aperture 1 : 4, $f = 58$ mm)

circle as a basis constitutes the surface of standard partial localization; indeed (4.85) becomes (4.107), if $\boldsymbol{Mw} = \Omega\boldsymbol{MEk} = \Omega\boldsymbol{Ek}_{\mathrm{N}} \cos\alpha$ is taken into account.

Finally, let us look at some very simple particular cases. If, in the vicinity of P, the *displacement gradient vanishes*, the condition of complete localization is reduced to

$$-\frac{1}{L_{\mathrm{S}}}\boldsymbol{NHu} + \frac{1}{L}\boldsymbol{NKu} = 0,$$

which implies, when $L_{\mathrm{S}} \to \infty$, that the fringes are localized at infinity. If, on the contrary, the *displacement* of P *is negligible*, we must have

$$L(\boldsymbol{\nabla}_{\mathrm{n}} \otimes \boldsymbol{u})\boldsymbol{g} = 0,$$

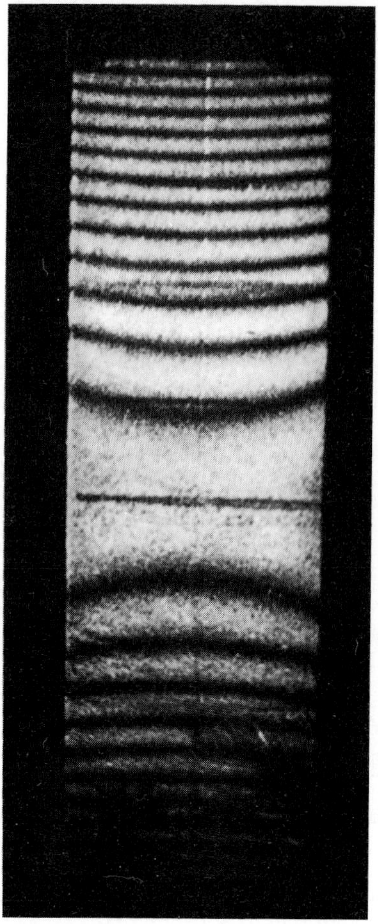

Fig. 4.29. Example of fringe pattern. The object is observed in a direction very near to the directions of the displacement vectors; the fringes appear on the object surface

i.e. the fringes are localized on the object surface (These two situations have been verified by a great number of authors, e.g. in [4.9, 120, 157, 169, 179, 180, 193, 198, 213]). It can also happen that simultaneously $\boldsymbol{Nw} = 0$ and $\boldsymbol{Ku} = 0$, in which case L is arbitrary; the fringes are thus *nonlocalized*, i.e. well visible everywhere in the direction $\boldsymbol{k}_{\mathrm{u}}$.

Photographs of actual fringes are shown in Figs. 4.27–29 [4.196, 197]. They concern the quite general case in which $\boldsymbol{u} \neq 0$, $\boldsymbol{V}_{\mathrm{n}} \otimes \boldsymbol{u}$ is composed of a strain tensor and of an inclination vector, and in which the vectors $\boldsymbol{h}$, $\boldsymbol{n}$ and $\boldsymbol{k}$ do not lie in the same plane.

In conclusion, we can assert that, owing to the foregoing examples and to the two general properties we have demonstrated, namely that the displacement vector as well as the object surface are tangent to the line of complete localization, the latter may generally be expected to form a butterfly centered in the point P to which it relates. In many cases, two loops of finite extension will be visible because $L \to \infty$, i.e. $Nw = 0$, may lead to the observer's positions located behind the object surface. Furthermore, these loops will be wide if the displacement is important, whereas they will remain close to the object surface if the displacement gradient, i.e. the strain or the rotation, is large.

4.2.5 Determination of Strain and Rotation by Means of the Derivatives of the Optical-Path Difference

In the preceding sections of this chapter, we have established relations that describe the interference phenomenon and contain derivatives of the optical-path difference D. In these derivatives, there appeared the displacement vector $\boldsymbol{u}$ and also the displacement "gradient" $\nabla_{\text{n}} \otimes \boldsymbol{u}$, which, we repeat, by virtue of the decomposition (2.109)

$$\nabla_{\text{n}} \otimes \boldsymbol{u} = \gamma + \Omega E + \boldsymbol{\omega} \otimes \boldsymbol{n},$$

consists of the symmetric tensor γ of surface strain, the pivot-motion scalar Ω and the inclination vector $\boldsymbol{\omega}$. We shall thus now examine how the fringes can be used to measure these mechanical quantities at some point P of the object surface. Fundamentally, we can use three kinds of results for that purpose: the kind that expresses the fringe spacing and direction, that which describes the fringe visibility or the partial localization and that given by the condition of complete localization.

First, let us speak of the only relation that contains only the displacement vector, namely (4.39). It assumes that the observation direction goes strictly through P and is in fact an application of the relative fringe-order method (see Sect. 4.1.2) in the case of neighboring fringes. If we measure the fringe spacing in three directions $\boldsymbol{m}_i$, we can write

$$\lambda \left[\frac{\Delta n_{\text{P}}}{\Delta \psi} \right]_i = \boldsymbol{m}_i \cdot \boldsymbol{u}, \qquad (i = 1, 2, 3) \tag{4.109}$$

where often $\Delta n_{\text{P}} = 1$ is chosen, i.e. we pass from one fringe to the next, or $\Delta n_{\text{P}} = 0$ is chosen, i.e. $\boldsymbol{m}_i$ is parallel to the fringe. From this, the displacement vector can be calculated in the way we described in Sect. 4.1.2 (see e.g. [4.9, 120, 126]). In order for the three $\boldsymbol{m}_i$ to form a complete basis, these must be non-coplanar; therefore, at least two observation directions $\boldsymbol{k}_1$, $\boldsymbol{k}_2$ must be used, because an $\boldsymbol{m}_i$ is perpendicular to either of these $\boldsymbol{k}_\beta$.

As for the relations containing both $\boldsymbol{u}$ and $\nabla_{\text{n}} \otimes \boldsymbol{u}$, let us begin with some general comments. Although, in principle, these relations allow us to find all unknowns, it is often easier to determine first the displacement vector by means of the optical-path difference, as we saw in Sect. 4.1, and to keep the more complicated equations with the derivatives of D exclusively for determination of the

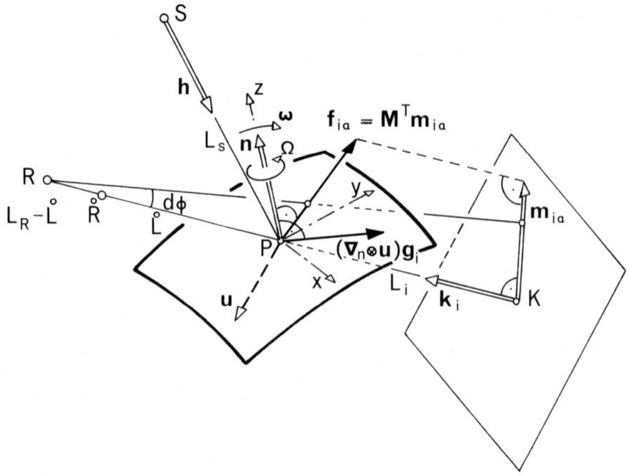

Fig. 4.30. The quantities that appear when $\nabla_n \otimes \boldsymbol{u}$ is measured

derivative of the displacement. For this reason and also for simplicity's sake, we shall suppose in the following that only $\nabla_n \otimes \boldsymbol{u}$ is sought. Because ∇_n lies in the plane tangent to the object surface, this tensor is semi-interior, i.e. $\boldsymbol{n}(\nabla_n \otimes \boldsymbol{u}) = 0$, and thus has six unknown components. In the established relations, $\nabla_n \otimes \boldsymbol{u}$ appears contracted with the sensitivity vector $\boldsymbol{g} = \boldsymbol{k} - \boldsymbol{h}$ (Fig. 4.30), which forms the vector $(\nabla_n \otimes \boldsymbol{u})\boldsymbol{g}$ that lies in the tangent plane. Moreover, the scalar product $\boldsymbol{m}\boldsymbol{M} \cdot (\nabla_n \otimes \boldsymbol{u})\boldsymbol{g}$ often appears, where $\boldsymbol{m}$ is a unit vector perpendicular to $\boldsymbol{k}$ and $\boldsymbol{M}^T\boldsymbol{m} = \boldsymbol{f}$, a vector perpendicular to $\boldsymbol{n}$. In a local cartesian coordinate system that has its origin in P and its axis z along $\boldsymbol{n}$, we have $\boldsymbol{n} \mathrel{\hat{=}} [0, 0, 1]$, and then

$$\boldsymbol{f} = \boldsymbol{M}^T\boldsymbol{m} \mathrel{\hat{=}} \begin{bmatrix} f_x \\ f_y \\ 0 \end{bmatrix} = \begin{bmatrix} 1 & 0 & -k_x/k_z \\ 0 & 1 & -k_y/k_z \\ 0 & 0 & 0 \end{bmatrix} \begin{bmatrix} m_x \\ m_y \\ m_z \end{bmatrix},$$

$$\boldsymbol{f} \cdot (\nabla_n \otimes \boldsymbol{u})\boldsymbol{g} \mathrel{\hat{=}} [f_x, f_y, 0] \begin{bmatrix} \varepsilon_x & \frac{1}{2}\gamma_{xy} + \Omega & \omega_x \\ \frac{1}{2}\gamma_{xy} - \Omega & \varepsilon_y & \omega_y \\ 0 & 0 & 0 \end{bmatrix} \begin{bmatrix} g_x \\ g_y \\ g_z \end{bmatrix}. \tag{4.110}$$

Hence, from the fact that $\nabla_n \otimes \boldsymbol{u}$ is semi-interior (the third row in the foregoing matrix has only zeros) it follows that three linearly independent vectors $\boldsymbol{g}_i$ ($i = 1, 2, 3$) and two nonparallel vectors $\boldsymbol{f}_\alpha$ ($\alpha = 1, 2$), are required. The three $\boldsymbol{g}_i$ are obtained by varying either the direction of observation (from $\boldsymbol{k}_1$ to $\boldsymbol{k}_2$, to $\boldsymbol{k}_3$) or that of illumination (from $\boldsymbol{h}_1$ to $\boldsymbol{h}_2$, to $\boldsymbol{h}_3$). However, if $\boldsymbol{k}$ is changed, two identical $\boldsymbol{m}_\alpha$ cannot be used in all measurements; therefore, with each $\boldsymbol{g}_i$, we shall generally use two vectors $\boldsymbol{m}_{i\alpha}$, which at last represents a set of six such unit vectors. Most commonly moreover, $\boldsymbol{h}$ is kept fixed and $\boldsymbol{k}$ only is changed.

Considering in this last case the expressions (4.34, 35) of the *fringe spacing* or direction, we write for example

$$\frac{\lambda}{L_{Ri}}\left[\frac{\Delta n_R}{\Delta\phi}\right]_{i\alpha} = \boldsymbol{m}_{i\alpha}\boldsymbol{M}_i\cdot\left[(\boldsymbol{V}_n \otimes \boldsymbol{u})\boldsymbol{g}_i - \left(\frac{1}{L_S}\boldsymbol{H} + \frac{1}{L_{Ri}}\boldsymbol{K}_i\right)\lambda n_k \boldsymbol{g}^k\right]$$

$$(i = 1, 2, 3, \text{ not summed}, \alpha = 1, 2), \qquad (4.111)$$

where again Δn_R often takes the value one or zero. For the displacement, we have used the expression (4.9), i.e. $\boldsymbol{u} = \lambda n_k \boldsymbol{g}^k$ with $\boldsymbol{g}_i \cdot \boldsymbol{g}^k = \delta_i^k$. In this way, if in each direction $\boldsymbol{k}_i$ the corresponding fringe-order n_i and two finite differences $[\Delta n_R/\Delta\phi]_{i\alpha}$ are measured, six linear equations are obtained, which determine entirely the components of $\boldsymbol{V}_n \otimes \boldsymbol{u}$ (see e.g. [4.145, 147, 166, 168, 174, 177, 213]). Contrary to the relations of Sect. 4.1.4, we need neither to know $\boldsymbol{u}$ at P explicitly, nor to form finite differences between $\boldsymbol{u}$ and neighboring $\boldsymbol{u}_1$ and $\boldsymbol{u}_2$. Let us also recall that the measurements are made independently of the localization or of the visibility of the fringes, the fringe-order being constant along an observation direction, and that these measurements can be performed, for example, in a plane perpendicular to $\boldsymbol{k}$ through P or in the plane tangent to the object surface (Sect. 4.2.1). Furthermore, the measurements can also be taken on an image of the object surface (Fig. 4.31). Then, the relations between the three possible vectorial increments are [see (4.26) and (2.70)]

$$\Delta\boldsymbol{r} = L_R \boldsymbol{M}^T \boldsymbol{m}\Delta\phi = -\frac{L_R}{\mathring{l}}\boldsymbol{M}^T\Delta\mathring{\boldsymbol{r}},$$

$$\Delta\mathring{\boldsymbol{r}} = -\mathring{l}\mathring{\boldsymbol{M}}^T\boldsymbol{m}\Delta\phi = -\frac{\mathring{l}}{L_R}\mathring{\boldsymbol{M}}^T\Delta\boldsymbol{r}, \qquad (4.112)$$

where $\mathring{\boldsymbol{M}} = \boldsymbol{I} - \mathring{\boldsymbol{n}} \otimes \boldsymbol{k}/\mathring{\boldsymbol{n}}\cdot\boldsymbol{k}$.

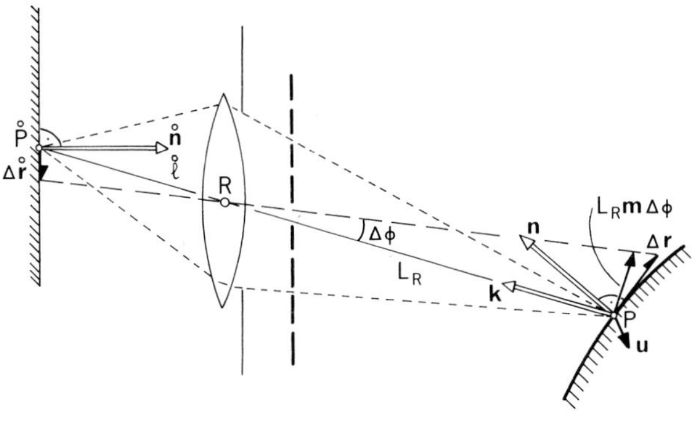

Fig. 4.31. Measurement of $\Delta D_R/\Delta s$ on an image $\mathring{P}$ of P, connections between the increments $\Delta\boldsymbol{r}$, $L_R\boldsymbol{m}\Delta\phi$ and $\Delta\boldsymbol{r}$

If now the results that pertain to the *partial localization* are used, not only the direction of observation but also the shape and the position of the aperture can be chosen. First, in the standard case of partial localization with a circular aperture, if the distance L_{p} of localization is measured, we have, from (4.85),

$$\left[(\boldsymbol{\nabla}_{\mathrm{n}} \otimes \boldsymbol{u})\boldsymbol{g}_i - \frac{1}{L_{\mathrm{S}}}\boldsymbol{H}\boldsymbol{u}\right] \cdot \boldsymbol{M}_i^{\mathrm{T}}\boldsymbol{M}_i\left[(\boldsymbol{\nabla}_{\mathrm{n}} \otimes \boldsymbol{u})\boldsymbol{g}_i - \frac{1}{L_{\mathrm{S}}}\boldsymbol{H}\boldsymbol{u} + \frac{1}{L_{\mathrm{p}i}}\boldsymbol{K}_{;i}\boldsymbol{u}\right] = 0$$

$$(i = 1, \ldots, 6, \text{ not summed}). \qquad (4.113)$$

With six such equations, which are unfortunately quadratic with respect to $\boldsymbol{\nabla}_{\mathrm{n}} \otimes \boldsymbol{u}$, the latter tensor could be calculated, at least in principle. Second, if a slit aperture is used, an additional degree of freedom is obtained. Indeed, in a given direction $\boldsymbol{k}_i$, we can either assign two slit directions $\boldsymbol{m}_a = \boldsymbol{m}_{i\alpha}$ ($\alpha = 1, 2$) and seek the distances $L_a = L_{i\alpha}$ for which the visibility becomes equal to one or, conversely, assign two distances $L_{i\alpha}$, and seek the directions $\boldsymbol{m}_{i\alpha}$ for which the visibility becomes equal to one. As suggested by *Stetson* [Ref. 4.164, p. 589; 4.176], (4.92) gives then the linear system of equations

$$\boldsymbol{m}_{i\alpha}\boldsymbol{M}_i \cdot \left[(\boldsymbol{\nabla}_{\mathrm{n}} \otimes \boldsymbol{u})\boldsymbol{g}_i - \frac{1}{L_{\mathrm{S}}}\boldsymbol{H}\boldsymbol{u} + \frac{1}{L_{i\alpha}}\boldsymbol{K}_{;i}\boldsymbol{u}\right] = 0$$

$$(i = 1, 2, 3, \ \alpha = 1, 2, \text{ not summed}). \qquad (4.114)$$

In a similar way, we could look for points where the visibility is zero. As indicated by *Charmet*, *Montel* and *Ebbeni* [4.201–203], by fixing, for instance, $\boldsymbol{m}_{i\alpha}$ and $L_{i\alpha}$, we can then determine the length $a_{i\alpha}$ of the slit. By means of the first zero $\xi = \pi$ of the function $\sin \xi/\xi$, we get from (4.90)

$$\frac{\lambda}{a_{i\alpha}} = \frac{L_{i\alpha}}{L_{i\alpha} + \mathring{L}}\boldsymbol{m}_{i\alpha}\boldsymbol{M}_i \cdot \left[(\boldsymbol{\nabla}_{\mathrm{n}} \otimes \boldsymbol{u})\boldsymbol{g}_i - \frac{1}{L_{\mathrm{S}}}\boldsymbol{H}\boldsymbol{u} + \frac{1}{L_{i\alpha}}\boldsymbol{K}_{;i}\boldsymbol{u}\right]$$

$$(i = 1, 2, 3, \ \alpha = 1, 2, \text{ not summed}). \qquad (4.115)$$

Finally, the condition of *complete localization* can be useful (see e.g. [4.9, 178, 195]), provided that we can be certain that this condition is actually fulfilled. For that purpose, we may measure the visibility, or verify that the point of complete localization does not move when the shape of the aperture is changed, for instance when a slit aperture is rotated. As proposed by *Monneret* [Ref. 4.184, p. 114], if the fringes are observed with a circular aperture perpendicular to $\boldsymbol{k}$, a point K can also be sought where the visibility is zero and, with (4.82), the length $t = |LM\boldsymbol{\nabla}_{\mathrm{n}}D_{\mathrm{K}}|$ between this point K and the point K* on the other homologous ray (Fig. 4.17) can be calculated. From this result, the distance

t_{min} corresponding to the point of localization and which should be zero, can be estimated. Another criterion lies in the normality theorem [see (4.94)] for the direction of the fringes [4.165]: with a special mechanical device based upon this theorem, one fiducial line of the reticule of the observation telescope is kept parallel to $\boldsymbol{Ku}$ while $\boldsymbol{k}$ is changed so that the observation direction for which the second fiducial line is tangent to the fringe is a direction of complete localization (Figs. 4.27–28) [4.196, 197]. If we measure three such $\boldsymbol{k}_i$ and the corresponding L_i, condition (4.93) gives the linear vectorial equations

$$(\boldsymbol{V}_{\mathrm{n}} \otimes \boldsymbol{u})\boldsymbol{g}_i - \frac{1}{L_{\mathrm{S}}}\boldsymbol{NHu} + \frac{1}{L_i}\boldsymbol{NK}_i\boldsymbol{u} = 0 \qquad (i = 1, 2, 3, \text{ not summed}). \tag{4.116}$$

Let us add here that, if the fringes are known to be completely localized, the expression (4.94) of the fringe direction can be used to determine the displacement, which appears alone therein [4.183, Ref. 4.184, p. 109].

Thus, holographic interferometry offers several possibilities for determining entirely the strain and the rotation at one point of an object surface. Each of these methods has advantages of its own, such as for example the large choice of observation directions in the partial-localization method, or the brightness and good visibility of the fringes in the complete-localization method, so that, in practice, combinations of them may be useful. Furthermore, as shown in (4.111), the manner of measuring the displacement may also be taken into account. Generally, numerous quantities are to be measured. Thus, $\Delta n_{\mathrm{R}}/\Delta\phi$, $\boldsymbol{m}$, $\boldsymbol{n}$, $\boldsymbol{h}$, $\boldsymbol{k}$, L_{S}, L_{R} or L and in most cases even $\boldsymbol{u}$ must be known before the system of six equations for $\boldsymbol{V}_{\mathrm{n}} \otimes \boldsymbol{u}$ can be written. This great number of measurements as well as the great number of unknowns may render the computations inaccurate and cumbersome. However, simplifications are possible: particular choices of the sensitivity vectors with respect to the coordinate system make some unknown components vanish in certain equations; especially, it is sometimes possible to avoid completely the determination of $\boldsymbol{u}$ by putting $L_{\mathrm{S}} \to \infty$ and L_{R} or $L_{\mathrm{p}} \to \infty$ (see e.g. [4.147, 174, 202]). As for the rotation tensor, as said in Sect. 2.2.1, it can be much larger than the strain tensor. In that case, its influence is important: a small error of determining the rotation then causes a large error of the strain. The degree of freedom in the choice of the conditions under which the measurements have to be performed varies with each method. Generally however, whatever technique is used, an arbitrary view of the fringe pattern does not depend on only one deformation component, but both the strain and the rotation tensors are involved; the only exception is that, if $\boldsymbol{g}$ is parallel to $\boldsymbol{n}$, the inclination appears alone. It remains therefore to find an appropriate criterion that would allow us to distinguish between the symmetric and the skew-symmetric part of $\boldsymbol{V}_{\mathrm{n}} \otimes \boldsymbol{u}$. More flexibility will perhaps be supplied by modifications of the optical arrangement at the reconstruction.

4.3 Modifications of the Optical Arrangement at the Reconstruction

We have not yet spoken in this chapter of the holographic process as such because we supposed it to be ideal, assuming the images to be identical to the object. The interference phenomenon was thus affected only by the object deformation and by the manner in which the object was illuminated and observed. Now, in this way, inconvenient fringe patterns may appear, particularly as far as the number of fringes or their contrast are concerned. To remedy these drawbacks and also to provide new means of measuring the displacement and its derivative, the fringe pattern can be changed in a known way by modifying the optical arrangement (For other methods, see e.g. [4.160, 214]). For this purpose, we must be able to act independently on the two interfering wave fields.

In the case of the *real-time technique*, these wave fields are clearly separated, so that no special equipment is required: we can, on the one hand, translate or rotate the object or its image, or, on the other hand, change the phase of the light source that illuminates the object, or also displace this source [4.215–220]. In the first case, the total object deformation is composed of the sought deformation and of a known rigid body motion; in the second, the optical-path difference contains an additional term, as indicated by (4.14). If on the contrary *both wave fields originate from a hologram*, a differentiated influence on them necessitates particular dispositions. First, we can record the wave fields with different reference sources and, during reconstruction, modify one of the sources. We can also record the wave fields on different hologram plates and, during reconstruction, move one of the plates or even both.

In the present section, we shall establish the counterpart of the principal preceding results in the mentioned cases of modification and, at the end, see how we can profit from these formulae for the determination of the deformation. Therefore, we shall use the concepts introduced in Sect. 3.2; especially, we shall speak of "optically displaced" images, which, as we saw, is not unwarranted, although these images suffer aberrations; but we shall need this representation provisionally and only for convenience. Let us add that the following results are also useful to estimate the errors made when part of the optical arrangement moves unintentionally, particularly in the real-time technique; in the double-exposure method with only one reference source, these errors are quite harmless because the fringe pattern is not affected, at least in our first-order approximations, as will be shown by the following relations.

4.3.1 Basic Relations when the Reconstruction Sources are Shifted and when the Wavelength is Changed

At present, we shall suppose that the images of the undeformed and of the deformed object are produced by *two reconstruction sources*. In order for each of these sources to reconstruct only one image, they must lie far away from each other, whereas the modifications they undergo are small (about this point, see Sect. 3.2). First, the modification of the optical arrangement will consist in a displacement and a phase change of one of the sources. We shall then examine the influence of wavelength changes that affect both sources.

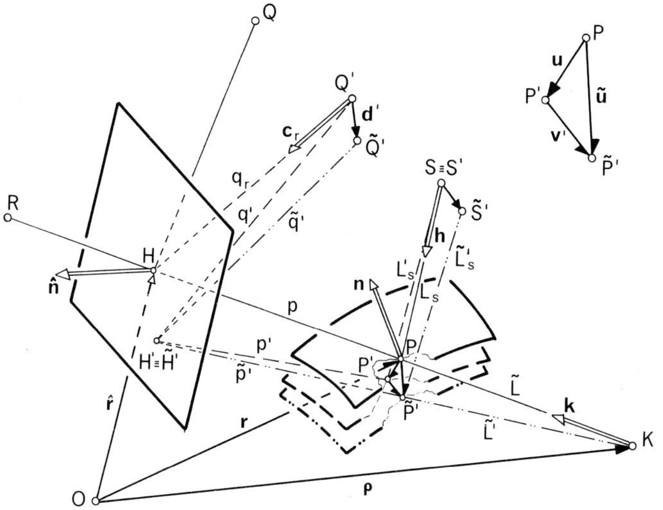

Fig. 4.32. Holographic interferometry with a fixed reference source Q for the undeformed object and a movable reference source Q′ for the deformed object

Let us first look at Fig. 4.32. The undeformed object, i.e. the set of points {P}, has been recorded and is reconstructed ideally by means of the same point source Q (Instead of its image, even the undeformed object itself could be present). On the other hand, the image $\{\tilde{\mathrm{P}}'\}$ of the deformed object does not coincide with the deformed object {P′} because the reconstruction source Q′ has been moved to $\tilde{\mathrm{Q}}'$ by the vector $\boldsymbol{d}'$. The two wave fields originate from the same fixed hologram with unit normal $\hat{\boldsymbol{n}}$. So, whereas a material point P has undergone the "*mechanical*" *displacement* $\boldsymbol{u}$ to P′, P and $\tilde{\mathrm{P}}'$ are separated by the *apparent total displacement* $\tilde{\boldsymbol{u}} = \boldsymbol{u} + \boldsymbol{v}'$, where $\boldsymbol{v}'$ is the "*optical*" *displacement* [4.178, 216–218, 221–229]. Now, let us calculate the difference between the optical path covered by two rays coming from corresponding points P and $\tilde{\mathrm{P}}'$ to some point K situated at the distance $\tilde{L}$ behind P, in which case $\tilde{L}$ is positive, as in the preceding sections. Whereas the optical path L_{S} gives us, in a simple way, the phase $\phi_{\mathrm{S}} = 2\pi L_{\mathrm{S}}/\lambda$ of the light "emitted" by P, that of $\tilde{\mathrm{P}}'$ must be calculated from the condition of interference identity at H′, which is the point on the hologram where the ray $\tilde{\mathrm{P}}'$K is formed. Similar to (3.12), we obtain here, because $\tilde{\lambda}' = \tilde{\lambda} = \lambda$,

$$\tilde{\phi}'_S = \phi'_S + \frac{2\pi}{\lambda}[(p' - \tilde{p}') - (q' - \tilde{q}')] - (\psi' - \tilde{\psi}'), \tag{4.117}$$

where ψ' and $\tilde{\psi}'$ represent the phases of Q′ and $\tilde{\mathrm{Q}}'$ and where $\phi'_S = 2\pi L'_S/\lambda$. The phase $\tilde{\phi}'_S$ at $\tilde{\mathrm{P}}'$ can be interpreted as follows: it defines the distance $\tilde{L}'_S = \lambda\tilde{\phi}'_S/2\pi$ covered by the light emitted by a fictitious source $\tilde{\mathrm{S}}'$ and falling on $\tilde{\mathrm{P}}'$ (Fig. 4.32). The optical-path difference at K is thus

$$D = \left(\frac{\lambda}{2\pi}\phi_S - \tilde{L}\right) - \left(\frac{\lambda}{2\pi}\tilde{\phi}'_S - \tilde{L}'\right) = (L_S - \tilde{L}) - (\tilde{L}'_S - \tilde{L}'). \tag{4.118}$$

Eliminating the unknown quantity $\tilde{\phi}'_S$ by means of (4.117), we obtain

$$D = (L_S - L'_S) - (\tilde{L} - \tilde{L}') - (p' - \tilde{p}') + (q' - \tilde{q}') + \frac{\lambda}{2\pi}(\psi' - \tilde{\psi}'). \tag{4.119}$$

Similar to what we did in Sects. 3.2.2 or 4.1.1, we can express D as a function of quantities that relate to a reference configuration, which we choose to take as the lines SP, PH and Q′H. Thus, D depends on the unit vectors $\boldsymbol{h}$, $\boldsymbol{k}$, $\boldsymbol{c}_r$ and also on the small displacement vectors $\boldsymbol{d}'$, $\boldsymbol{u}$. We have, then,

$$D = -\boldsymbol{u}\cdot\boldsymbol{h} + \tilde{\boldsymbol{u}}\cdot\boldsymbol{k} - \boldsymbol{v}'\cdot\boldsymbol{k} + \boldsymbol{d}'\cdot\boldsymbol{c}_r + \frac{\lambda}{2\pi}(\psi' - \tilde{\psi}'),$$

that is, with $\tilde{\boldsymbol{u}} = \boldsymbol{u} + \boldsymbol{v}'$ and $\boldsymbol{k} - \boldsymbol{h} = \boldsymbol{g}$,

$$D = \boldsymbol{u}\cdot\boldsymbol{g} + \boldsymbol{d}'\cdot\boldsymbol{c}_r + \frac{\lambda}{2\pi}(\psi' - \tilde{\psi}'). \tag{4.120}$$

The first term is the usual one that appears alone in standard holographic interferometry [see (4.6)], whereas the two last terms show the influence of the modification of the optical arrangement and will, therefore, be called the *fringe-control terms.* These terms depend on only the parameters of the fringe modification and the exact location of $\tilde{\mathrm{P}}'$ does not play any role in this result.

Starting from this relation, we can further calculate the *fringe spacing and direction,* i.e. the derivative of D with respect to a change $d\phi$ of the direction along which the observer looks at the fringes (Fig. 4.33). In the same way as in Sect. 4.2.1, this derivative can be expressed as a function of several increments that are related through the collineation with center R. However, we shall restrict

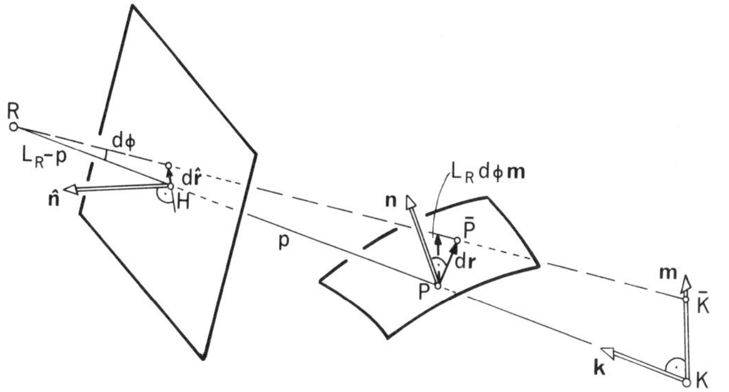

Fig. 4.33. Connections between the increments dr, $L_R m d\phi$ and $d\hat{r}$ established by the collineation with center R

ourselves to the form [see (4.27)]

$$\frac{dD_R}{d\phi} = L_R \boldsymbol{m} \cdot \boldsymbol{M} \boldsymbol{\nabla}_n D_R = L_R \boldsymbol{m} \cdot \boldsymbol{M}[\boldsymbol{\nabla}_n(\boldsymbol{u} \cdot \boldsymbol{g})|_R + (\boldsymbol{\nabla}_n \otimes \boldsymbol{c}_r)|_R \boldsymbol{d}'],$$

where we have taken into account that ψ', $\tilde{\psi}'$ and $\boldsymbol{d}'$ are constant. The derivative of $\boldsymbol{u} \cdot \boldsymbol{g}$ is the same as previously in (4.31), but that of $\boldsymbol{c}_r$ is more awkward because $\boldsymbol{\nabla}_n$ is on the object surface. However, using once more the collineation, we can transform it into a derivation on the hologram plane. Similar to (2.70, 72), we can write

$$d\boldsymbol{r} = \frac{L_R}{L_R - p} \boldsymbol{M}^T d\hat{\boldsymbol{r}}, \qquad \boldsymbol{\nabla}_n = \frac{L_R - p}{L_R} \boldsymbol{N} \hat{\boldsymbol{M}} \boldsymbol{\nabla}_{\hat{n}}. \tag{4.121}$$

The two-dimensional derivative operator in the hologram plane $\boldsymbol{\nabla}_{\hat{n}}$ and the oblique projection along $\hat{\boldsymbol{n}}$ onto the plane perpendicular to $\boldsymbol{k}$, $\hat{\boldsymbol{M}} = \boldsymbol{I} - \hat{\boldsymbol{n}} \otimes \boldsymbol{k}/\hat{\boldsymbol{n}} \cdot \boldsymbol{k}$ (see Fig. 2.11), were already used in Chap. 3 for the image formation. Like (2.73) and like in Sects. 3.2.1, 2, we have

$$\boldsymbol{\nabla}_{\hat{n}} \otimes \boldsymbol{c}_r = \frac{1}{q_r} \hat{\boldsymbol{N}} \boldsymbol{C}_r,$$

with the reference distance q_r between Q′ and H and the normal projection $\boldsymbol{C}_r = \boldsymbol{I} - \boldsymbol{c}_r \otimes \boldsymbol{c}_r$. Then, since $\hat{\boldsymbol{M}} \hat{\boldsymbol{N}} = \hat{\boldsymbol{M}}$ [see (2.71)], we get, by superposed linear transformation

$$\boldsymbol{\nabla}_n \otimes \boldsymbol{c}_r|_R = \frac{L_R - p}{L_R q_r} \boldsymbol{N} \hat{\boldsymbol{M}} \boldsymbol{C}_r. \tag{4.122}$$

So, instead of (4.34), we find now with $\boldsymbol{M}\hat{\boldsymbol{M}} = \hat{\boldsymbol{M}}$

$$\frac{dD_{\mathrm{R}}}{d\phi} = \lambda \frac{dn_{\mathrm{R}}}{d\phi} = \boldsymbol{m} \cdot \left(L_{\mathrm{R}} \boldsymbol{M} \boldsymbol{w} - \boldsymbol{K} \boldsymbol{u} + \frac{L_{\mathrm{R}} - p}{q_{\mathrm{r}}} \hat{\boldsymbol{M}} \boldsymbol{C}_{\mathrm{r}} \boldsymbol{d}' \right), \tag{4.123}$$

where we recall that, as previously, the strain and rotation are contained in $\boldsymbol{w}$ because of (4.32). The last term in (4.123) on the contrary is new and marks the influence of the fringe control. Otherwise, the same reasoning as in Sect. 4.2.1 is valid; in particular, the fringes are perpendicular to the vector between the parentheses. Let us remark also that, like D, this result does not depend on the position of $\tilde{\mathrm{P}}'$. However, remembering (3.46), we can make the lateral part $\boldsymbol{K}\boldsymbol{v}'$ of the displacement of P′ appear, i.e.

$$\frac{dD_{\mathrm{R}}}{d\phi} = \boldsymbol{m} \cdot \left(L_{\mathrm{R}} \boldsymbol{M} \boldsymbol{w} - \boldsymbol{K} \boldsymbol{u} + \frac{L_{\mathrm{R}} - p}{p} \boldsymbol{K} \boldsymbol{v}' \right). \tag{4.124}$$

Now, as far as the fringe contrast and localization are concerned, we need not repeat the calculations of Sects. 4.2.2, 3, 4. Indeed, because the wave field of the deformed object was supposed to differ from that of the undeformed object by only the optical-path difference D [see (4.50)], and because we could express D as a function of the same reference configuration, it suffices to insert in the preceding relations the value of D given by (4.120) and to replace its derivative (4.58) by

$$\nabla_{\mathrm{n}} D_{\mathrm{K}} = \boldsymbol{N} \left(\boldsymbol{w} + \frac{1}{\tilde{L}} \boldsymbol{K} \boldsymbol{u} + \frac{p + \tilde{L}}{\tilde{L} q_{\mathrm{r}}} \hat{\boldsymbol{M}} \boldsymbol{C}_{\mathrm{r}} \boldsymbol{d}' \right), \tag{4.125}$$

where, compared to (4.123), L_{R} has been replaced by $-\tilde{L}$ because K is, in this case, the collineation center. In this way, in the standard case of *partial localization*, we arrive at the condition

$$\frac{d}{d\tilde{L}} \left| \tilde{L} \boldsymbol{M} \boldsymbol{w} + \boldsymbol{K} \boldsymbol{u} + \frac{p + \tilde{L}}{q_{\mathrm{r}}} \hat{\boldsymbol{M}} \boldsymbol{C}_{\mathrm{r}} \boldsymbol{d}' \right|_{\tilde{L} = \tilde{L}_{\mathrm{p}}} = 0,$$

i.e. that again the length t between the pair of points K, K* on the homologous rays must be minimal. This condition is fulfilled when

$$\tilde{L}_{\mathrm{p}} = - \frac{\left(\boldsymbol{M} \boldsymbol{w} + \frac{1}{q_{\mathrm{r}}} \hat{\boldsymbol{M}} \boldsymbol{C}_{\mathrm{r}} \boldsymbol{d}' \right) \cdot \left(\boldsymbol{K} \boldsymbol{u} + \frac{p}{q_{\mathrm{r}}} \hat{\boldsymbol{M}} \boldsymbol{C}_{\mathrm{r}} \boldsymbol{d}' \right)}{\left(\boldsymbol{M} \boldsymbol{w} + \frac{1}{q_{\mathrm{r}}} \hat{\boldsymbol{M}} \boldsymbol{C}_{\mathrm{r}} \boldsymbol{d}' \right)^2}, \tag{4.126}$$

which can be compared to (4.85). Furthermore, equating (4.125) to zero yields the condition of *complete localization*, i.e.

$$\left[N w + \frac{1}{\tilde{L}} N K u + \frac{p + \tilde{L}}{\tilde{L} q_r} N \hat{M} C_r d'\right]_{\substack{\tilde{L}=\tilde{L}_c \\ \tilde{k}=\tilde{k}_c}} = 0 \tag{4.127}$$

or with the lateral displacement of P′

$$N\left[w + \frac{1}{\tilde{L}} K u + \left(\frac{1}{\tilde{L}} + \frac{1}{p}\right) K v'\right]_{\substack{\tilde{L}=\tilde{L}_c \\ \tilde{k}=\tilde{k}_c}} = 0. \tag{4.128}$$

If this condition is satisfied, we can write for the fringe spacing

$$\frac{dD_R}{d\phi} = -\frac{L_R + \tilde{L}}{\tilde{L}} m \cdot K(u + v'). \tag{4.129}$$

We obtain thus again a normality theorem: *if and only if the fringes are observed in a direction of complete localization, they are perpendicular to the projection* $K(u + v') = K\tilde{u}$ *of the total displacement.* Finally, let us say that, in the particular case $u = 0$, $V_n \otimes u = 0$, the fringes are localized on the hologram for any k, because $\tilde{L} \to -p$ [Ref. 4.178, p. 1095; Ref. 4.227, p. 403], and that the partial and the complete localization loci are confounded.

To investigate now the influence of a *change of the wavelength*, let us first look at a situation similar to that of the heterodyne holographic interferometry we spoke about in Sect. 4.1.3. With this method, the two wave fields are recorded with the same wavelength λ, but reconstructed with different wavelengths $\tilde{\lambda} = \lambda - \Delta\lambda$ and $\tilde{\lambda}' = \lambda + \Delta\lambda$, for the undeformed and the deformed object, respectively. On the other hand, the positions of the reference and reconstruction sources are not changed (Fig. 4.34). Using again the concept of an image $\tilde{\mathrm{P}}$ of P, we write for the vibration that it emits, with the frequency $\tilde{\nu} = \nu + \Delta\nu$, to the point K

$$\tilde{U}(\rho, t) = \frac{|A_{\tilde{\mathrm{P}}}|}{|\tilde{L}|} \exp\left[\mathrm{i}\left(\tilde{\phi}_S - \frac{2\pi}{\tilde{\lambda}}\tilde{L} - 2\pi\Delta\nu t\right)\right] \exp(-2\pi \mathrm{i}\nu t).$$

For the amplitude $\tilde{U}'(\rho, t)$ produced by the image $\tilde{\mathrm{P}}'$ we have a similar expression with the frequency $\tilde{\nu}' = \nu - \Delta\nu$. Considering $\exp(-2\pi\mathrm{i}\nu t)$ as the common time-harmonic factor, we obtain for the resulting intensity at K

$$\frac{1}{2}\left\{\frac{|A_{\tilde{\mathrm{P}}}|^2}{\tilde{L}^2} + \frac{|A_{\tilde{\mathrm{P}}'}|^2}{\tilde{L}'^2} + 2\frac{|A_{\tilde{\mathrm{P}}}||A_{\tilde{\mathrm{P}}'}|}{\tilde{L}\tilde{L}'} \cos\left[\left(\tilde{\phi}_S - \frac{2\pi}{\tilde{\lambda}}\tilde{L}\right) - \left(\tilde{\phi}'_S - \frac{2\pi}{\tilde{\lambda}'}\tilde{L}'\right) - 4\pi\Delta\nu t\right]\right\}. \tag{4.130}$$

The time-dependent term indicates that the fringe-order oscillates with the beat frequency $2\Delta\nu$. The interesting quantity is the phase of this vibration, which can

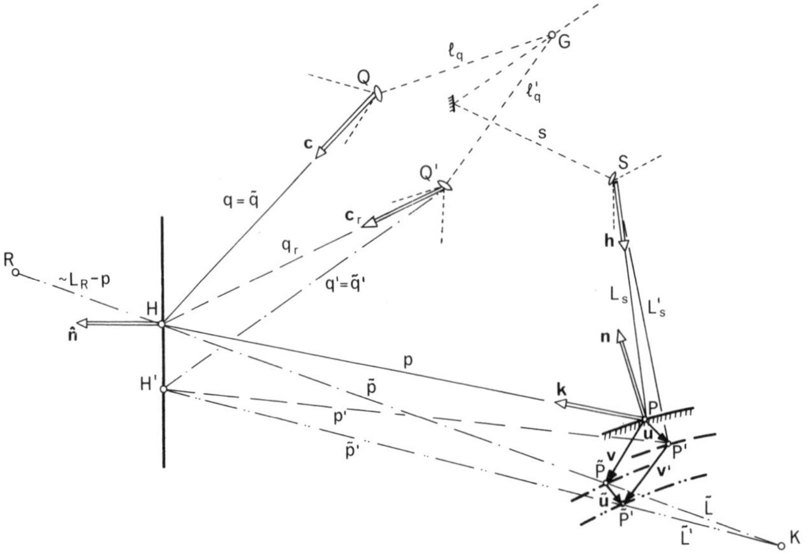

Fig. 4.34. Holographic interferometry with two fixed reference sources and with wavelength changes

be expressed as an optical-path difference referred to λ

$$D = \frac{\lambda}{2\pi}\left[\left(\tilde{\Phi}_S - \frac{2\pi}{\tilde{\lambda}}\tilde{L}\right) - \left(\tilde{\Phi}'_S - \frac{2\pi}{\tilde{\lambda}'}\tilde{L}'\right)\right].$$

To calculate it, we proceed as before; however, to evaluate the phases of the point sources Q, Q′ and S, we take into account the optical paths l_q, l'_q and s traversed by the light between the point G, where it is split up into three beams, and the points Q, Q′ and S (Fig. 4.34). Consequently, observing also that $\tilde{q} = q$ $\tilde{q}' = q'$, we get from the conditions of interference identity at H and H′ [see (3.12)] the phases of the "sources" $\tilde{\mathrm{P}}$ and $\tilde{\mathrm{P}}'$

$$\tilde{\Phi}_S = \frac{2\pi}{\lambda}(s + L_S) + 2\pi\left[\left(\frac{p}{\lambda} - \frac{\tilde{p}}{\tilde{\lambda}}\right) - \left(\frac{1}{\lambda} - \frac{1}{\tilde{\lambda}}\right)(q + l_q)\right],$$

$$\tilde{\Phi}'_S = \frac{2\pi}{\lambda}(s + L'_S) + 2\pi\left[\left(\frac{p'}{\lambda} - \frac{\tilde{p}'}{\tilde{\lambda}'}\right) - \left(\frac{1}{\lambda} - \frac{1}{\tilde{\lambda}'}\right)(q' + l'_q)\right].$$

The optical-path difference is then

$$D = (L_S - L'_S) + \lambda\left[-\left(\frac{\tilde{L}}{\tilde{\lambda}} - \frac{\tilde{L}'}{\tilde{\lambda}'}\right) + \left(\frac{p}{\lambda} - \frac{\tilde{p}}{\tilde{\lambda}}\right) - \left(\frac{p'}{\lambda} - \frac{\tilde{p}'}{\tilde{\lambda}'}\right) - \left(\frac{1}{\lambda} - \frac{1}{\tilde{\lambda}}\right)(q + l_q) + \left(\frac{1}{\lambda} - \frac{1}{\tilde{\lambda}'}\right)(q' + l'_q)\right]. \quad (4.131)$$

Supposing that $u \ll p, L_S, q, \ldots, \Delta\lambda \ll \lambda$ and moreover that

$$\frac{\Delta\lambda}{\lambda} = O\left(\frac{u}{p}\right), \tag{4.132}$$

we can use first-order approximations of the form

$$\frac{1}{\tilde{\lambda}} = \frac{1}{\lambda - \Delta\lambda} \simeq \frac{1}{\lambda}\left(1 + \frac{\Delta\lambda}{\lambda}\right), \qquad L_S' \simeq L_S + \boldsymbol{u}\cdot\boldsymbol{h}, \ldots$$

so that we obtain the linearized expression

$$D = -\boldsymbol{u}\cdot\boldsymbol{h} + \tilde{\boldsymbol{u}}\cdot\boldsymbol{k} + \boldsymbol{v}\cdot\boldsymbol{k} - \boldsymbol{v}'\cdot\boldsymbol{k} - \frac{2\Delta\lambda}{\lambda}\tilde{L} - \frac{\Delta\lambda}{\lambda}(p + p') + \frac{\Delta\lambda}{\lambda}(q + l_q) + \frac{\Delta\lambda}{\lambda}(q' + l_q').$$

With the decomposition of the total displacement $\tilde{\boldsymbol{u}} = \boldsymbol{u} + \boldsymbol{v}' - \boldsymbol{v}$ and with $\Delta\lambda|p' - p|/\lambda p = O(u^2/p^2), \ldots$ the optical-path difference becomes finally

$$D = \boldsymbol{u}\cdot\boldsymbol{g} + \frac{\Delta\lambda}{\lambda}[q + l_q + q_r + l_q' - 2(\tilde{p} + \tilde{L})]. \tag{4.133}$$

In practice however, if $\Delta\lambda/\lambda$ is much smaller than u/p, the assumption (4.132) does not hold. Then, D and also the phase of the fringe-order vibration contains only $\boldsymbol{u}\cdot\boldsymbol{g}$, as we assumed in (4.13).

Let us now look at another case of wavelength change, where we suppose that the recordings are made with two wavelengths λ and λ', for the undeformed and the deformed object, respectively, but, in turn, that the reconstruction is made with only one wavelength $\tilde{\lambda}$ [4.230, 231]. Here, the fringe-order does not oscillate any more. Proceeding as above, with the hypothesis $|\lambda - \tilde{\lambda}| \ll \tilde{\lambda}$, $|\lambda' - \tilde{\lambda}| \ll \tilde{\lambda}$, we obtain for the linearized expression of the optical-path difference

$$D = \boldsymbol{u}\cdot\boldsymbol{g} + \frac{\lambda - \tilde{\lambda}}{\tilde{\lambda}}[q + l_q - (p + L_S + s)] - \frac{\lambda' - \tilde{\lambda}}{\tilde{\lambda}}[q_r + l_q' - (p + L_S + s)], \tag{4.134}$$

or, if we separate the given difference $\lambda - \lambda'$ from the modifiable one $\lambda - \tilde{\lambda}$,

$$D = \boldsymbol{u}\cdot\boldsymbol{g} + \frac{\lambda - \lambda'}{\lambda}[q_r + l_q' - (p + L_S + s)] + \frac{\lambda - \tilde{\lambda}}{\lambda}[q + l_q - (q_r + l_q')]. \tag{4.135}$$

In (4.134) as well as in (4.133), the fringe-control terms thus consist of optical paths relative to H and H′ multiplied by the wavelength change that they undergo.

As for the fringe spacing, using the connection (4.121) between ∇_{n} and $\nabla_{\hat{\mathrm{n}}}$, we have first, similar to (2.35),

$$\nabla_{\mathrm{n}} L_{\mathrm{S}} = \boldsymbol{N}\boldsymbol{h},$$

$$\nabla_{\mathrm{n}} q|_{\mathrm{R}} = \frac{L_{\mathrm{R}} - p}{L_{\mathrm{R}}} \boldsymbol{N}\hat{\boldsymbol{M}}\nabla_{\hat{\mathrm{n}}} q = \frac{L_{\mathrm{R}} - p}{L_{\mathrm{R}}} \boldsymbol{N}\hat{\boldsymbol{M}}\boldsymbol{c},$$

$$\nabla_{\mathrm{n}} p|_{\mathrm{R}} = \nabla_{\mathrm{n}}[L_{\mathrm{R}} - (L_{\mathrm{R}} - p)]_{\mathrm{R}} = \nabla_{\mathrm{n}} L_{\mathrm{R}} - \frac{L_{\mathrm{R}} - p}{L_{\mathrm{R}}} \boldsymbol{N}\hat{\boldsymbol{M}}\nabla_{\hat{\mathrm{n}}}(L_{\mathrm{R}} - p)$$

$$= -\boldsymbol{N}\boldsymbol{k} + \frac{L_{\mathrm{R}} - p}{L_{\mathrm{R}}} \boldsymbol{N}\hat{\boldsymbol{M}}\boldsymbol{k},$$

so that, from (4.135), we obtain

$$\frac{dD_{\mathrm{R}}}{d\phi} = L_{\mathrm{R}}\boldsymbol{m}\cdot\boldsymbol{M}\nabla_{\mathrm{n}} D_{\mathrm{R}} = \boldsymbol{m}\cdot\left\{L_{\mathrm{R}}\boldsymbol{M}\boldsymbol{w} - \boldsymbol{K}\boldsymbol{u} + \frac{\lambda - \lambda'}{\lambda}[(L_{\mathrm{R}} - p)\hat{\boldsymbol{M}}(\boldsymbol{c}_{\mathrm{r}} - \boldsymbol{k}) + L_{\mathrm{R}}\boldsymbol{M}(\boldsymbol{k} - \boldsymbol{h})] + \frac{\lambda - \tilde{\lambda}}{\lambda}(L_{\mathrm{R}} - p)\hat{\boldsymbol{M}}(\boldsymbol{c} - \boldsymbol{c}_{\mathrm{r}})\right\}. \quad (4.136)$$

Let us add, in conclusion, that, because we deal with linear approximations, the general relation for the case in which the wavelengths are changed and the sources are shifted can be calculated simply by adding the fringe-control terms that correspond to each considered modification. Of course D is still a function of the position vector $\boldsymbol{r}$ of the point P on the object surface and a function of the observation direction $\boldsymbol{k}$ [see (4.7)]; in addition, it now also depends on the position vector $\hat{\boldsymbol{r}}$ of the small area of the hologram around H (Fig. 4.32) from which the observed rays originate. Moreover, in the case in which two wavelengths are used in the reconstruction, D is also a function of the position vector $\boldsymbol{\rho}$ of K, which means that here D is not constant along a direction $\boldsymbol{k}$.

4.3.2 Principal Relations that Result from the Movement of One Hologram or of Two Sandwiched Holograms

Early in the history of holographic interferometry, it was suggested that the two wave fields that are to interfere can originate from two holograms [4.232, 233]. We can profit from this arrangement to modify the fringe appearance by moving one of the holograms. The other hologram is supposed to give an ideal image of the object, which, in the real-time technique, can of course be replaced by the object itself. It is also possible to move both holograms together, the distance between them causing different modifications of each image. In both cases, only one source is necessary for the recording and for the reconstruction of the two holograms. The existence of the glass plates will be neglected here, which is warrantable as long as these plates are thin relative to p, L_S, . . . and they lie "outside the interferometer", i.e. if the emulsions face each other. Furthermore, during the exposure of the hologram on the observer's side, a glass plate must be put in front of it to compensate the additional optical path involved in the other hologram (the thicknesses of both the glass plate and the hologram plate should of course be accurately the same).

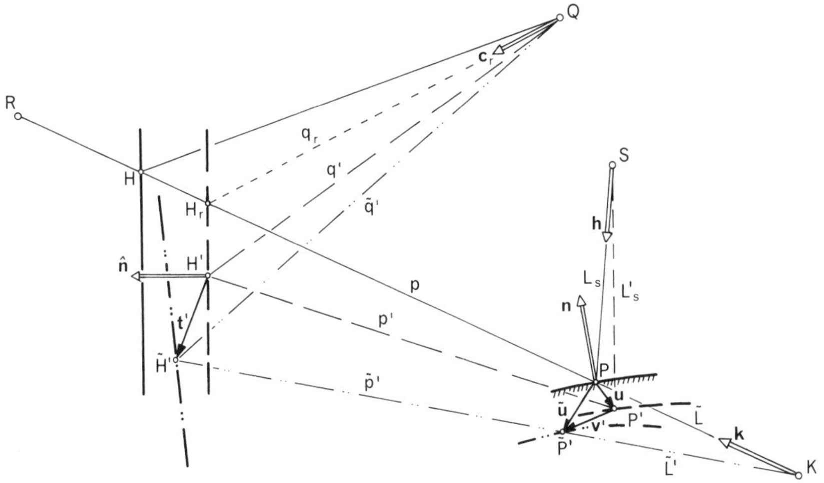

Fig. 4.35. Holographic interferometry with a fixed hologram for the undeformed object and a movable hologram for the deformed object

To start with, we thus assume (Fig. 4.35) [3.82; Ref. 4.184, p. 73; 4.219, 220, 226, 234, 235] that the wave field produced by the undeformed object undergoes no modification but that the *hologram of the deformed object is displaced and rotated* at the reconstruction, one point H′ on it being displaced to $\tilde{\mathrm{H}}'$ by the vector $\boldsymbol{t}'$, which induces the "optical" displacement $\boldsymbol{v}'$ of P′ to $\tilde{\mathrm{P}}'$. Again, we look first at the optical-path difference D at a point K. Formally, the calculation is exactly the same as in Sect. 4.3.1; the distances are however those shown in Fig. 4.35 and the condition of interference identity relates to the points H′ and $\tilde{\mathrm{H}}'$. We thus arrive at (4.119), i.e., with $\psi' = \tilde{\psi}' = \psi$ because the source Q is alone,

$$D = (L_S - L_S') - (\tilde{L} - \tilde{L}') - (p' - \tilde{p}') + (q' - \tilde{q}').$$

By use of the assumption t', u, $v' \ll p$, q, L_S, . . . we obtain a linear function in terms of unit vectors relative to the reference frame, i.e. the lines SP, $\mathrm{PH_r}$ and

QH_r, and of the displacement vectors

$$D = -\boldsymbol{u}\cdot\boldsymbol{h} + \tilde{\boldsymbol{u}}\cdot\boldsymbol{k} + (\boldsymbol{t}' - \boldsymbol{v}')\cdot\boldsymbol{k} - \boldsymbol{t}'\cdot\boldsymbol{c}_r.$$

Because of $\tilde{\boldsymbol{u}} = \boldsymbol{u} + \boldsymbol{v}'$, the location of the image $\tilde{P}'$ vanishes, so that the *optical-path difference* becomes

$$D = \boldsymbol{u}\cdot(\boldsymbol{k} - \boldsymbol{h}) + \boldsymbol{t}'\cdot(\boldsymbol{k} - \boldsymbol{c}_r) = \boldsymbol{u}\cdot\boldsymbol{g} + \boldsymbol{t}'\cdot\boldsymbol{f}_r. \tag{4.137}$$

The fringe-control term thus contains the shift $\boldsymbol{t}'$ of the hologram area around H′, or H_r, from which one of the two interfering rays comes, and the "*sensitivity vector of the fringe control*" $\boldsymbol{f}_r$.

Then, to calculate the derivatives of D, in addition to the results encountered in the preceding section, we need the derivative of the hologram shift. In the absence of a strain of the hologram, we obtain, similar to (2.104) and (3.77),

$$\nabla_{\hat{n}} \otimes \boldsymbol{t}' = \hat{\boldsymbol{N}}\boldsymbol{\Psi}', \tag{4.138}$$

where $\boldsymbol{\Psi}'$ is the skew-symmetric tensor that describes the *small rotation of the hologram.* In the same way as we decomposed the rotation of the object surface [see (2.109)], we can also distinguish a pivot motion Ψ' and an inclination $\boldsymbol{\psi}'$ of the hologram, i.e. write

$$\hat{\boldsymbol{N}}\boldsymbol{\Psi}' = \Psi'\hat{\boldsymbol{E}} + \boldsymbol{\psi}' \otimes \hat{\boldsymbol{n}}. \tag{4.139}$$

The derivative of D, when R is fixed, is then

$$\nabla_n D_R = \boldsymbol{N}\boldsymbol{w} - \frac{1}{L_R}\boldsymbol{N}\boldsymbol{K}(\boldsymbol{u} + \boldsymbol{t}') + \frac{L_R - p}{L_R}\boldsymbol{N}\hat{\boldsymbol{M}}\left(\boldsymbol{\Psi}'\boldsymbol{f}_r - \frac{1}{q_r}\boldsymbol{C}_r\boldsymbol{t}'\right),$$

so that we have for the *fringe spacing*

$$\begin{aligned}\frac{dD_R}{d\phi} &= L_R\boldsymbol{m}\cdot\boldsymbol{M}\nabla_n D_R \\ &= \boldsymbol{m}\cdot\left[L_R\boldsymbol{M}\boldsymbol{w} - \boldsymbol{K}(\boldsymbol{u} + \boldsymbol{t}') + (L_R - p)\hat{\boldsymbol{M}}\left(\boldsymbol{\Psi}'\boldsymbol{f}_r - \frac{1}{q_r}\boldsymbol{C}_r\boldsymbol{t}'\right)\right].\end{aligned} \tag{4.140}$$

Here we may recall formula (3.80), which allows us to replace the last term in the bracket of (4.140) by the relative lateral shift of the image $\tilde{P}'$, i.e.

$$\boldsymbol{K}\boldsymbol{v}' - \boldsymbol{K}\boldsymbol{t}' = p\hat{\boldsymbol{M}}\left(\boldsymbol{\Psi}'\boldsymbol{f}_r - \frac{1}{q_r}\boldsymbol{C}_r\boldsymbol{t}'\right).$$

In addition, the derivative $\nabla_n D_K$ required for the formulae concerning the visibility and the localization would be obtained simply by replacing L_{R} by $-\tilde{L}$ in the expression of $V_{\mathrm{n}}D_{\mathrm{R}}$ and all the reasoning of Sect. 4.3.1 could be repeated.

To conclude this passage on the motion of one hologram, let us add that such a movement can also be produced by the object itself if the hologram is bound to it [4.236–239]. Then, the quantities $\boldsymbol{t}'$ and $\boldsymbol{\Psi}'$ are determined by the deformation of the object-surface element where the hologram is fixed. In this arrangement, because the hologram is close to the object, the reference source will often be placed on the observer's side and illuminate also the object through the hologram.

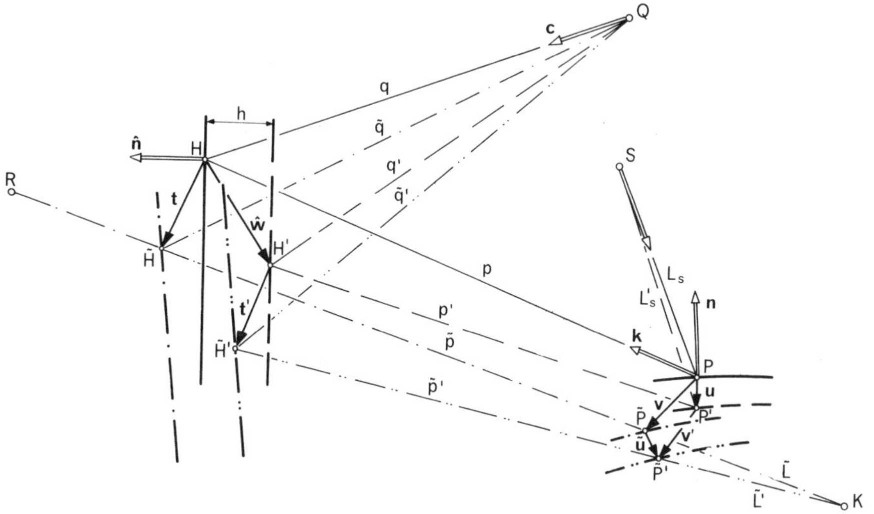

Fig. 4.36. Sandwich holography

Let us now pass to the so-called *sandwich holography technique* due to *Abramson* [3.83, 4.240–243] (see also [3.81, 4.244–247]). Here (Fig. 4.36), the two holograms are separated by a *small distance h* but are rigidly linked to each other and thus shifted and rotated parallel to one another. Consequently, both images $\tilde{\mathrm{P}}$ and $\tilde{\mathrm{P}}'$ are "*optically*" displaced with respect to P and P′. As we saw in Sect. 3.2.5 where we calculated the difference $\boldsymbol{K}(\boldsymbol{v}' - \boldsymbol{v})$ of the lateral displacements, we must now pay attention to three orders of magnitude:

1° the lengths p, q, L_{S}, $\tilde{L}$, ..., in the optical arrangement are of order 1 compared to a characteristic length $l \gg h$,

2° the kinematic quantities of the hologram's motion, $\boldsymbol{t}/l$, $\boldsymbol{\Psi}$, ..., are of order $\delta = h/l$,

3° the quantities of the object deformation, $\boldsymbol{u}/l$, γ, Ω, ω, are of order $\delta^2 = (h/l)^2$.

Therefore, we cannot simply make use of the results we have just obtained, but must start again with the definition of the optical-path difference in K, i.e.

$$D = \left(\frac{\lambda}{2\pi}\tilde{\Phi}_S - \tilde{L}\right) - \left(\frac{\lambda}{2\pi}\tilde{\Phi}'_S - \tilde{L}'\right).$$

By use of two conditions of interference identity similar to (3.12) in H and $\tilde{H}$ on the one hand and in H′ and $\tilde{H}'$ on the other hand, we eliminate the unknown phases of $\tilde{P}$ and $\tilde{P}'$, so that

$$D = (L_S - L'_S) - (\tilde{L} - \tilde{L}') + (p - \tilde{p}) - (p' - \tilde{p}') - (q - \tilde{q}) + (q' - \tilde{q}'),$$

or with the definition (3.86)

$$D = (L_S - L'_S) - (\tilde{L} - \tilde{L}') + (\Theta' - \Theta).$$

The three length differences that appear here are all of the order δ^2. Choosing now as a reference frame the lines SP, PH and QH, we can write a linear expression of D; for $\Theta' - \Theta$ we shall use the approximation (3.89) of Sect. 3.2.5. However, this last equation relates to rays that intersect at the observer's position R (Fig. 3.15), whereas we are now applying it to rays that intersect at K (Fig. 4.36). This is nevertheless possible because the term neglected by such a change is of order δ^3 (it concerns only a very small modification of the location of H′). We have then

$$D = -\boldsymbol{u}\cdot\boldsymbol{h} + \tilde{\boldsymbol{u}}\cdot\boldsymbol{k} + (\boldsymbol{v} - \boldsymbol{v}')\cdot\boldsymbol{k} + \frac{h}{\hat{\boldsymbol{n}}\cdot\boldsymbol{k}}\boldsymbol{k}\cdot\left(\boldsymbol{\Psi}\boldsymbol{c} + \frac{1}{q}\boldsymbol{C}\boldsymbol{t}\right),$$

i.e. with $\tilde{\boldsymbol{u}} = \boldsymbol{u} + \boldsymbol{v}' - \boldsymbol{v}$

$$D = \boldsymbol{u}\cdot\boldsymbol{g} + \frac{h}{\hat{\boldsymbol{n}}\cdot\boldsymbol{k}}\boldsymbol{k}\cdot\left(\boldsymbol{\Psi}\boldsymbol{c} + \frac{1}{q}\boldsymbol{C}\boldsymbol{t}\right). \tag{4.141}$$

Thus, in sandwich holography, the optical-path difference consists again of the sum of the usual term and of a fringe-control term, which now no longer contains only unit direction vectors and displacement vectors; on the contrary, the influence of $\boldsymbol{t}$ depends on the "*sensitivity vector for translation*" $h\boldsymbol{C}\boldsymbol{k}/q\hat{\boldsymbol{n}}\cdot\boldsymbol{k}$ whereas $\boldsymbol{\Psi}$ is combined with the "*sensitivity tensor for rotation*" $h(\boldsymbol{k} \otimes \boldsymbol{c})/\hat{\boldsymbol{n}}\cdot\boldsymbol{k}$. In contrast with the case in which only one hologram is moved, the rotation tensor $\boldsymbol{\Psi}$ and the projection $\boldsymbol{C}$ are thus present already in D and not only in $dD_R/d\phi$.

Having in this way expressed the optical-path difference as a function of a reference position, we can suppose the points R and K to lie on the line PH and use again Fig. 4.33 for the calculation of the derivatives. Keeping in mind that $\boldsymbol{\Psi}$ is a constant tensor along the holograms, because they behave like a rigid body, we obtain, according to the rules for formation of derivatives (2.30),

$$\nabla_{\mathrm{n}} D = \nabla_{\mathrm{n}}(\boldsymbol{u}\cdot\boldsymbol{g}) + h\left(\nabla_{\mathrm{n}} \otimes \frac{\boldsymbol{k}}{\hat{\boldsymbol{n}}\cdot\boldsymbol{k}}\right)\left(\boldsymbol{\Psi}\boldsymbol{c} + \frac{1}{q}\boldsymbol{C}\boldsymbol{t}\right)$$
$$+ \left[(\nabla_{\mathrm{n}} \otimes \boldsymbol{c})\boldsymbol{\Psi}^{\mathrm{T}} + \left(\nabla_{\mathrm{n}} \otimes \frac{1}{q}\boldsymbol{C}\right)\boldsymbol{t} + \frac{1}{q}(\nabla_{\mathrm{n}} \otimes \boldsymbol{t})\boldsymbol{C}^{\mathrm{T}}\right]\frac{h}{\hat{\boldsymbol{n}}\cdot\boldsymbol{k}}\boldsymbol{k}.$$

If we fix the point R, we know from Sect. 4.2.1 the derivative of the usual term (4.33). For the derivative of the fringe-control term we can use the connection (4.121) between ∇_{n} and $\nabla_{\hat{\mathrm{n}}}$; we have also the derivatives of $\boldsymbol{c}$ (4.122), of $\boldsymbol{t}$ (4.138) and, similar to (2.37), the derivative of the projection $\boldsymbol{C}/q$ makes appear the superprojection $\mathscr{C}$.

$$\nabla_{\hat{\mathrm{n}}} \otimes \left(\frac{1}{q}\boldsymbol{C}\right) = -\frac{1}{q^2}\hat{\boldsymbol{N}}\mathscr{C} = -\frac{1}{q^2}\hat{\boldsymbol{N}}[\boldsymbol{c} \otimes \boldsymbol{C} + \boldsymbol{C} \otimes \boldsymbol{c} + \boldsymbol{C} \otimes \boldsymbol{c})^{\mathrm{T}}].$$

Finally, as in (3.91), we get

$$\nabla_{\mathrm{n}} \otimes \left(\frac{\boldsymbol{k}}{\hat{\boldsymbol{n}}\cdot\boldsymbol{k}}\right)\bigg|_{\mathrm{R}} = -\frac{1}{L_{\mathrm{R}}\hat{\boldsymbol{n}}\cdot\boldsymbol{k}}\boldsymbol{N}\hat{\boldsymbol{M}}.$$

Observing furthermore that $\boldsymbol{\Psi}^{\mathrm{T}} = -\boldsymbol{\Psi}$, $\boldsymbol{C}^{\mathrm{T}} = \boldsymbol{C}$ and taking into account the combinations of successive projections (2.71), we can write for the *fringe spacing*

$$\frac{dD_{\mathrm{R}}}{d\phi} = L_{\mathrm{R}}\boldsymbol{m}\cdot\boldsymbol{M}\nabla_{\mathrm{n}}D_{\mathrm{R}} = \boldsymbol{m}\cdot\left\{L_{\mathrm{R}}\boldsymbol{M}\boldsymbol{w} - \boldsymbol{K}\boldsymbol{u} - \frac{h}{\hat{\boldsymbol{n}}\cdot\boldsymbol{k}}\hat{\boldsymbol{M}}\left[\left(\boldsymbol{\Psi}\boldsymbol{c} + \frac{1}{q}\boldsymbol{C}\boldsymbol{t}\right)\right.\right.$$
$$\left.\left. + \frac{L_{\mathrm{R}} - p}{q}(\boldsymbol{\Psi}\boldsymbol{C} - \boldsymbol{C}\boldsymbol{\Psi})\boldsymbol{k} - \frac{L_{\mathrm{R}} - p}{q^2}\boldsymbol{k}\mathscr{C}\boldsymbol{t}\right]\right\}. \tag{4.142}$$

As in the preceding, we can also calculate the derivative $\nabla_{\mathrm{n}}D_{\mathrm{K}}$ by replacing L_{R} by $-\tilde{L}$ in $\nabla_{\mathrm{n}}D_{\mathrm{R}}$ and write the formulae for the fringe visibility and localization [3.81].

4.3.3 Generalization and Application of the Results Concerning the Modification of the Optical Arrangement

Before explaining how the foregoing results help determine the displacement vector and the rotation and strain tensors, we shall summarize them and show their common features as well as their relation to the problem of image formation that we discussed in Chap. 3.

If we leave aside the case of two wavelengths used together in the reconstruction, where the fringe-order oscillates and is not constant along an observation direction, we can take as a common and general starting point the *optical-path difference*

$$D = \left(\frac{\tilde{\lambda}}{2\pi}\tilde{\phi}_S - \tilde{L}\right) - \left(\frac{\tilde{\lambda}}{2\pi}\tilde{\phi}'_S - \tilde{L}'\right). \tag{4.143}$$

This applies to the paths covered by two light rays that come from points $\tilde{P}$ and $\tilde{P}'$, which are the images of the pair of points P and P′ on the undeformed and deformed object, respectively. These rays intersect at some point K in space at the distance $\tilde{L}$ from $\tilde{P}$ and are reconstructed with light of wavelength $\tilde{\lambda}$. The unknown phases $\tilde{\phi}_S$ and $\tilde{\phi}'_S$ of the "sources" $\tilde{P}$ and $\tilde{P}'$ are determined by two conditions of interference identity of type (3.12) at the points of the holograms where the considered rays are formed. D can thus be expressed by known quantities

$$D = (\tilde{q} - \tilde{p} - \tilde{L}) - (\tilde{q}' - \tilde{p}' - \tilde{L}') - \frac{\tilde{\lambda}}{\lambda}(q - p - L_S) + \frac{\tilde{\lambda}}{\lambda'}(q' - p' - L'_S) - \frac{\tilde{\lambda}}{2\pi}[(\psi - \tilde{\psi}) - (\psi' - \tilde{\psi}')]. \tag{4.144}$$

If we define as in (3.13) the optical-path difference functions

$$\Theta = \frac{\tilde{\lambda}}{2\pi}\theta = (\tilde{p} - \tilde{q}) - \frac{\tilde{\lambda}}{\lambda}(p - q), \qquad \Theta' = \cdots, \tag{4.145}$$

we can also write

$$D = \left(\frac{\tilde{\lambda}}{\lambda}L_S - \frac{\tilde{\lambda}}{\lambda'}L'_S\right) - (\tilde{L} - \tilde{L}') - (\Theta - \Theta') - \frac{\tilde{\lambda}}{2\pi}[(\psi - \psi') - (\tilde{\psi} - \tilde{\psi}')]. \tag{4.146}$$

The concepts of Sect. 3.2 that pertain to image formation are thus utilized in the present problem of fringe formation. They allow us to speak, as in standard holography, of "image points" $\tilde{P}$ and $\tilde{P}'$ and to refer to them the optical-path difference D (4.143). However, the position of these images can be made to disappear, because, in (4.144), we need only the total distances $\tilde{p} + \tilde{L}$ and $\tilde{p}' + \tilde{L}'$. The problem of fringe formation is thus independent of that of image formation; in other words, fringes are formed whatever aberrations the images may suffer. Nevertheless, because we are interested in the deformation of the object surface, i.e. in the ensembles {P} and {P′}, it is sometimes convenient to refer to the images of these surfaces, as in (4.146) or in the following.

The next step consists in expressing D as a linear function of unit vectors and lengths in the reference configuration, which we choose to be that of the undeformed object during recording, and as a function of the small changes, i.e. the displacement vectors, the rotation tensors and the wavelength changes. D thus generally takes the form

$$D = \boldsymbol{u} \cdot \boldsymbol{g} + \tilde{D} + \Lambda, \tag{4.147}$$

where the first term is the usual one of standard holographic interferometry, whereas $\tilde{D}$ takes into account the geometric modifications and Λ, the wavelength changes. Because this expression is linear, the influence of each modification is represented by a separate term. In the preceding, we have established the formulae:

Type of modification	Additional terms
a) Standard holography (no modification)	0
b) Shift and phase change of one reference source (Q′)	$\tilde{D}=\boldsymbol{d}'\cdot\boldsymbol{c}_r+\frac{\lambda}{2\pi}(\psi'-\tilde{\psi}')$
c) Movement of one hologram {H′}	$\tilde{D}=\boldsymbol{t}'\cdot(\boldsymbol{k}-\boldsymbol{c}_r)$
d) Movement of two sandwiched holograms	$\tilde{D}=\frac{h}{\hat{\boldsymbol{n}}\cdot\boldsymbol{k}}\boldsymbol{k}\cdot\left(\boldsymbol{\Psi}\boldsymbol{c}+\frac{1}{q}\boldsymbol{C}\boldsymbol{t}\right)$
e) wavelength change	$\Lambda=\frac{\lambda-\lambda'}{\lambda}[q_r+l'_q-(p+L_S+s)] + \frac{\lambda-\tilde{\lambda}}{\lambda}[q+l_q-(q_r+l'_q)]$

Two remarks may be added here. First, the linearized form of (4.145) is

$$\Theta' - \Theta = \tilde{D} + \Lambda - (\boldsymbol{v}' - \boldsymbol{v}) \cdot \boldsymbol{k}, \tag{4.148}$$

which proves that the "longitudinal" displacements $\boldsymbol{v}' \cdot \boldsymbol{k}$ and $\boldsymbol{v} \cdot \boldsymbol{k}$ are contained in Θ' and Θ; but, as shown previously, they are not contained in $\tilde{D}$ and Λ. Second, as we said in Sect. 4.1.1, $\boldsymbol{u} \cdot \boldsymbol{g}$ is solely a function of the position vector $\boldsymbol{r}$ of the object point P and of the observation direction $\boldsymbol{k}$. As for the additional terms, $\tilde{D}$ may conveniently be expressed as a function of $\boldsymbol{k}$ and of the position vector $\hat{\boldsymbol{r}}$ of the point H on the hologram where the interfering rays are produced, i.e. as $\tilde{D}(\hat{\boldsymbol{r}}, \boldsymbol{k})$. Then, for Λ, the most suitable form is $\Lambda(\hat{\boldsymbol{r}}, \boldsymbol{r})$. $\boldsymbol{k}$, $\boldsymbol{r}$ and $\hat{\boldsymbol{r}}$ have only two independent components, since $\boldsymbol{k}$ is a unit vector and since $\boldsymbol{r}$ and $\hat{\boldsymbol{r}}$ vary only on surfaces. Moreover, we must remember that these three vectors are linked by

$$\hat{\boldsymbol{r}} = \boldsymbol{r} + p\boldsymbol{k}.$$

If we now calculate the derivatives of optical-path differences, we have to consider the ensembles of rays that go through a given point. So, we encountered two derivatives of D, first, the derivative $\nabla_{\mathrm{n}} D_{\mathrm{R}}$ when the position R of the observer is fixed, which led us to the *fringe spacing and direction*, and, second, the derivative $\nabla_{\mathrm{n}} D_{\mathrm{K}}$ when the position K of the fringes is fixed, which determined the *visibility and the localization*. In these cases, in addition to the preceding one, there exist the connections

$$\boldsymbol{r} = \boldsymbol{\rho}_{\mathrm{R}} - L_{\mathrm{R}} \boldsymbol{k} \qquad \text{when R is fixed, and}$$

$$\boldsymbol{r} = \boldsymbol{\rho} + \tilde{L} \boldsymbol{k} \qquad \text{when K is fixed.}$$

Consequently, D can be expressed as a function of $\boldsymbol{r}$ alone, i.e. as

$$D(\boldsymbol{r}) = \boldsymbol{u} \cdot \boldsymbol{g} + \tilde{D}[\hat{\boldsymbol{r}}(\boldsymbol{r}), \boldsymbol{k}(\boldsymbol{r})] + \Lambda[\hat{\boldsymbol{r}}(\boldsymbol{r}), \boldsymbol{r}]. \tag{4.149}$$

On the other hand, in Chap. 3, the "*lateral*" *displacement* $\boldsymbol{K v}$ was determined by the condition $\nabla_{\hat{\mathrm{n}}} \Theta = 0$ where P was fixed and $\nabla_{\hat{\mathrm{n}}}(\Theta' - \Theta) = 0$ gave us the difference $\boldsymbol{K}(\boldsymbol{v}' - \boldsymbol{v})$. In these derivatives, $\boldsymbol{r}$ is a constant, so that Θ' and Θ can be said to depend only on $\hat{\boldsymbol{r}}$, i.e.

$$\Theta'(\hat{\boldsymbol{r}}) - \Theta(\hat{\boldsymbol{r}}) = \tilde{D}[\hat{\boldsymbol{r}}, \boldsymbol{k}(\hat{\boldsymbol{r}})] + \Lambda(\hat{\boldsymbol{r}}) - (\boldsymbol{v}' - \boldsymbol{v}) \cdot \boldsymbol{k}(\hat{\boldsymbol{r}}). \tag{4.150}$$

With ∂_r, $\partial_{\hat{r}}$ and ∂_k, the partial-derivative operators with respect to $\boldsymbol{r}$, $\hat{\boldsymbol{r}}$ and $\boldsymbol{k}$ [see (2.78)], the derivatives of D and $\Theta' - \Theta$ then become

$$\begin{aligned} \nabla_{\mathrm{n}} D &= \partial_r(\boldsymbol{u} \cdot \boldsymbol{g}) + (\nabla_{\mathrm{n}} \otimes \boldsymbol{k}) \partial_k(\boldsymbol{u} \cdot \boldsymbol{g}) + (\nabla_{\mathrm{n}} \otimes \hat{\boldsymbol{r}}) \partial_{\hat{r}} \tilde{D} \\ &\quad + (\nabla_{\mathrm{n}} \otimes \boldsymbol{k}) \partial_k \tilde{D} + (\nabla_{\mathrm{n}} \otimes \hat{\boldsymbol{r}}) \partial_{\hat{r}} \Lambda + \partial_r \Lambda, \\ \nabla_{\hat{\mathrm{n}}}(\Theta' - \Theta) &= \partial_{\hat{r}} \tilde{D} + (\nabla_{\hat{\mathrm{n}}} \otimes \boldsymbol{k}) \partial_k \tilde{D} + \partial_{\hat{r}} \Lambda - (\nabla_{\hat{\mathrm{n}}} \otimes \boldsymbol{k})(\boldsymbol{v}' - \boldsymbol{v}) = 0. \end{aligned}$$

The partial derivatives of the usual term $\boldsymbol{u} \cdot \boldsymbol{g} = \boldsymbol{u}(\boldsymbol{r}) \cdot [\boldsymbol{k} - \boldsymbol{h}(\boldsymbol{r})]$ are [see (4.31)]

$$\partial_r(\boldsymbol{u} \cdot \boldsymbol{g}) = (\nabla_{\mathrm{n}} \otimes \boldsymbol{u}) \boldsymbol{g} - \frac{1}{L_{\mathrm{S}}} \boldsymbol{N H u} = \boldsymbol{N w}, \qquad \partial_k(\boldsymbol{u} \cdot \boldsymbol{g}) = \boldsymbol{K u}.$$

For those of the additional terms in $\tilde{D}$, we have found

Type of modification	$\partial_{\hat{r}} \tilde{D}$	$\partial_k \tilde{D}$
a)	0	0
b)	$\frac{1}{q_{\mathrm{r}}} \hat{\boldsymbol{N}} \boldsymbol{C}_{\mathrm{r}} \boldsymbol{d}'$	0
c)	$\hat{\boldsymbol{N}}\left[\boldsymbol{\Psi}'(\boldsymbol{k} - \boldsymbol{c}_{\mathrm{r}}) - \frac{1}{q_{\mathrm{r}}} \boldsymbol{C}_{\mathrm{r}} \boldsymbol{t}'\right]$	$\boldsymbol{K t}'$
d)	$\frac{h}{(\hat{\boldsymbol{n}} \cdot \boldsymbol{k}) q} \hat{\boldsymbol{N}}\left[(\boldsymbol{\Psi C} - \boldsymbol{C \Psi}) \boldsymbol{k} - \frac{1}{q} \boldsymbol{k} \mathcal{C} \boldsymbol{t}\right]$	$\frac{h}{\hat{\boldsymbol{n}} \cdot \boldsymbol{k}} \hat{\boldsymbol{M}}\left(\boldsymbol{\Psi c} + \frac{1}{q} \boldsymbol{C t}\right)$

whereas, for those of the terms in Λ, we have (case *e*)

$$\partial_{\hat{r}}\Lambda = \frac{\lambda - \lambda'}{\lambda}\hat{N}(c_r - k) + \frac{\lambda - \tilde{\lambda}}{\lambda}\hat{N}(c - c_r), \qquad \partial_r\Lambda = \frac{\lambda - \lambda'}{\lambda}N(k - h).$$

Finally, summarizing the derivatives with respect to the several collineations, we know that

$$\left.\begin{aligned} \nabla_n \otimes k &= \frac{1}{l}NK \\ \nabla_n \otimes \hat{r} &= \frac{p + l}{l}N\hat{M} \end{aligned}\right\} \text{ with } \begin{cases} l = -L_R, \text{ when R is fixed,} \\ l = \tilde{L}, \text{ when K is fixed.} \end{cases}$$

$$\nabla_{\hat{n}} \otimes k = \frac{1}{p}\hat{N}K, \quad \text{where P is fixed.}$$

So, for any case of modification, we can write

$$\nabla_n D = Nw + \frac{1}{l}NKu + \frac{p + l}{l}N\hat{M}\partial_{\hat{r}}\tilde{D} + \frac{1}{l}N\partial_k\tilde{D} + \frac{p + l}{l}N\hat{M}\partial_{\hat{r}}\Lambda + \partial_r\Lambda, \tag{4.151}$$

$$K(v' - v) = p\hat{M}\partial_{\hat{r}}\tilde{D} + \partial_k\tilde{D} + p\hat{M}\partial_{\hat{r}}\Lambda. \tag{4.152}$$

At this point, the two phenomena of fringe formation and of image formation are still independent. If, however, we define

$$\tilde{w} = w + \hat{M}(\partial_{\hat{r}}\tilde{D} + \partial_{\hat{r}}\Lambda) + \partial_r\Lambda, \tag{4.153}$$

and use again the total displacement $\tilde{u} = u + v' - v$, we can associate the two problems and write

$$\nabla_n D = N\left(\tilde{w} + \frac{1}{l}K\tilde{u}\right). \tag{4.154}$$

Using this form of the derivative, we are able to obtain in a simple way the counterparts of all results for standard holographic interferometry, as, for example, the *fringe spacing* that replaces (4.34)

$$\frac{dD_R}{d\phi} = m \cdot (L_R M\tilde{w} - K\tilde{u}), \tag{4.155}$$

the condition of *standard partial localization* that replaces (4.85)

$$\tilde{L}_p = -\frac{M\tilde{w} \cdot K\tilde{u}}{(M\tilde{w})^2}, \tag{4.156}$$

or the condition of *complete localization* that replaces (4.93)

$$N\left(\tilde{w} + \frac{1}{\tilde{L}}K\tilde{u}\right) = 0. \tag{4.157}$$

If (4.157) is fulfilled, (4.155) becomes

$$\frac{dD_R}{d\phi} = -\frac{L_R + \tilde{L}}{\tilde{L}}\, m \cdot K\tilde{u}, \tag{4.158}$$

which means that the *normality theorem* (4.94) can be generalized: if and only if the fringes are observed in a direction of complete localization, they are normal to the projection of the *total* displacement vector onto the plane perpendicular to the observation direction. Thus, although image formation and fringe formation are two different phenomena, their association can contribute to a simpler and more expressive representation.

Let us finally briefly point out some of the advantages and disadvantages that result from modification of the optical arrangement in measurement of deformation. As we observed, the displacement vector $\boldsymbol{u}$ and its derivative $\boldsymbol{\nabla}_n \otimes \boldsymbol{u}$ appear in the same terms as in standard holographic interferometry. Therefore, the same methods as in that case can be applied to determine them. Particularly, the same minimum number of base vectors $\boldsymbol{g}_i$ or $\boldsymbol{m}_{i\alpha}$ is required when all components of $\boldsymbol{u}$ or of $\boldsymbol{\nabla}_n \otimes \boldsymbol{u}$ are sought. The first advantage of being able to modify the fringe pattern is thus that the usual methods can be applied under *convenient observation conditions*: the fringe-order, the fringe spacing or the localization points can, if they are not convenient, be changed to facilitate measurement. For that purpose, in case *e* we have at our disposal one parameter, namely the wavelength change $\lambda - \tilde{\lambda}$, in case *b* four parameters, namely the components of the shift $\boldsymbol{d}'$ and the phase difference $\psi' - \tilde{\psi}'$; and in cases *c* and *d* six parameters, namely the components of the displacement vector $\boldsymbol{t}$ and of the rotation tensor $\boldsymbol{\Psi}$. If, moreover, one hologram were strained [Ref. 4.184, Sect. III.4.2], three parameters more would be offered. However, the actual degree of freedom depends on how these parameters appear in the equations. According to an idea of *Abramson* [3.83], an interesting modification consists in making the fringes vanish around the investigated object point, i.e. making $dD_R/d\phi = 0\ \forall\ \boldsymbol{m}$, as at the singularity in Fig. 4.29. In this way, we need not measure any fringe spacing; the remaining equation then relates the optical modification (e.g. the rotation of the hologram) directly to the object deformation. Two possibilities for practical application should be examined. First, the object deformation may be *transferred and amplified* into motions of the optical arrangement: in sandwich holography for example, the movement of the holograms is of the order of magnitude l/h greater than that of the object. Second, part of the object deformation, especially the symmetric strain tensor might be *isolated and measured separately* from the displacement and the rotation.

As for the disadvantages, let us mention that we supposed an ideal reconstruction. This requires, on the one hand, that the shrinkage, the change of the refractive index and the nonlinearity of the emulsions can be neglected (see Chap. 3), and, on the other hand, that the elements of the optical arrangement can be exactly repositioned (see in particular the technique of *Hariharan* and *Hegedus* [4.246, 247]) before being displaced and that their movements are accurately known. As shown by the formulae, these movements are often of the same order of magnitude as the object deformation so that each uncontrolled or ignored modification can alter the results.

4.4 Conclusion

In this chapter, we have studied the fringes formed in holographic interferometry; we have seen how they depend on the object deformation and conversely how this deformation can be measured through them. We have utilized the results of Chap. 2 about continuum kinematics as well as those of Chap. 3 about holographic recording and reconstruction.

First, we have investigated a simple model of the interference phenomenon, considering only two light rays and the difference between the optical paths they traverse. The mechanical quantity that appeared there was the displacement vector. Next, we examined the first derivatives of this optical-path difference, either to calculate the fringe spacing and direction, or to take into account, in a more complete model of the fringe formation, all superimposed light rays that emanate from the vicinity of the object point under consideration. Thus, whereas only the displacement vector could be determined from the optical-path difference, the derivatives of the latter quantities allowed us, on the one hand, to measure directly the derivative of the displacement, i.e. the strain and the rotation tensors, and, on the other hand, to explain the contrast or visibility of the fringes.

We could ignore the holographic process as such as long as the reconstructed wave fields were supposed to reproduce exactly the recorded ones, i.e. in what we called standard holographic interferometry. On the contrary, we had to deal with the disposition of the optical arrangement when this, and consequently the fringe pattern, is modified during reconstruction.

Let us remark also that the basic concepts introduced at the beginning first appeared only in form of vectors and some simple scalar products, but along the way it turned out that the essential features consisted in a combined application of differential operators, linear transformations and projections in particular.

Chapter 5 Second Derivatives of the Displacement and of the Optical-Path Difference

When we spoke about the formation of holographic images in Chap. 3, we used first as well as second derivatives of phase differences. In Chap. 4, we described the fringe formation by looking at the optical-path difference, and then, in a more complete treatment, at its first derivative. At the same time, we saw how the displacement vector and its first derivative, i.e. the strain and rotation tensors, are related to the optical quantities and can thus be measured at the surface of an opaque body. Therefore, because each additional order of derivation leads to more information, we now intend to have a look at the second derivative of the optical-path difference, which will make the second derivative of the displacement appear. Thus, we must first briefly explain what are the mechanical quantities that depend on this derivative and what relations we have at our disposal. Then we shall calculate the second derivative of the optical-path difference and outline some of its possible applications.

5.1 Some Additional Equations of Continuum Mechanics, Particularly Relations Containing the Second Derivative of the Displacement

In Sect. 2.2, we established the kinematic relations that link the strain and the rotation to the first derivative of the displacement vector. In addition, we shall introduce here, on the one hand, the constitutive equations for an elastic body, which relate the strain tensor to the stress tensor, and, on the other hand, the (differential) equations of motion or of equilibrium, which relate the gradient of the stress tensor to the acceleration of an element; in the latter, the second derivative of the displacement is thus in fact implicitly contained. Before doing so however, we shall stay in the domain of kinematics and calculate the curvature change of the object surface, where the second derivative of the displacement will be present explicitly. We also have to consider the conditions of compatibility, i.e. of integrability, that these successive derivatives must fulfil. Besides the mentioned relations that are valid in any case, supplementary ones, such as the normality condition, come from the smallness of some parameters that are related to the particular geometry of the body.

As holographic interferometry is concerned with the deformation that affects the surface (generally curved) of a nontransparent body, we must take into account the presence of this surface and distinguish, as before, interior, i.e. tangential, and exterior, i.e. normal, quantities. Therefore, in addition to the concepts of differential geometry that we presented in Chap. 2 (see also [2.8, 9, 5.1]), we shall use some concepts of shell theory [5.2–19].

5.1.1 Curvature Change of the Outer Surface of an Arbitrary Body or of the Middle Surface of a Shell

When we calculated the strain in Sect. 2.2, we compared the length ds of an elemental vector $\boldsymbol{dr}$ on the undeformed surface to its counterpart ds' on the deformed surface. We used then the difference of the squares $ds'^2 - ds^2$, which constitutes the difference $\mathrm{I}' - \mathrm{I}$ of the first fundamental forms [see (2.42)] of the two surfaces. Likewise, we now consider the difference of the normal curvatures B and B' (i.e. the reciprocals of the curvature radii R and R') in the direction of the unit vectors $\boldsymbol{e}$ and $\boldsymbol{e}'$ parallel to $\boldsymbol{dr}$ and $\boldsymbol{dr}'$ respectively

$$B' - B = \boldsymbol{e}' \cdot \boldsymbol{B}'\boldsymbol{e}' - \boldsymbol{e} \cdot \boldsymbol{B}\boldsymbol{e} = \frac{\boldsymbol{dr}' \cdot \boldsymbol{B}'\boldsymbol{dr}'}{ds'^2} - \frac{\boldsymbol{dr} \cdot \boldsymbol{B}\boldsymbol{dr}}{ds^2} = \frac{\mathrm{II}'}{\mathrm{I}'} - \frac{\mathrm{II}}{\mathrm{I}}. \tag{5.1}$$

The tensors $\boldsymbol{B} = -\boldsymbol{\nabla}_{\mathrm{n}} \otimes \boldsymbol{n}$ and $\boldsymbol{B}' = -\boldsymbol{\nabla}'_{\mathrm{n}} \otimes \boldsymbol{n}'$ [see (2.51)] describe the curvature of the two surfaces and II, II′ denote the second fundamental forms of these surfaces [see (2.52)].

In order to express the quantities that characterize the deformed surface as functions of those that characterize the undeformed surface, we use first the transformation (2.97), i.e.

$$\boldsymbol{\nabla}' = (\boldsymbol{F}^{\mathrm{T}})^{-1}\boldsymbol{\nabla},$$

between the three-dimensional derivative operators in each configuration. Let us recall that $\boldsymbol{F} = \boldsymbol{I} + (\boldsymbol{\nabla} \otimes \boldsymbol{u})^{\mathrm{T}}$ [see (2.83)] is the three-dimensional deformation gradient and that, in the case of small deformations, we can write $(\boldsymbol{F}^{\mathrm{T}})^{-1} \simeq \boldsymbol{I} - \boldsymbol{\nabla} \otimes \boldsymbol{u}$. With $\boldsymbol{n}' \simeq \boldsymbol{n} - \boldsymbol{\omega}$ [see (2.110)], we thus have

$$\begin{aligned}\boldsymbol{\nabla}'_{\mathrm{n}} &= \boldsymbol{N}'\boldsymbol{\nabla}' \simeq [\boldsymbol{I} - (\boldsymbol{n} - \boldsymbol{\omega}) \otimes (\boldsymbol{n} - \boldsymbol{\omega})](\boldsymbol{I} - \boldsymbol{\nabla} \otimes \boldsymbol{u})\boldsymbol{\nabla} \\ &\simeq (\boldsymbol{N} + \boldsymbol{n} \otimes \boldsymbol{\omega} + \boldsymbol{\omega} \otimes \boldsymbol{n} - \boldsymbol{\nabla}_{\mathrm{n}} \otimes \boldsymbol{u})\boldsymbol{\nabla}.\end{aligned}$$

Using the decomposition (2.109) of $\boldsymbol{\nabla}_{\mathrm{n}} \otimes \boldsymbol{u}$, we see that only the interior part $\boldsymbol{\nabla}_{\mathrm{n}}$ of $\boldsymbol{\nabla}$ remains, so that we obtain

$$\boldsymbol{\nabla}'_{\mathrm{n}} \simeq (\boldsymbol{N} + \boldsymbol{n} \otimes \boldsymbol{\omega} - \Omega\boldsymbol{E} - \boldsymbol{\gamma})\boldsymbol{\nabla}_{\mathrm{n}}, \tag{5.2}$$

and then

$$\boldsymbol{B}' \simeq (\boldsymbol{N} + \boldsymbol{n} \otimes \boldsymbol{\omega} - \Omega\boldsymbol{E} - \boldsymbol{\gamma})(\boldsymbol{\nabla}_{\mathrm{n}} \otimes \boldsymbol{\omega} - \boldsymbol{\nabla}_{\mathrm{n}} \otimes \boldsymbol{n}).$$

If we distinguish further in the derivative of the (interior) vector $\boldsymbol{\omega}$ [see (2.55)] an interior, nonsymmetric part

$$\boldsymbol{\kappa} = -(\boldsymbol{\nabla}_{\mathrm{n}} \otimes \boldsymbol{\omega})\boldsymbol{N}, \tag{5.3}$$

and an exterior part $\boldsymbol{B\omega} \otimes \boldsymbol{n}$, we find

$$\boldsymbol{B}' \simeq \boldsymbol{B} - \boldsymbol{\kappa} - \Omega \boldsymbol{EB} - \boldsymbol{\gamma B} + \boldsymbol{n} \otimes \boldsymbol{B\omega} + \boldsymbol{B\omega} \otimes \boldsymbol{n},$$

or, because both $\boldsymbol{B}'$ and $\boldsymbol{B}$ are symmetric,

$$\boldsymbol{B}' \simeq \boldsymbol{B} - \frac{1}{2}[\boldsymbol{\kappa} + \boldsymbol{\kappa}^{\mathrm{T}} - \Omega(\boldsymbol{BE} - \boldsymbol{EB}) + (\boldsymbol{B\gamma} + \boldsymbol{\gamma B})] + \boldsymbol{n} \otimes \boldsymbol{B\omega} + \boldsymbol{B\omega} \otimes \boldsymbol{n}. \tag{5.4}$$

On the other hand, we have with (2.98) for the elemental vectors

$$d\boldsymbol{r}' = ds'\boldsymbol{e}' = ds\boldsymbol{e}\boldsymbol{N}\boldsymbol{F}^{\mathrm{T}} = ds\boldsymbol{e}(\boldsymbol{N} + \boldsymbol{\gamma} + \Omega\boldsymbol{E} + \boldsymbol{\omega} \otimes \boldsymbol{n}), \tag{5.5}$$

and for the squares of the elemental lengths

$$ds'^2 = d\boldsymbol{r}' \cdot d\boldsymbol{r}' \simeq (1 + 2\boldsymbol{e}\cdot\boldsymbol{\gamma e})ds^2.$$

Thus, we obtain for the difference of the normal curvatures [Ref. 5.19, Eq. IV.12] with (5.4, 5)

$$\begin{aligned} B' - B &= \left(\frac{ds}{ds'}\right)^2 \boldsymbol{e}\cdot\boldsymbol{N}\boldsymbol{F}^{\mathrm{T}}\boldsymbol{B}'\boldsymbol{F}\boldsymbol{N}\boldsymbol{e} - \boldsymbol{e}\cdot\boldsymbol{B}\boldsymbol{e} \\ &\simeq -\frac{1}{2}\boldsymbol{e}\cdot[\boldsymbol{\kappa} + \boldsymbol{\kappa}^{\mathrm{T}} + \Omega(\boldsymbol{BE} - \boldsymbol{EB}) - (\boldsymbol{B\gamma} + \boldsymbol{\gamma B})]\boldsymbol{e} - 2(\boldsymbol{e}\cdot\boldsymbol{\gamma e})(\boldsymbol{e}\cdot\boldsymbol{Be}). \end{aligned} \tag{5.6}$$

We may thus define a first *tensor describing the curvature change*

$$\boldsymbol{\kappa}_{\mathrm{c}} = \frac{1}{2}[(\boldsymbol{\kappa} + \boldsymbol{\kappa}^{\mathrm{T}}) + \Omega(\boldsymbol{BE} - \boldsymbol{EB}) - (\boldsymbol{B\gamma} + \boldsymbol{\gamma B})], \tag{5.7}$$

so that we can abbreviate

$$B' - B = -\boldsymbol{e}\cdot\boldsymbol{\kappa}_{\mathrm{c}}\boldsymbol{e} - 2(\boldsymbol{e}\cdot\boldsymbol{\gamma e})(\boldsymbol{e}\cdot\boldsymbol{Be}). \tag{5.8}$$

If we consider the special case in which the principal directions $\boldsymbol{e}_1$, $\boldsymbol{e}_2$ of $\boldsymbol{\gamma}$ and $\boldsymbol{B}$ coincide, we have

$$B_1' - B_1 = -\boldsymbol{e}_1\cdot\boldsymbol{\kappa}_{\mathrm{s}}\boldsymbol{e}_1, \qquad B_2' - B_2 = -\boldsymbol{e}_2\cdot\boldsymbol{\kappa}_{\mathrm{s}}\boldsymbol{e}_2, \tag{5.9}$$

where

$$\boldsymbol{\kappa}_s = \frac{1}{2}[(\boldsymbol{\kappa} + \boldsymbol{\kappa}^T) + \Omega(\boldsymbol{BE} - \boldsymbol{EB}) + (\boldsymbol{B}\boldsymbol{\gamma} + \boldsymbol{\gamma}\boldsymbol{B})] \tag{5.10}$$

is a second tensor that describes the curvature change. Third, if we leave aside the terms in γ, we get the so-called *reduced tensor describing the curvature change* introduced by *Sanders* [5.13] and *Koiter* [5.14] in shell theory

$$\boldsymbol{\kappa}_r = \frac{1}{2}[\boldsymbol{\kappa} + \boldsymbol{\kappa}^T + \Omega(\boldsymbol{BE} - \boldsymbol{EB})], \tag{5.11}$$

depending only on the rotations $\boldsymbol{\omega}$, Ω. It must be noticed that $\boldsymbol{\kappa}_c$, $\boldsymbol{\kappa}_s$ and $\boldsymbol{\kappa}_r$ are all *symmetric* and that they become identical in case the surface is *not* strained. In particular, they are equal to zero in case the surface is neither strained nor bent (rigid surface), i.e. when $\gamma = 0$ and $B' = B \;\forall\; \boldsymbol{e}$.

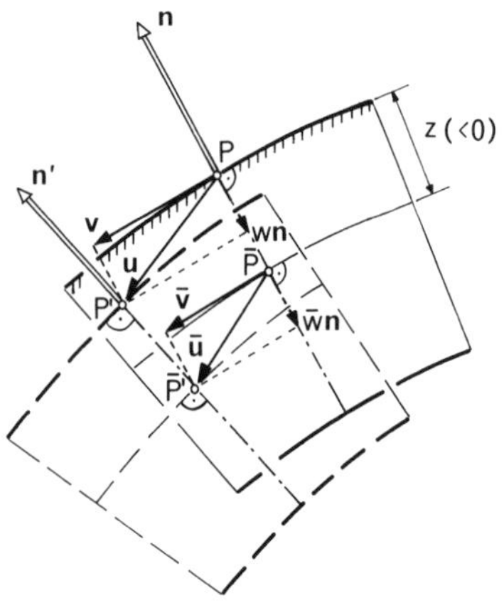

Fig. 5.1. Deformation of a shell element according to the normality condition

Let us now look at another role of the tensor $\boldsymbol{\kappa}_s$ besides the one in (5.9). In the particular case of a *thin shell* with parallel free surfaces, the so-called *normality condition* of Kirchhoff–Love (see e.g. [Ref. 5.6, p. 430; Ref. 5.11, p. 38]) is approximately valid (in a rigorous treatment, it is replaced by an asymptotic analysis, see e.g. [5.10]). This condition states that, during the deformation, straight lines normal to the free surfaces remain normal to them, that consequently parallel inner surfaces remain parallel, and moreover that the distance between them does not change (Fig. 5.1). Then, the displacement vector $\bar{\boldsymbol{u}} = \bar{\boldsymbol{v}} + \bar{w}\boldsymbol{n}$ of a point $\bar{\mathrm{P}}$ situated in the interior of the shell, at the distance $z (z < 0)$ from the outer surface, fulfils the relations

$$\bar{\boldsymbol{u}} = \boldsymbol{u} - z\boldsymbol{\omega} \quad \text{and} \quad \bar{w} = w. \tag{5.12}$$

Consequently, the surface strain tensor in $\bar{\mathrm{P}}$, i.e. with (2.105)

$$\bar{\gamma} = \frac{1}{2}N[\bar{\boldsymbol{V}} \otimes \bar{\boldsymbol{u}} + (\bar{\boldsymbol{V}} \otimes \bar{\boldsymbol{u}})^{\mathrm{T}}]N,$$

can be expressed as

$$\bar{\gamma} = \gamma + z\boldsymbol{\kappa}_{\mathrm{s}}, \qquad (5.13)$$

provided that we take into account the transformation (2.80) between $\bar{\boldsymbol{V}}_{\mathrm{n}}$ and $\boldsymbol{V}_{\mathrm{n}}$ and neglect the terms in z^2. Since (5.13) can also be written in the form

$$\gamma = \bar{\gamma} - z\boldsymbol{\kappa}_{\mathrm{s}},$$

we may state that, in this approximation, $\boldsymbol{\kappa}_{\mathrm{s}}$ describes either the curvature change of the outer free surface or also the curvature change of an inner (e.g. the middle) surface of the shell. Equation (5.13) will be useful for an extrapolation of strain from the outer surface, where it is measured, to the interior of the body.

5.1.2 Integrability and Compatibility Conditions on a Curved Surface

The gradient fields defined on surfaces that are embedded in the three-dimensional Euclidean space $\mathbb{R}^3$ must satisfy certain conditions that can be useful in holographic interferometry because they afford information about the quantities to be measured. In order to establish them, we start from the well-known *Stokes' theorem*

$$\iint_{A_{\mathrm{n}}} (\boldsymbol{V} \times \boldsymbol{t})\cdot \boldsymbol{n}\, dA_{\mathrm{n}} = \oint_{\partial A_{\mathrm{n}}} \boldsymbol{t}\cdot d\boldsymbol{r}, \qquad (5.14)$$

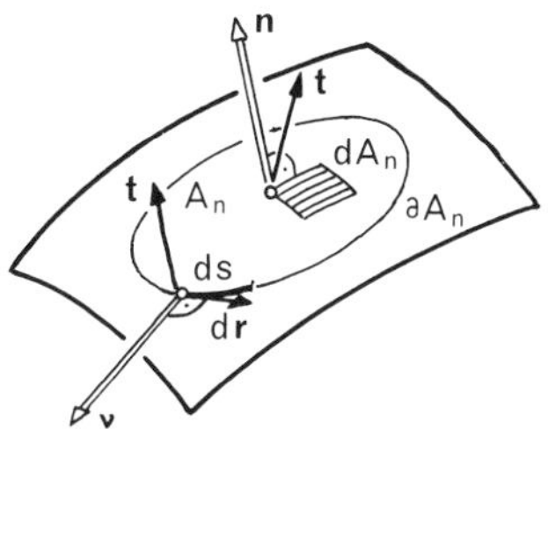

Fig. 5.2. Definition of quantities that appear in the integral theorems on a curved surface

which is valid for any differentiable vector field $\boldsymbol{t}$ in $\mathbb{R}^3$ and any surface A_{n} with a regular boundary ∂A_{n} (Fig. 5.2). $\boldsymbol{n}$ denotes the unit normal to a surface element dA_n and $d\boldsymbol{r} = -\boldsymbol{E}\boldsymbol{v}ds$, a directed arc element on ∂A_n, ds being the length of this element and $\boldsymbol{v}$, the outer unit normal to ∂A_n. Because, with reference to (2.28, 30, 45),

$$(\nabla \times \boldsymbol{t})\cdot\boldsymbol{n} = (\nabla_n \times \boldsymbol{t})\cdot\boldsymbol{n} + \left(\boldsymbol{n} \times \frac{\partial \boldsymbol{t}}{\partial z}\right)\cdot\boldsymbol{n} = -(\nabla_n \mathscr{E}\boldsymbol{t})\cdot\boldsymbol{n}$$

$$= -\nabla_n\cdot(\boldsymbol{t}\mathscr{E}\boldsymbol{n}) + \boldsymbol{B}\cdot(\mathscr{E}\boldsymbol{t}) = \nabla_n\cdot(\boldsymbol{E}\boldsymbol{t}),$$

(5.14) is equivalent to

$$\iint\limits_{A_n} \nabla_n\cdot(\boldsymbol{E}\boldsymbol{t})dA_n = \oint\limits_{\partial A_n} \boldsymbol{v}\cdot\boldsymbol{E}\boldsymbol{t}ds, \tag{5.15}$$

which constitutes an expression of the *Gauss' theorem* on a curved surface. This theorem is valid not only for vectors but also for tensors, the scalar products being then replaced by linear transformations. For an arbitrary tensor $\boldsymbol{T}$, we have thus

$$\iint\limits_{A_n} \nabla_n(\boldsymbol{E}\boldsymbol{T})dA_n = \oint\limits_{\partial A_n} \boldsymbol{v}\boldsymbol{E}\boldsymbol{T}ds. \tag{5.16}$$

With the help of this theorem, we can demonstrate that

$$\nabla_n(\boldsymbol{E}\boldsymbol{T}) = 0 \tag{5.17}$$

is a necessary and sufficient *condition of integrability* for a semi-interior tensor in a simply connected two-dimensional domain A, in other words that, when (5.17) is fulfilled, the tensor $\boldsymbol{T} \equiv \boldsymbol{N}\boldsymbol{T}$ can be expressed as the gradient of a vector, let us say $\boldsymbol{N}\boldsymbol{T} = \nabla_n \otimes \boldsymbol{t}$. That this condition is necessary follows from

$$\nabla_n(\boldsymbol{E}\nabla_n \otimes \boldsymbol{t}) = (\nabla_n\boldsymbol{E})\nabla_n \otimes \boldsymbol{t} - \boldsymbol{E}\cdot(\nabla_n \otimes \nabla_n \otimes \boldsymbol{t}) = 0,$$

where the second term vanishes because $\boldsymbol{E}$ is skew symmetric and the second derivative of $\boldsymbol{t}$ is symmetric with respect to the two ∇_n, and where $\nabla_n\boldsymbol{E} = 0$. Indeed, with (2.45),

$$\nabla_n \otimes \boldsymbol{E} = \nabla_n \otimes (\boldsymbol{n}\mathscr{E}) = (\nabla_n \otimes \boldsymbol{n})\mathscr{E} = \boldsymbol{B}\boldsymbol{E} \otimes \boldsymbol{n})^{\mathrm{T}} - \boldsymbol{B}\boldsymbol{E} \otimes \boldsymbol{n}, \tag{5.18}$$

a contraction of which yields $\nabla_n\boldsymbol{E} = \boldsymbol{B}\boldsymbol{E}\boldsymbol{n} - (\boldsymbol{B}\cdot\boldsymbol{E})\boldsymbol{n} = 0$. To prove that (5.17) is a sufficient condition, we shall use (5.16) in the form

$$\oint\limits_{\partial A_n} \boldsymbol{v}\boldsymbol{E}\boldsymbol{T}ds = \int_0^{\mathrm{P}} \boldsymbol{v}_1\boldsymbol{E}\boldsymbol{T}ds_1 - \int_0^{\mathrm{P}} \boldsymbol{v}_2\boldsymbol{E}\boldsymbol{T}ds_2 = 0,$$

where O is a fixed and P, a variable point in A, these points being joined by two different paths s_1 and s_2. Because A is simply connected, the line OPO can be regarded as the boundary of a part A_n of A. We thus conclude that the integral

$$\boldsymbol{t}(\boldsymbol{r}) = \int_0^P \boldsymbol{v}\boldsymbol{E}\boldsymbol{T}\,ds = \int_0^P \boldsymbol{T}^T d\boldsymbol{r},$$

depends on only the position vector $\boldsymbol{r}$ of P. Writing the increment of $\boldsymbol{t}$ from P to a neighboring point $\bar{\text{P}}$, i.e.

$$\boldsymbol{t}(\bar{\boldsymbol{r}}) - \boldsymbol{t}(\boldsymbol{r}) = d\boldsymbol{t} = d\boldsymbol{r}(\boldsymbol{\nabla}_n \otimes \boldsymbol{t}) = d\boldsymbol{r}\boldsymbol{T},$$

shows that $\boldsymbol{T}$ is the gradient of $\boldsymbol{t}$. We notice that the condition (5.17) as well as its proof are fully analogous to the case of the gradient of a scalar function in three-dimensional space (see e.g. [Ref. 2.11, pp. 40, 71]).

As far as surfaces are concerned, the quantities that pertain to the first and the second fundamental forms must fulfil, among others, integrability conditions; these are expressed by the *Gauss' equation*

$$K\boldsymbol{N} = -\boldsymbol{E}\boldsymbol{B}\boldsymbol{E}\boldsymbol{B} \tag{5.19}$$

for the Gaussian curvature K (which, as can be demonstrated, depends on only the first fundamental form) and the equations of *Codazzi–Mainardi*, i.e.

$$\boldsymbol{\nabla}_n(\boldsymbol{E}\boldsymbol{B}) = 0 \quad \text{and also} \quad \boldsymbol{\nabla}_n(\boldsymbol{E}\boldsymbol{B})\boldsymbol{N} = 0, \tag{5.20}$$

for the tensor $\boldsymbol{B}$ (which is, as we know, the gradient of the normal $\boldsymbol{n}$).

Let us now establish the conditions that the strain and the rotation tensors must fulfil because they form the decomposition of the gradient of the displacement field. We could write (5.19, 20) on the deformed surface and relate them to their counterparts on the undeformed surface. However, a more direct way, which yields useful intermediate results and which we choose here for that reason, consists in applying the criterion (5.17), i.e. in developing

$$\boldsymbol{\nabla}_n[\boldsymbol{E}(\boldsymbol{\gamma} + \Omega\boldsymbol{E} + \boldsymbol{\omega} \otimes \boldsymbol{n})] = \boldsymbol{\nabla}_n(\boldsymbol{E}\boldsymbol{\gamma} - \Omega\boldsymbol{N} + \boldsymbol{E}\boldsymbol{\omega} \otimes \boldsymbol{n}) = 0.$$

According to (2.57), we have first

$$\boldsymbol{\nabla}_n(\boldsymbol{E}\boldsymbol{\gamma} - \Omega\boldsymbol{N})\boldsymbol{N} - \boldsymbol{B}\boldsymbol{E}\boldsymbol{\omega} + [\boldsymbol{B}\cdot(\boldsymbol{E}\boldsymbol{\gamma} - \Omega\boldsymbol{N}) + \boldsymbol{\nabla}_n\cdot(\boldsymbol{E}\boldsymbol{\omega})]\boldsymbol{n} = 0.$$

Observing that, by contraction of (2.54),

$$\boldsymbol{\nabla}_n\boldsymbol{N} = (\text{tr}\,\boldsymbol{B})\boldsymbol{n} = 2H\boldsymbol{n},$$

and that, according to the rules (2.30) and with (5.18),

$$\boldsymbol{\nabla}_n(\boldsymbol{E}\boldsymbol{\gamma}\boldsymbol{E}) = \boldsymbol{\nabla}_n(\boldsymbol{E}\boldsymbol{\gamma})\boldsymbol{E} + (\boldsymbol{\nabla}_n \otimes \boldsymbol{E})^T\cdot\boldsymbol{E}\boldsymbol{\gamma} = \boldsymbol{\nabla}_n(\boldsymbol{E}\boldsymbol{\gamma})\boldsymbol{E} + (\boldsymbol{E}\boldsymbol{B}\cdot\boldsymbol{E}\boldsymbol{\gamma})\boldsymbol{n},$$

setting $\boldsymbol{N} = -\boldsymbol{E}^2$, we obtain from the interior part

$$\nabla_{\mathrm{n}}(\boldsymbol{E}\gamma\boldsymbol{E})\boldsymbol{N} = (\nabla_{\mathrm{n}}\Omega)\boldsymbol{E} - \boldsymbol{E}\boldsymbol{B}\boldsymbol{E}\boldsymbol{\omega}. \tag{5.21}$$

The exterior part, which, with (5.3), becomes

$$[\boldsymbol{\kappa} + (\gamma + \Omega\boldsymbol{E})\boldsymbol{B}]\cdot\boldsymbol{E} = 0, \tag{5.22}$$

shows that $\boldsymbol{\kappa} + (\gamma + \Omega\boldsymbol{E})\boldsymbol{B}$ is a symmetric tensor. By the way, the transformation $\boldsymbol{E}(\ldots)\boldsymbol{E}$, which appears often in the calculations, is an *involution* (see e.g. [Ref. 5.18, Eq. (4)]) because we have, for any interior tensor $\boldsymbol{D}$,

$$\boldsymbol{E}(\boldsymbol{E}\boldsymbol{D}\boldsymbol{E})\boldsymbol{E} = \boldsymbol{N}\boldsymbol{D}\boldsymbol{N} = \boldsymbol{D}.$$

Equation (5.21) does not yet constitute a proper integrability condition for γ, because it contains the rotation quantities that do not vanish when the body moves rigidly, i.e. when $\gamma = 0$. In this case, in contrast, it rather expresses the relation that exists between $\boldsymbol{\omega}$ and Ω. Therefore, we attempt to find another equation by using the trivial condition

$$\nabla_{\mathrm{n}}[\boldsymbol{E}\nabla_{\mathrm{n}} \otimes (\boldsymbol{E}\boldsymbol{\omega})] = 0. \tag{5.23}$$

With (2.30) and (5.18), we have

$$\nabla_{\mathrm{n}} \otimes (\boldsymbol{E}\boldsymbol{\omega}) = (\nabla_{\mathrm{n}} \otimes \boldsymbol{E})\boldsymbol{\omega} - (\nabla_{\mathrm{n}} \otimes \boldsymbol{\omega})\boldsymbol{E} = \boldsymbol{B}\boldsymbol{E}\boldsymbol{\omega} \otimes \boldsymbol{n} + \boldsymbol{\kappa}\boldsymbol{E},$$

so that, if we insert (5.21), we obtain

$$\nabla_{\mathrm{n}}\{[(\nabla_{\mathrm{n}}\Omega)\boldsymbol{E} - \nabla_{\mathrm{n}}(\boldsymbol{E}\gamma\boldsymbol{E})\boldsymbol{N}] \otimes \boldsymbol{n} + \boldsymbol{E}\boldsymbol{\kappa}\boldsymbol{E}\} = 0.$$

Thanks to the symmetry of $\boldsymbol{\kappa} + (\gamma + \Omega\boldsymbol{E})\boldsymbol{B}$, we get from (5.11)

$$\boldsymbol{\kappa} = \boldsymbol{\kappa}_{\mathrm{r}} - \Omega\boldsymbol{B}\boldsymbol{E} + \frac{1}{2}(\boldsymbol{B}\gamma - \gamma\boldsymbol{B}) = \boldsymbol{\kappa}_{\mathrm{r}} - \Omega\boldsymbol{B}\boldsymbol{E} + \frac{1}{4}[(\boldsymbol{B}\boldsymbol{E} - \boldsymbol{E}\boldsymbol{B})\cdot\boldsymbol{E}\gamma\boldsymbol{E}]\boldsymbol{E}. \tag{5.24}$$

Inserting this result into the preceding relation and developing as before, with (5.20), we obtain from the interior part the vector equation

$$\nabla_{\mathrm{n}}(\boldsymbol{E}\boldsymbol{\kappa}_{\mathrm{r}}\boldsymbol{E})\boldsymbol{N} - \frac{1}{4}\nabla_{\mathrm{n}}[(\boldsymbol{B}\boldsymbol{E} - \boldsymbol{E}\boldsymbol{B})\cdot\boldsymbol{E}\gamma\boldsymbol{E}]\boldsymbol{E} + \nabla_{\mathrm{n}}(\boldsymbol{E}\gamma\boldsymbol{E})\boldsymbol{B} = 0, \tag{5.25}$$

and from the exterior part the scalar equation

$$\boldsymbol{B}\cdot\boldsymbol{E}\boldsymbol{\kappa}_{\mathrm{r}}\boldsymbol{E} - [(\nabla_{\mathrm{n}} \otimes \nabla_{\mathrm{n}})\cdot(\boldsymbol{E}\gamma\boldsymbol{E})] = 0. \tag{5.26}$$

These integrability conditions, which are valid for the deformation of surfaces of any curvature and any shape, are called the *compatibility equations* for the strain tensor. Thanks to them, the deformation of a surface can be specified by the two symmetric interior tensors γ and $\boldsymbol{\kappa}_r$ instead of the displacement vector $\boldsymbol{u}$. Finally, we add that, in the particular case of a plate where $\boldsymbol{B} \equiv 0$, (5.25) becomes trivial, whereas (5.26) is reduced to

$$-(\boldsymbol{V}_n \otimes \boldsymbol{V}_n)\cdot(\boldsymbol{E}\gamma\boldsymbol{E}) = 0,$$

i.e., expressed in cartesian components,

$$-\begin{pmatrix} \dfrac{\partial^2}{\partial x^2} & \dfrac{\partial^2}{\partial x \partial y} \\ \dfrac{\partial^2}{\partial y \partial x} & \dfrac{\partial^2}{\partial y^2} \end{pmatrix} \cdot \begin{pmatrix} -\varepsilon_y & \dfrac{1}{2}\gamma_{xy} \\ \dfrac{1}{2}\gamma_{xy} & -\varepsilon_x \end{pmatrix} = \frac{\partial^2 \varepsilon_y}{\partial x^2} + \frac{\partial^2 \varepsilon_x}{\partial y^2} - \frac{\partial^2 \gamma_{xy}}{\partial x \partial y} = 0.$$

5.1.3 Equations of Motion or of Equilibrium and Kinematic Relations in the Neighborhood of a Free Surface. Constitutive Equations for an Elastic Body

Up to now, among the equations of continuum mechanics, we have dealt only with those that relate to the kinematics of deformation. Because the other basic relations can prove useful in holographic interferometry, especially when seeking the strain in the interior of a body or also the constitutive equations, we briefly recall them here. In the next sections, we shall then see how they can help us to find all of the characteristics of the deformation in the neighborhood of a free surface.

First, we examine the statics in a point $\bar{\mathrm{P}}$ in the interior of the object (Fig. 5.3). An elemental "force" $\Delta \boldsymbol{f}$ acts on an elemental surface $\Delta \bar{A}_e$, with unit normal $\bar{\boldsymbol{e}}$; the *stress vector* $\bar{\boldsymbol{t}} = \lim \Delta \boldsymbol{f}/\Delta \bar{A}_e (\Delta \bar{A}_e \to 0)$ is determined by the linear transformation

$$\bar{\boldsymbol{t}} = \bar{\boldsymbol{T}}\bar{\boldsymbol{e}} = \bar{\boldsymbol{e}}\bar{\boldsymbol{T}} \tag{5.27}$$

produced by the symmetric three-dimensional stress tensor $\bar{\boldsymbol{T}}$. It must be noted that $\bar{\boldsymbol{T}}$ is *Kirchhoff's stress tensor*, i.e. it describes the stress present in the deformed object but is referred to the undeformed configuration (see e.g. [Ref. 2.11, Sect. IX.4]). If, instead of $\bar{\mathrm{P}}$, we consider a point P on the free surface, the *static boundary conditions* imply that

$$\boldsymbol{t} = \boldsymbol{T}\boldsymbol{n} = 0. \tag{5.28}$$

To take into account the vicinity of the free surface, let us decompose, similarly to (2.40), the stress tensor $\bar{\boldsymbol{T}}$ with respect to the unit normal $\boldsymbol{n}$ at the point P situated above $\bar{\mathrm{P}}$, i.e.

$$\bar{\boldsymbol{T}} = \bar{\boldsymbol{\tau}} + \bar{\boldsymbol{\tau}}_z \otimes \boldsymbol{n} + \boldsymbol{n} \otimes \bar{\boldsymbol{\tau}}_z + \bar{\sigma}_z(\boldsymbol{n} \otimes \boldsymbol{n}), \tag{5.29}$$

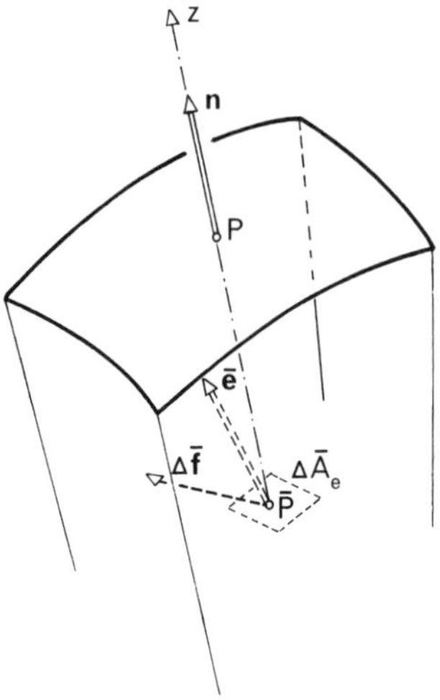

Fig. 5.3. Definition of the elemental "force" $\Delta \boldsymbol{f}$ that acts on a surface element $\Delta \bar{A}_e$ in the interior of the undeformed body

or in matrix notation

$$\bar{\boldsymbol{T}} \triangleq \left[\begin{array}{c|c} \bar{\tau} & \bar{\tau}_z \\ \hline \bar{\tau}_z & \bar{\sigma}_z \end{array}\right]$$

with the two-dimensional tensor $\bar{\tau}$ of in-plane stress (e.g. with cartesian components $\bar{\sigma}_x$, $\bar{\sigma}_y$, $\bar{\tau}_{xy}$), the shear stress vector $\bar{\tau}_z$(e.g. $\bar{\tau}_{xz}$, $\bar{\tau}_{yz}$) and the normal stress $\bar{\sigma}_z$ along $\boldsymbol{n}$. The condition (5.28) can then be expressed as

$$\tau_z = 0, \qquad \sigma_z = 0. \tag{5.30}$$

With the stress tensor, we are now able to write the *equations of motion* (*or of equilibrium*) for the material point $\bar{\mathrm{P}}$; in the case of *small* deformations, they are

$$\bar{\nabla}\bar{\boldsymbol{T}} = \rho\ddot{\boldsymbol{u}}(=0), \tag{5.31}$$

where ρ denotes the specific mass and $\ddot{\boldsymbol{u}}$, the acceleration at $\bar{\mathrm{P}}$. With the decompositions (5.29), $\bar{\boldsymbol{u}} = \bar{\boldsymbol{v}} + \bar{w}\boldsymbol{n}$, $\bar{\nabla} = \bar{\nabla}_{\mathrm{n}} + \boldsymbol{n}\partial_z$, where ∂_z is an abbreviation for ∂/∂_z, and according to (2.57), we can split (5.31) into a tangential and a normal part, i.e. into

$$\begin{aligned} &\bar{\nabla}_{\mathrm{n}}\bar{\tau}N - \bar{\boldsymbol{B}}\bar{\tau}_z - 2\bar{H}\bar{\tau}_z + \partial_z\bar{\tau}_z = \rho\ddot{\bar{v}}, \\ &\bar{\boldsymbol{B}}\cdot\bar{\tau} + \bar{\nabla}_{\mathrm{n}}\cdot\bar{\tau}_z - 2\bar{H}\bar{\sigma}_z + \partial_z\bar{\sigma}_z = \rho\ddot{\bar{w}}. \end{aligned} \tag{5.32}$$

On the free surface, because (5.30) is valid everywhere, the equation of motion is reduced to

$$\begin{aligned} \bar{\nabla}_{\mathrm{n}}\boldsymbol{\tau}\boldsymbol{N} + \partial_z\boldsymbol{\tau}_z &= \rho\ddot{\boldsymbol{v}}, \\ \boldsymbol{B}\cdot\boldsymbol{\tau} + \partial_z\sigma_z &= \rho\ddot{w}. \end{aligned} \tag{5.33}$$

Second, let us complete the relations that we deduced in Sect. 2.2.1 and, similar to the above procedure, decompose the three-dimensional kinematic relations with respect to $\boldsymbol{n}$. So, we write for the three-dimensional strain tensor in $\bar{\mathrm{P}}$

$$\bar{\boldsymbol{\varepsilon}} = \bar{\boldsymbol{\gamma}} + \bar{\boldsymbol{\gamma}}_z \otimes \boldsymbol{n} + \boldsymbol{n} \otimes \bar{\boldsymbol{\gamma}}_z + \bar{\varepsilon}_z \boldsymbol{n} \otimes \boldsymbol{n}, \tag{5.34}$$

i.e. in matrix notation

$$\bar{\boldsymbol{\varepsilon}} \mathrel{\hat{=}} \left[\begin{array}{c:c} \bar{\boldsymbol{\gamma}} & \bar{\boldsymbol{\gamma}}_z \\ \hdashline \bar{\boldsymbol{\gamma}}_z & \bar{\varepsilon}_z \end{array}\right].$$

With the linear kinematic relation (2.91), i.e.

$$\bar{\boldsymbol{\varepsilon}} = \frac{1}{2}[\bar{\nabla} \otimes \bar{\boldsymbol{u}} + (\bar{\nabla} \otimes \bar{\boldsymbol{u}})^{\mathrm{T}}],$$

and with the decompositions of $\bar{\boldsymbol{u}}$, $\bar{\nabla}$ and $\bar{\nabla}_{\mathrm{n}} \otimes \bar{\boldsymbol{u}}$ [see (2.55)], we obtain the *kinematic relations*

$$\begin{aligned} \bar{\boldsymbol{\gamma}} &= \frac{1}{2}[(\bar{\nabla}_{\mathrm{n}} \otimes \bar{\boldsymbol{v}})\boldsymbol{N} + \boldsymbol{N}(\bar{\nabla}_{\mathrm{n}} \otimes \bar{\boldsymbol{v}})^{\mathrm{T}}] - \bar{\boldsymbol{B}}\bar{w}, \\ \bar{\boldsymbol{\gamma}}_z &= \frac{1}{2}(\bar{\boldsymbol{B}}\bar{\boldsymbol{v}} + \bar{\nabla}_{\mathrm{n}}\bar{w} + \partial_z\bar{\boldsymbol{v}}), \\ \bar{\varepsilon}_z &= \partial_z\bar{w}. \end{aligned} \tag{5.35}$$

Let us remark here that we already obtained the relation for $\bar{\boldsymbol{\gamma}} = \boldsymbol{N}\bar{\boldsymbol{\varepsilon}}\boldsymbol{N}$ in (2.106) and that the present decomposition of the semiexterior terms of $\bar{\boldsymbol{\varepsilon}}$ is connected with that of the semiexterior term of $\bar{\nabla}_{\mathrm{n}} \otimes \bar{\boldsymbol{u}}$ in (2.109) by

$$\begin{aligned} \bar{\boldsymbol{\gamma}}_z &= \boldsymbol{N}\bar{\boldsymbol{\varepsilon}}\boldsymbol{n} = \frac{1}{2}(\bar{\boldsymbol{\omega}} + \partial_z\bar{\boldsymbol{v}}), \\ \bar{\boldsymbol{\omega}}_{\mathrm{i}} &= \boldsymbol{N}\bar{\boldsymbol{\Omega}}\boldsymbol{n} = \frac{1}{2}(\bar{\boldsymbol{\omega}} - \partial_z\bar{\boldsymbol{v}}), \end{aligned} \tag{5.36}$$

because with (2.111) $\bar{\boldsymbol{\omega}} = \bar{\boldsymbol{B}}\bar{\boldsymbol{v}} + \bar{\nabla}_{\mathrm{n}}\bar{w}$.

Finally, we shall have to relate the dynamic to the kinematic quantities; this must be done with the *constitutive equations* that link the stress tensor to the strain tensor and depend on the kind of material of which the body is made.

Here, we restrict our attention to the most simple case, namely that of an isotropic elastic body where these equations follow *Hooke's law*

$$\bar{\boldsymbol{\mathcal{E}}} = \frac{1}{2G}\left[\bar{\boldsymbol{T}} - \frac{\nu}{1+\nu}(\operatorname{tr}\bar{\boldsymbol{T}})\boldsymbol{I}\right]; \tag{5.37}$$

$G = E/2(1+\nu)$ represents here the shear modulus, ν, Poisson's ratio and E, the modulus of elasticity or Young's modulus. Also this result can be decomposed; the interior part is then

$$\bar{\gamma} = \frac{1}{2G}\left[\bar{\tau} - \frac{\nu}{1+\nu}(\operatorname{tr}\bar{\tau} + \bar{\sigma}_z)\boldsymbol{N}\right],$$

where we can use $(\operatorname{tr}\bar{\tau})\boldsymbol{N} = \bar{\tau} - \boldsymbol{E}\bar{\tau}\boldsymbol{E}$. The whole decomposition is thus

$$\begin{aligned}
\bar{\gamma} &= \frac{1}{E}(\bar{\tau} + \nu\boldsymbol{E}\bar{\tau}\boldsymbol{E}) - \frac{\nu}{E}\bar{\sigma}_z\boldsymbol{N},\\
\bar{\gamma}_z &= \frac{1}{2G}\bar{\tau}_z,\\
\bar{\varepsilon}_z &= \frac{1}{E}(\bar{\sigma}_z - \nu\operatorname{tr}\bar{\tau}).
\end{aligned} \tag{5.38}$$

Now let us finally calculate the inverse of the first of these relations. Applying the operation $\boldsymbol{N}(\ldots)\boldsymbol{N} - \nu\boldsymbol{E}(\ldots)\boldsymbol{E}$, we find that

$$\bar{\tau} = \frac{E}{1-\nu^2}(\bar{\gamma} - \nu\boldsymbol{E}\bar{\gamma}\boldsymbol{E}) + \frac{\nu}{1-\nu}\bar{\sigma}_z\boldsymbol{N}. \tag{5.39}$$

5.2 The Second Derivatives of the Optical-Path Difference and Some of Their Possible Applications

In the preceding chapters, we already twice encountered second derivatives of optical quantities: in Sect. 3.2, when we investigated the aberrations of holographically reconstructed images and, in Sect. 4.2.4, when we spoke of the absolute localization. In Chap. 4 on the other hand, we took into account only first derivatives (about the validity of this approximation, see e.g. [Ref. 4.192, Chap. 12]). Now, we shall first calculate in detail the second derivative of the optical-path difference, which relates to the curvature of the interference fringes; this will be done for a fixed position of the observer. We shall then indicate how the mechanical quantities underneath the object surface can be determined by using the information given on the surface and the basic relations of continuum mechanics. There, the second derivatives of the optical quantities will be present only implicitly, since the starting point will be optically measured mechanical quantities on the surface.

5.2.1 Derivative of the Fringe Spacing, Curvature of a Fringe, Singularities in the Fringe Pattern

In this section, we refer again to Fig. 4.9, i.e. we consider the changes that affect the optical-path difference D (as defined in Sect. 4.1.1) when the observer looks from a fixed position R at the fringes that correspond to several object points P. $D = D_{\mathrm{R}}$ is thus a function of only one independent vector variable, namely of the position vector $\boldsymbol{r}$ of P or of the observation direction $\boldsymbol{k}$, as expressed in (4.23, 24). It must also be recalled that, because of the linearization, D_{R} is constant along a straight line RP; it is consequently defined on the object surface as well as on the unit sphere centered in R on which $\boldsymbol{k}$ varies. So, for $\bar{D}_{\mathrm{R}}$, which belongs to a line R$\bar{\mathrm{P}}$ near RP, we can envisage the two following equivalent series developments, similar to (2.60),

$$\bar{D}_{\mathrm{R}} = D_{\mathrm{R}} + d\boldsymbol{r}\cdot\boldsymbol{\nabla}_{\mathrm{n}}D_{\mathrm{R}} + \frac{1}{2}[d\boldsymbol{r}\cdot(\boldsymbol{\nabla}_{\mathrm{n}} \otimes \boldsymbol{\nabla}_{\mathrm{n}}D_{\mathrm{R}})d\boldsymbol{r} + ds^2\boldsymbol{b}\cdot\boldsymbol{\nabla}_{\mathrm{n}}D_{\mathrm{R}}] + \cdots, \tag{5.40}$$

$$\bar{D}_{\mathrm{R}} = D_{\mathrm{R}} + d\boldsymbol{k}\cdot\boldsymbol{\nabla}_{\mathrm{k}}D_{\mathrm{R}} + \frac{1}{2}[d\boldsymbol{k}\cdot(\boldsymbol{\nabla}_{\mathrm{k}} \otimes \boldsymbol{\nabla}_{\mathrm{k}}D_{\mathrm{R}})d\boldsymbol{k} + d\phi^2\boldsymbol{b}_{\mathrm{k}}\cdot\boldsymbol{\nabla}_{\mathrm{k}}D_{\mathrm{R}}] + \cdots. \tag{5.41}$$

$d\boldsymbol{r}$ is an increment on the object surface, to which corresponds the increment $d\boldsymbol{k}$ on the unit sphere, on account of the collineation with center R. The unit vectors $\boldsymbol{e}$ and $\boldsymbol{m}$ parallel to these increments are so associated to them that $d\boldsymbol{r} = \boldsymbol{e}ds$ and $d\boldsymbol{k} = -\boldsymbol{m}d\phi$. $\boldsymbol{b}$ and $\boldsymbol{b}_{\mathrm{k}}$ are the geodesic curvatures of the curves, on the object surface and on the unit sphere, respectively, along which D_{R} and $\bar{D}_{\mathrm{R}}$ are examined.

Two special cases are of particular importance for the application of these relations. First, if we go along a fringe such as it appears on the object surface or on the unit sphere, we have

$$\begin{aligned} &\bar{D}_{\mathrm{R}} = D_{\mathrm{R}}, \\ &d\boldsymbol{r}\cdot\boldsymbol{\nabla}_{\mathrm{n}}D_{\mathrm{R}} = d\boldsymbol{k}\cdot\boldsymbol{\nabla}_{\mathrm{k}}D_{\mathrm{R}} = 0, \\ &\boldsymbol{e}\cdot(\boldsymbol{\nabla}_{\mathrm{n}} \otimes \boldsymbol{\nabla}_{\mathrm{n}}D_{\mathrm{R}})\boldsymbol{e} = -\boldsymbol{b}\cdot\boldsymbol{\nabla}_{\mathrm{n}}D_{\mathrm{R}} \quad \text{or} \\ &\boldsymbol{m}\cdot(\boldsymbol{\nabla}_{\mathrm{k}} \otimes \boldsymbol{\nabla}_{\mathrm{k}}D_{\mathrm{R}})\boldsymbol{m} = -\boldsymbol{b}_{\mathrm{k}}\cdot\boldsymbol{\nabla}_{\mathrm{k}}D_{\mathrm{R}}. \end{aligned} \tag{5.42}$$

$\boldsymbol{b}$ and $\boldsymbol{b}_{\mathrm{k}}$ are now the (non-equal) geodesic curvatures of the fringes that appear on the two surfaces. Except for a magnification factor, $\boldsymbol{b}_{\mathrm{k}}$ is also the *curvature of the fringe observed on a screen*, e.g. a photographic plate, perpendicular to $\boldsymbol{k}$. Because $\boldsymbol{b}_{\mathrm{k}}$ and $\boldsymbol{\nabla}_{\mathrm{k}}D_{\mathrm{R}}$ are both normal to the fringe, we can replace the gradient of D_R by the derivative $dD_R/d\psi$ in a direction perpendicular to $\boldsymbol{m}$, where, if $\Delta D_R = \lambda$, $\Delta\psi$ represents the distance between two fringes on the unit sphere. With the curvature radius $\rho_k = 1/b_k$, we thus have

$$\boldsymbol{m}\cdot(\boldsymbol{V}_k \otimes \boldsymbol{V}_k D_R)\boldsymbol{m} = -\frac{1}{\rho_k}\frac{dD_R}{d\psi}, \tag{5.43}$$

which means that, by measuring the distance between two fringes and the curvature radius (see e.g. [Ref. 4.240, p. 983]), we are able to calculate the component of the tensor $(\boldsymbol{V}_k \otimes \boldsymbol{V}_k D_R)\boldsymbol{K}$ in the direction $\boldsymbol{m}$ tangent to the fringe.

Second, if the fringe pattern shows a singularity around the direction RP (possibly after a modification of the optical arrangement), i.e. if the fringes disappear in the neighborhood of P, as in the middle of Fig. 4.29, we have

$$\boldsymbol{V}_n D_R = \boldsymbol{V}_k D_R = 0,$$

and if $\bar{\mathrm{P}}$ is situated on either asymptote to the approximating indicatrix of the fringe pattern, then also

$$\boldsymbol{e}\cdot(\boldsymbol{V}_n \otimes \boldsymbol{V}_n D_R)\boldsymbol{e} = \boldsymbol{m}\cdot(\boldsymbol{V}_k \otimes \boldsymbol{V}_k D_R)\boldsymbol{m} = 0, \tag{5.44}$$

which again provides information about the second derivative of D_R. For instance, in the particular deformation of a beam or a plate, measuring the direction of these asymptotes and using the relations of elasticity of the preceding sections makes it possible to determine Poisson's ratio [4.60, 64; Ref. 4.184, Sect. III.3.2]. More generally, if we take into account the differential field equations, it should be possible to find the material coefficients for more complicated constitutive equations (e.g. with anisotropy) in one point of the object surface by measuring higher-order derivatives of the optical-path difference.

Let us thus develop the second derivative of D_R and see in general in which way it relates the deformation to the optical quantities. Because the measurement of the curvature $\boldsymbol{b}_k$ seems to be easier than that of $\boldsymbol{b}$ on the object surface, which mostly consists in a virtual image, we shall calculate the derivative $(\boldsymbol{V}_k \otimes \boldsymbol{V}_k D_R)\boldsymbol{K}$ [Ref. 4.197, Sect. VI.3]. Because of the relation (4.26) between $d\boldsymbol{r}$ and $d\boldsymbol{k}$ due to the collineation, we know that, similar to (2.75), the two-dimensional derivative operators are connected by

$$\boldsymbol{V}_k = -L_R \boldsymbol{M}\boldsymbol{V}_n, \tag{5.45}$$

so that, with $\boldsymbol{MN} = \boldsymbol{M}$, $\boldsymbol{MK} = \boldsymbol{K}$ [see (2.71)], we rewrite (4.33) as

$$\boldsymbol{V}_k D_R = -L_R \boldsymbol{M}\boldsymbol{w} + \boldsymbol{K}\boldsymbol{u}, \tag{5.46}$$

where again [see (4.32)]

$$\boldsymbol{w} = (\boldsymbol{V} \otimes \boldsymbol{u})\boldsymbol{g} - \frac{1}{L_S}\boldsymbol{H}\boldsymbol{u}.$$

From there, applying the rules for forming derivatives (2.30), we get

$$\nabla_k \otimes \nabla_k D_R = -\nabla_k L_R \otimes Mw - L_R(\nabla_k \otimes M)Nw - L_R[\nabla_k \otimes (Nw)]M^T + (\nabla_k \otimes K)u + (\nabla_k \otimes u)K. \quad (5.47)$$

Similar to (2.35, 36), and the results of Sect. 2.1.5, we have

$$\nabla_k L_R = -L_R M \nabla_n L_R = L_R M k,$$

$$\nabla_k \otimes K = -K \otimes k - K \otimes k)^T,$$

$$\nabla_k \otimes M = -\frac{1}{n\cdot k}\left[(\nabla_k \otimes n) \otimes k + (\nabla_k \otimes k)|_R \otimes n)^T - \frac{1}{n\cdot k}\nabla_k(n\cdot k)|_R \otimes n \otimes k\right]$$

$$= -\frac{1}{n\cdot k}[L_R MBM^T \otimes k + M \otimes n)^T].$$

With these derivatives, $Kn/n\cdot k = -Mk$ and $K = KM^T$, we obtain

$$(\nabla_k \otimes \nabla_k D_R)K = L_R^2 M\left[\nabla_n \otimes (Nw) + \frac{1}{n\cdot k}B(k\cdot Nw)\right]M^T - K(k\cdot u) - L_R(Mk \otimes Mw + Mw \otimes Mk) - L_R M(\nabla_n \otimes u)KM^T. \quad (5.48)$$

In this form, this result still contains the possibility of generalization through an optical modification (see Sect. 4.3); we simply have to replace u, w by $\tilde{u}$, $\tilde{w}$. In the case of standard holography, we need for the explicit calculation of $\nabla_n \otimes (Nw)$, in addition, the formulae

$$\nabla_n \otimes N = B \otimes n + B \otimes n)^T, \qquad \text{[see (2.54)]}$$

$$\nabla_n \otimes \left(\frac{1}{L_S}H\right) = -\frac{1}{L_S^2}N[h \otimes H + H \otimes h + H \otimes h)^T] = -\frac{1}{L_S^2}N\mathscr{H},$$

so that (5.48) becomes

$$(\nabla_k \otimes \nabla_k D_R)K = M\left\{L_R^2\left[(\nabla_n \otimes \nabla_n \otimes u)g + \frac{1}{n\cdot k}Bk\cdot(\nabla_n \otimes u)g\right] - L_R[k \otimes (\nabla_n \otimes u)g + (\nabla_n \otimes u)g \otimes k + (\nabla_n \otimes u)K + K(\nabla_n \otimes u)^T] + \frac{L_R}{L_S}[k \otimes Hu + Hu \otimes k] - \frac{L_R^2}{L_S}\left[(\nabla_n \otimes u)H + H(\nabla_n \otimes u)^T + \frac{k\cdot Hu}{n\cdot k}B\right] + \frac{L_R^2}{L_S^2}\mathscr{H}u - (k\cdot u)K\right\}M^T.$$

With the decomposition (2.109) of $\boldsymbol{V}_{\mathrm{n}} \otimes \boldsymbol{u}$, the second derivative of $\boldsymbol{u}$ can be expressed by the sum of

$$\begin{aligned}
&\boldsymbol{V}_{\mathrm{n}} \otimes \gamma = \boldsymbol{V}_{\mathrm{n}} \otimes (N\gamma N) = \boldsymbol{V}_{\mathrm{n}} \otimes \gamma)^{\mathrm{T}} N]^{\mathrm{T}} N + \boldsymbol{B}\gamma \otimes \boldsymbol{n})^{\mathrm{T}} + \boldsymbol{B}\gamma \otimes \boldsymbol{n}, \\
&\boldsymbol{V}_{\mathrm{n}} \otimes (\Omega E) = \boldsymbol{V}_{\mathrm{n}}\Omega \otimes E + \Omega \boldsymbol{B}E \otimes \boldsymbol{n})^{\mathrm{T}} - \Omega \boldsymbol{B}E \otimes \boldsymbol{n}, \\
&\boldsymbol{V}_{\mathrm{n}} \otimes (\boldsymbol{\omega} \otimes \boldsymbol{n}) = -\boldsymbol{\kappa} \otimes \boldsymbol{n} + \boldsymbol{B}\boldsymbol{\omega} \otimes \boldsymbol{n} \otimes \boldsymbol{n} - \boldsymbol{B} \otimes \boldsymbol{\omega})^{\mathrm{T}},
\end{aligned}$$

where (2.54, 55), (5.3, 18) have been used. Because the tensor $(\boldsymbol{V}_{\mathrm{n}} \otimes \boldsymbol{V}_{\mathrm{n}} \otimes \boldsymbol{u})\boldsymbol{g}$ is symmetric, we can also write its decomposition in a purely symmetric form; observing furthermore that $\boldsymbol{n}\boldsymbol{M}^{\mathrm{T}} = 0$, we finally obtain

$$\begin{aligned}
(\boldsymbol{V}_{\mathrm{k}} \otimes \boldsymbol{V}_{\mathrm{k}} D_{\mathrm{R}})\boldsymbol{K} = \boldsymbol{M}\Big\{ &-L_{\mathrm{R}}^2(\boldsymbol{n}\cdot\boldsymbol{g})\boldsymbol{\kappa}_{\mathrm{c}} \\
&+ \frac{L_{\mathrm{R}}^2}{2}[[(\boldsymbol{V}_{\mathrm{n}} \otimes \gamma)N\boldsymbol{g}]N + N[(\boldsymbol{V}_{\mathrm{n}} \otimes \gamma)N\boldsymbol{g}]^{\mathrm{T}} \\
&+ \boldsymbol{V}_{\mathrm{n}}\Omega \otimes E\boldsymbol{g} + E\boldsymbol{g} \otimes \boldsymbol{V}_{\mathrm{n}}\Omega - \boldsymbol{B}\boldsymbol{g} \otimes \boldsymbol{\omega} - \boldsymbol{\omega} \otimes \boldsymbol{B}\boldsymbol{g}\big] \\
&- L_{\mathrm{R}}[2\gamma - \gamma\boldsymbol{h} \otimes \boldsymbol{k} - \boldsymbol{k} \otimes \gamma\boldsymbol{h} \\
&- \Omega(E\boldsymbol{h} \otimes \boldsymbol{k} + \boldsymbol{k} \otimes E\boldsymbol{h}) - (\boldsymbol{n}\cdot\boldsymbol{h})(\boldsymbol{\omega} \otimes \boldsymbol{k} + \boldsymbol{k} \otimes \boldsymbol{\omega})] \\
&- \frac{L_{\mathrm{R}}^2}{L_{\mathrm{S}}}[\gamma H + H\gamma + \Omega(EH - HE) + \boldsymbol{\omega} \otimes H\boldsymbol{n} + H\boldsymbol{n} \otimes \boldsymbol{\omega}] \\
&+ \frac{L_{\mathrm{R}}^2}{L_{\mathrm{S}}^2}\mathscr{H}\boldsymbol{u} + \frac{L_{\mathrm{R}}}{L_{\mathrm{S}}}(H\boldsymbol{u} \otimes \boldsymbol{k} + \boldsymbol{k} \otimes H\boldsymbol{u}) - (\boldsymbol{k}\cdot\boldsymbol{u})\boldsymbol{K} \\
&+ \frac{L_{\mathrm{R}}^2}{\boldsymbol{n}\cdot\boldsymbol{k}}\left[\boldsymbol{k}\cdot(\gamma + \Omega E + \boldsymbol{\omega} \otimes \boldsymbol{n})\boldsymbol{g} - \frac{1}{L_{\mathrm{S}}}\boldsymbol{k}\cdot H\boldsymbol{u}\right]\boldsymbol{B}\Big\}\boldsymbol{M}^{\mathrm{T}}. \qquad (5.49)
\end{aligned}$$

The change of curvature of the object surface is thus described by $\boldsymbol{\kappa}_{\mathrm{c}}$, which has been defined by (5.7). To simplify this expression, let us suppose that the object is illuminated with collimated light, i.e. $L_{\mathrm{S}} \to \infty$, and that $\boldsymbol{g}$ is parallel to $\boldsymbol{n}$ at the considered point P. Although $D = \boldsymbol{u}\cdot\boldsymbol{g}$ is then constant at P, the fringe pattern and, in particular, the direction and the curvature of the fringe in P vary if $\boldsymbol{h}$ and $\boldsymbol{k}$ are changed, whereas $\boldsymbol{g} = g\boldsymbol{n}$. With $N\boldsymbol{g} = E\boldsymbol{g} = \boldsymbol{B}\boldsymbol{g} = 0$, we have in this case

$$\begin{aligned}
(\boldsymbol{V}_{\mathrm{k}} \otimes \boldsymbol{V}_{\mathrm{k}} D_{\mathrm{R}})\boldsymbol{K} = \boldsymbol{M}\Big\{ &-L_{\mathrm{R}}^2 g\boldsymbol{\kappa}_{\mathrm{c}} - L_{\mathrm{R}}[2\gamma - \gamma\boldsymbol{h} \otimes \boldsymbol{k} - \boldsymbol{k} \otimes \gamma\boldsymbol{h} \\
&- \Omega(E\boldsymbol{h} \otimes \boldsymbol{k} + \boldsymbol{k} \otimes E\boldsymbol{h}) - (\boldsymbol{n}\cdot\boldsymbol{h})(\boldsymbol{\omega} \otimes \boldsymbol{k} + \boldsymbol{k} \otimes \boldsymbol{\omega})] \\
&- (\boldsymbol{k}\cdot\boldsymbol{u})\boldsymbol{K} + \frac{L_{\mathrm{R}}^2 g}{\boldsymbol{n}\cdot\boldsymbol{k}}(\boldsymbol{k}\cdot\boldsymbol{\omega})\boldsymbol{B}\Big\}\boldsymbol{M}^{\mathrm{T}}. \qquad (5.50)
\end{aligned}$$

If $\boldsymbol{u}$, γ, Ω and ω are known, this together with (5.43) allows us to determine the change of curvature of the object surface by measuring the fringe spacing and the curvature of the fringes with several directions $\boldsymbol{h}$ and $\boldsymbol{k}$.

5.2.2 Extrapolation of Mechanical Quantities into the Interior of a Nontransparent Body

In the case of a transparent body, apart from strain measurements by photoelasticity for example, measurements of the displacement by holographic interferometry can be undertaken directly in the interior of the body (see e.g. [5.20, 21]). In the case of a nontransparent body however, where the displacement field is determined only on its surface, we can utilize the relations of continuum mechanics to extrapolate this information underneath the surface. Two situations must be distinguished here.

First, if certain equations that describe the deformation of the investigated object are available separately from others, for instance in an integrated form and under simplifying assumptions, we shall revert to them. Especially, when the body is thin (plate or shell), the normality condition is often valid so that the strain tensor $\bar{\gamma}$ for an interior surface parallel to the exterior one can be calculated by (5.13), provided that the tensor $\boldsymbol{\kappa}_s$ that describes the curvature change is known (see e.g. [4.144, 145]). In some instances, it can also be advisable to make measurements on the two exterior surfaces of the body (in case they are both attainable by the laser beam) and to interpolate between the obtained values (see also [4.74]).

Second, if no assumption about the deformation is available, we must use the basic field relations of Sect. 5.1 in their differential form and also add the constitutive law, e.g. the relations of elasticity. In an analytical or in a numerical approach, these equations are usually solved for the whole object, i.e. with a closed contour on which either static or kinematic boundary conditions are given. In an experimental approach, we will often want to consider only part of the object surface (e.g. an edge of a very complicated machine). In such a case we have an *open* contour on which the static boundary conditions are given *and* the kinematic quantities are measured. The problem is thus improperly formulated. Nevertheless, an approximate solution can be found by means of finite differences; the error increases however with increased distance from the exterior surface. In addition to the equilibrium equations and Hooke's law, we can use either the three-dimensional compatibility conditions, as proposed by *Dändliker* [Ref. 4.148, Chap. 6], or, in order to keep the distinction between tangential and normal parts, the decomposed kinematic relations. In the latter case, with (5.32, 35, 38, 39), we have the scheme

$$\partial_z \bar{\tau}_z = -\bar{\boldsymbol{V}}_n \bar{\tau} N + \bar{\boldsymbol{B}} \bar{\tau}_z + 2\bar{H} \bar{\tau}_z,$$

$$\partial_z \bar{\sigma}_z = -\bar{\boldsymbol{B}} \cdot \bar{\tau} - \bar{\boldsymbol{V}}_n \cdot \bar{\tau}_z + 2\bar{H} \bar{\sigma}_z,$$

$$\begin{aligned}
\partial_z \bar{v} &= -\bar{\boldsymbol{B}}\bar{v} - \bar{\boldsymbol{V}}_{\mathrm{n}}\bar{w} + \frac{1}{G}\bar{\tau}_z, \\
\partial_z \bar{w} &= \frac{1}{E}(\bar{\sigma}_z - \nu \operatorname{tr} \bar{\tau}), \\
\bar{\gamma} &= \frac{1}{2}[(\bar{\boldsymbol{V}}_{\mathrm{n}} \otimes \bar{v})\boldsymbol{N} + \boldsymbol{N}(\bar{\boldsymbol{V}}_{\mathrm{n}} \otimes \bar{v})^{\mathrm{T}}] - \bar{\boldsymbol{B}}\bar{w}, \\
\bar{\tau} &= \frac{E}{1 - \nu^2}(\bar{\gamma} - \nu \boldsymbol{E}\bar{\gamma}\boldsymbol{E}) + \frac{\nu}{1 - \nu}\bar{\sigma}_z \boldsymbol{N}.
\end{aligned} \tag{5.51}$$

By calculating $\bar{\tau}_z + \Delta\bar{\tau}_z$, $\bar{\sigma}_z + \Delta\bar{\sigma}_z$, $\bar{v} + \Delta\bar{v}$, $\bar{w} + \Delta\bar{w}$ from the first four relations, we can thus pass from a surface with coordinate z to a neighboring one with the coordinate $z + \Delta z$. It must be noted that the tangential derivatives (with $\bar{\boldsymbol{V}}_{\mathrm{n}}$) are also approximated by finite differences, so that, because of the limitation of the open contour, we can enter into the object only within a conical domain, the aperture of which depends on the steps chosen for the calculation of the derivatives with $\bar{\boldsymbol{V}}_{\mathrm{n}} \ldots$ and with $\partial_z \ldots$

5.3 Conclusion

In the present chapter, we first saw the relations of continuum mechanics that we have at our disposal in order to obtain more knowledge about the deformation. Let us recall that those relations are valid only as long as the strains, and also the rotations, are small and if the derivative of the displacement vector exists, i.e. if there is no dislocation. Then, in the last section, we sketched how holographic interferometry enables us to measure not only the displacement vector and the strain tensor, but also second derivatives of the displacement and particularly curvature changes and material coefficients. However, in this domain, as in others, much remains still to be discovered.

References

Chapter 2. Some Basic Concepts of Differential Geometry and Continuum Mechanics

2.1 R. Abraham: *Linear and Multilinear Algebra* (W. A. Benjamin, New York 1966)
2.2 D. C. Leigh: *Nonlinear Continuum Mechanics* (McGraw-Hill, New York 1968)
2.3 C. C. Wang, C. Truesdell: *Introduction to Rational Elasticity* (Noordhoff, Leyden 1973)
2.4 P. Villagio: *Qualitative Methods in Elasticity* (Noordhoff, Leyden 1977)
2.5 A. E. Green, W. Zerna: *Theoretical Elasticity,* 2nd ed. (Clarendon Press, Oxford 1968)
2.6 L. Brillouin: *Les tenseurs en mécanique et en élasticité,* 2nd ed. (Masson, Paris 1949)
2.7 I. S. Sokolnikoff: *Tensor Analysis: Theory and Applications* (John Wiley, New York 1951)
2.8 H. Guggenheimer: *Differential Geometry* (McGraw-Hill, New York 1963)
2.9 D. Laugwitz: *Differentialgeometrie,* 2nd ed. (Teubner, Stuttgart 1968)
2.10 A. E. Green, J. E. Adkins: *Large Elastic Deformations* (Clarendon Press, Oxford 1960)
2.11 W. Prager: *Einführung in die Kontinuumsmechanik* (Birkhäuser, Basel 1961)
2.12 A. C. Eringen: *Nonlinear Theory of Continuous Media* (McGraw-Hill, New York 1962)
2.13 A. C. Eringen: *Mechanics of Continua* (John Wiley, New York 1967)
2.14 C. Truesdell, W. Noll: "The Non-Linear Field Theories of Mechanics", in *Encyclopedia of Physics,* ed. by S. Flügge, Vol. 3/3 (Springer, Berlin, Heidelberg, New York 1965)
2.15 C. Truesdell: *The Elements of Continuum Mechanics* (Springer, Berlin, Heidelberg, New York 1966)

Chapter 3. Principles of Image Formation in Holography

3.1 D. Gabor: "Twenty-five Years of Holography", in *Holography in Medicine,* Proc. Int. Symp. on Holography in Biomedical Sciences, New York 1973, ed. by P. Greguss (IPC Science and Technology Press 1975) pp. 11–16
3.2 D. Gabor: A new microscopic principle. Nature *161,* 777–778 (1948)
3.3 D. Gabor: Microscopy by reconstructed wave-fronts. Proc. Roy. Soc. (London) *A197,* 454–487 (1949)
3.4 D. Gabor: Microscopy by reconstructed wave-fronts II. Proc. Phys. Soc. (London) *B64,* 449–469 (1951)
3.5 M. Born, E. Wolf: *Principles of Optics,* 5th ed. (Pergamon Press, Oxford 1975)
3.6 G. W. Stroke: *An Introduction to Coherent Optics and Holography* (Academic Press, New York, London 1966)
3.7 J. B. de Velis, G. O. Reynolds: *Theory and Applications of Holography* (Addison-Wesley, Reading, Mass. 1967)
3.8 J. W. Goodman: *Introduction to Fourier Optics* (McGraw-Hill, San Francisco 1968)
3.9 M. Françon: *Holographie* (Masson, Paris 1969)
3.10 H. Kiemle, D. Röss: *Einführung in die Technik der Holographie,* Technisch-Physikalische Sammlung, Bd. 8 (Akad. Verlagsgesellschaft, Frankfurt am Main 1969)
3.11 H. Lenk: *Holographie,* Fortschr. Exp. Theor. Biophys., Heft 9 (VEB Georg Thieme, Leipzig 1969)
3.12 H. J. Caulfield, S. Lu: *The Applications of Holography* (John Wiley, New York 1970)
3.13 J. N. Butters: *Holography and Its Technology* (Institution of Electrical Engineers, Peregrinus Ltd, London 1971)
3.14 R. J. Collier, C. B. Burckhardt, L. H. Lin: *Optical Holography* (Academic Press, New York, London 1971)
3.15 J.-Ch. Viénot, P. Smigielski, H. Royer: *Holographie optique* (Dunod, Paris 1971)

3.16 G. Groh: *Holographie, physikalische Grundlagen und Anwendungen* (Berliner Union, Stuttgart; W. Kohlhammer, Stuttgart 1973)
3.17 E. Menzel, W. Mirandé, I. Weingärtner: *Fourier-Optik und Holographie* (Springer, Wien, New York 1973)
3.18 F. T. S. Yu: *Introduction to Diffraction, Information Processing, and Holography* (MIT Press, Cambridge 1973)
3.19 W. E. Kock: *Engineering Applications of Lasers and Holography,* Optical Physics and Engineering, ed. by W. L. Wolfe, Vol. 3 (Plenum Press, New York 1975)
3.20 H. M. Smith: *Principles of Holography,* 2nd ed. (John Wiley, New York 1975)
3.21 R. K. Erf (ed.): *Holographic Nondestructive Testing* (Academic Press, New York, London 1974)
3.22 K. Biedermann: Information storage materials for holography and optical data processing. Opt. Acta *22,* 103–124 (1975)
3.23 H. M. Smith (ed.): *Holographic Recording Materials,* Topics in Applied Physics, Vol. 20 (Springer, Berlin, Heidelberg, New York 1977)
3.24 R. Dändliker: *Laser-Kurzlehrgang* (Fachschriftenverlag Aargauer Tagblatt, Aarau 1971)
3.25 M. Young: *Optics and Lasers. An Engineering Physics Approach,* Springer Series in Optical Sciences, ed. by D. L. MacAdam, Vol. 5 (Springer, Berlin, Heidelberg, New York 1977)
3.26 W. Lukosz: Equivalent-lens theory of holographic imaging. J. Opt. Soc. Am. *58,* 1084–1091 (1968)
3.27 W. Lukosz: "Physikalische Optik I und II. Vorlesung an der ETH Zürich", Lecture notes (1973)
3.28 E. N. Leith, J. Upatnieks: Reconstructed wavefronts and communication theory. J. Opt. Soc. Am. *52,* 1123–1130 (1962)
3.29 E. N. Leith, J. Upatnieks: Wavefront reconstruction with continuous-tone objects. J. Opt. Soc. Am. *53,* 1377–1381 (1963)
3.30 E. N. Leith, J. Upatnieks: Wavefront reconstruction with diffused illumination and three-dimensional objects. J. Opt. Soc. Am. *54,* 1295–1301 (1964)
3.31 A. Kozma: Photographic recording of spatially modulated coherent light. J. Opt. Soc. Am. *56,* 428–432 (1966)
3.32 R. F. VanLigten: Influence of photographic film on wavefront reconstruction. I: plane wavefronts. J. Opt. Soc. Am. *56,* 1–9 (1966)
3.33 R. F. VanLigten: Influence of photographic film on wavefront reconstruction. II: "cylindrical" wavefronts. J. Opt. Soc. Am. *56,* 1009–1014 (1966)
3.34 R. Dändliker, B. Ineichen: Nonlinear cross-talk in two-reference-beam holographic interferometry. Opt. Commun. *19,* 365–369 (1976)
3.35 D. Dameron, C. M. Vest: Fringe sharpening and diffraction in nonlinear two-exposure holographic interferometry. J. Opt. Soc. Am. *66,* 1418–1421 (1976)
3.36 J. Katz: "Non-Linear Effects in Holographic Interferometry and Their Influence on Measurements Accuracy", M. S. thesis, Tel Aviv Univ. (1976)
3.37 E. Marom, J. Katz: Unconventional interferometric realizations based on holographic nonlinear effects. Appl. Opt. *16,* 1400–1403 (1977)
3.38 R. Dändliker, B. Ineichen, J. Katz, E. Marom: Effects of nonlinear recording in holographic interferometry. (to be published)
3.39 I. Přikryl: A contribution to hologram imagery. Opt. Acta *21,* 517–528 (1974)
3.40 Y. Belvaux: Influence de divers paramètres d'enregistrement lors de la restitution d'un hologramme. Nouv. Rev. Opt. *6,* 137–147 (1975)
3.41 C. Durou, J.-P. Hot, R. Lefèvre: Effets simultanés du tassement et de la diminution d'indice de réfraction de l'émulsion photographique en holographie. Nouv. Rev. Opt. *7,* 87–94 (1976)
3.42 M. Lurie: Effects of partial coherence on holography with diffuse illumination. J. Opt. Soc. Am. *56,* 1369–1372 (1966)
3.43 F. M. Mottier, R. Dändliker, B. Ineichen: Relaxation of the coherence requirements in holography. Appl. Opt. *12,* 243–248 (1973)

3.44 D. Gabor, G. W. Stroke, R. Restrick, A. Funkhouser, D. Brumm: Optical image synthesis (complex amplitude addition and substraction) by holographic Fourier transformation. Phys. Lett. *18*, 116–118 (1965)
3.45 G. M. Brown, R. M. Grant, G. W. Stroke: Theory of holographic interferometry. J. Acoust. Soc. Am. *45*, 1166–1179 (1969)
3.46 J. A. Armstrong: Fresnel holograms: their imaging properties and aberrations. IBM J. Res. Dev. *9*, 171–178 (1965)
3.47 E. N. Leith, J. Upatnieks, K. A. Haines: Microscopy by wavefront reconstruction. J. Opt. Soc. Am. *55*, 981–986 (1965)
3.48 M. Marquet, H. Royer: Etude des aberrations géométriques des images reconstituées par holographie. C. R. Acad. Sci. Paris *260*, 6051–6053 (1965)
3.49 R. W. Meier: Magnification and third-order aberrations in holography. J. Opt. Soc. Am. *55*, 987–992 (1965)
3.50 R. W. Meier: Cardinal points and the novel imaging properties of a holographic system. J. Opt. Soc. Am. *56*, 219–223 (1966)
3.51 D. B. Neumann: Geometrical relationships between the original object and the two images of a hologram reconstruction. J. Opt. Soc. Am. *56*, 858–861 (1966)
3.52 E. B. Champagne: Nonparaxial imaging, magnification and aberration properties in holography. J. Opt. Soc. Am. *57*, 51–55 (1967)
3.53 E. B. Champagne, N. G. Massey: Resolution in holography. Appl. Opt. *8*, 1879–1885 (1969)
3.54 K. A. Stetson: An analysis of the properties of total internal reflection holograms. Optik *29*, 520–536 (1969)
3.55 W. Richter: Theorie der Queraberrationen holographischer Abbildungen. Opt. Acta *17*, 285–298 (1970)
3.56 E. Storck: "Exakte Abbildungsgleichungen für ausgedehnte Flächenhologramme", in *Applications de l'Holographie*, Proc. Symp. Besançon 1970, ed. by J.-Ch. Viénot, J. Bulabois, J. Pasteur (Univ. de Besançon 1970) paper 1.3
3.57 J. N. Latta: Computer-based analysis of hologram imagery and aberrations. I: hologram types and their nonchromatic aberrations. II: aberrations induced by a wavelength shift. Appl. Opt. *10*, 599–608, 609–618 (1971)
3.58 M. Miler: Projective properties of holographic imaging. Opt. Acta *19*, 555–568 (1972)
3.59 J. F. Miles: Imaging and magnification properties in holography. Opt. Acta *19*, 165–186 (1972)
3.60 J. F. Miles: Evaluation of the wavefront aberration in holography. Opt. Acta *20*, 19–31 (1973)
3.61 G. N. Buinov, R. K. Gizatullin, K. S. Mustafin: Effect of hologram aberrations on image quality. Opt. Spectrosc. *34*, 443–446 (1973)
3.62 G. N. Buinov, K. S. Mustafin: Method of compensating for the spherical aberration of axial hologram lenses. Opt. Spectrosc. *41*, 196–197 (1976)
3.63 P. Hariharan: Longitudinal distortion in images reconstructed by reflection holograms. Opt. Commun. *17*, 52–54 (1976)
3.64 W. Schumann, M. Dubas: On the motion of holographic images caused by movements of the reconstruction light source, with the aim of application to deformation analysis. Optik *46*, 377–392 (1976)
3.65 R. W. Smith: Astigmatism free holographic lens elements. Opt. Commun. *21*, 102–105 (1977)
3.66 R. W. Smith: The s and t formulae for holographic lens elements. Opt. Commun. *21*, 106–109 (1977)
3.67 H. H. Hopkins: *Wave Theory of Aberrations* (Clarendon Press, Oxford 1950)
3.68 A. Maréchal: "Optique géométrique générale", in *Encyclopedia of Physics*, ed. by S. Flügge, Vol. 24 (Springer, Berlin, Göttingen, Heidelberg 1956) pp. 44–170
3.69 W. Lukosz, A. Wüthrich: Holography with evanescent waves. I: theory of the diffraction efficiency for s-polarized light. Optik *41*, 191–211 (1974)

3.70 A. Wüthrich, W. Lukosz: Holographie mit quergedämpften Wellen. II: experimentelle Untersuchungen der Beugungswirkungsgrade. Optik *42*, 315–334 (1975)
3.71 C. W. Helstrom: Image luminance and ray tracing in holography. J. Opt. Soc. Am. *56*, 433–441 (1966)
3.72 W. C. Sweatt: Describing holographic optical elements as lenses. J. Opt. Soc. Am. *67*, 803–808 (1977)
3.73 A. Offner: Ray tracing through a holographic system. J. Opt. Soc. Am. *56*, 1509–1512 (1966)
3.74 I. A. Abramowitz, J. M. Ballantyne: Evaluation of hologram aberrations by ray tracing. J. Opt. Soc. Am. *57*, 1522–1526 (1967)
3.75 I. A. Abramowitz: Evaluation of hologram imaging by ray tracing. Appl. Opt. *8*, 403–410 (1969)
3.76 J. N. Latta: Computer-based analysis of holography using ray tracing. Appl. Opt. *10*, 2698–2710 (1971)
3.77 I. Přikryl: Studying hologram imagery by a ray-tracing method. Opt. Acta *19*, 623–631 (1972)
3.78 H. H. Arsenault: Geometrical optics of holograms. J. Opt. Soc. Am. *65*, 903–908 (1975)
3.79 A. Maréchal: "Etude des effets combinés de la diffraction et des aberrations géométriques sur l'image d'un point lumineux", Thesis, Université de Paris (1948)
3.80 M. Françon: "Interférences, diffraction et polarisation", in *Encyclopedia of Physics*, ed. by S. Flügge, Vol. 24 (Springer, Berlin, Göttingen, Heidelberg 1956) pp. 171–460
3.81 M. Dubas, W. Schumann: Contribution à l'étude théorique des images et des franges produites par deux hologrammes en sandwich. Opt. Acta *24*, 1193–1209 (1977)
3.82 W. Schumann, M. Dubas: "Holographic Interferometry with the Possibility of Modifying the Fringes During Reconstruction", in *Proc. 1st Europ. Conf. on Opt. Appl. to Metrol.*, Strasbourg 1977, ed. by M. Grosmann, P. Meyrueis, SPIE, Vol. 136 (SPIE, Washington 1978) pp. 174–180
3.83 N. Abramson: Sandwich hologram interferometry: a new dimension in holographic comparison. Appl. Opt. *13*, 2019–2025 (1974)

Chapter 4. Fringe Interpretation in Holographic Interferometry

4.1 R. E. Brooks, L. O. Heflinger, R. F. Wuerker: Interferometry with a holographically reconstructed comparison beam. Appl. Phys. Lett. *7*, 248–249 (1965)
4.2 J. M. Burch: The application of lasers in production engineering. Prod. Eng. *44*, 431–443 (1965)
4.3 R. J. Collier, E. T. Doherty, K. S. Pennington: Application of moiré techniques to holography. Appl. Phys. Lett. *7*, 223–225 (1965)
4.4 D. Gabor, G. W. Stroke, D. Brumm, A. Funkhouser, A. Labeyrie: Reconstruction of phase objects by holography. Nature *208*, 1159–1162 (1965)
4.5 M. H. Horman: An application of wavefront reconstruction to interferometry. Appl. Opt. *4*, 333–336 (1965)
4.6 R. L. Powell, K. A. Stetson: Interferometric vibration analysis by wavefront reconstruction. J. Opt. Soc. Am. *55*, 1593–1598 (1965)
4.7 K. A. Stetson, R. L. Powell: Interferometric hologram evaluation and real-time vibration analysis of diffuse objects. J. Opt. Soc. Am. *55*, 1694–1695 (1965)
4.8 J. M. Burch, A. E. Ennos, R. J. Wilton: Dual- and multiple-beam interferometry by wavefront reconstruction. Nature *209*, 1015–1016 (1966)
4.9 K. A. Haines, B. P. Hildebrand: Surface-deformation measurement using the wavefront reconstruction technique. Appl. Opt. *5*, 595–602 (1966)
4.10 L. O. Heflinger, R. F. Wuerker, R. E. Brooks: Holographic interferometry. J. Appl. Phys. *37*, 642–649 (1966)
4.11 B. P. Hildebrand, K. A. Haines: Interferometric measurements using the wavefront reconstruction technique. Appl. Opt. *5*, 172–173 (1966)

4.12 H. Nassenstein: Holographische Interferometrie diffus reflektierender Objekte. Phys. Lett. *21*, 290–291 (1966)

4.13 K. A. Stetson, R. L. Powell: Hologram interferometry. J. Opt. Soc. Am. *56*, 1161–1166 (1966)

4.14 G. W. Stroke, A. E. Labeyrie: Two-beam interferometry by successive recording of intensities in a single hologram. Appl. Phys. Lett. *8*, 42–44 (1966)

4.15 E. Archbold, A. E. Ennos: "Techniques of Hologram Interferometry for Engineering Inspection and Vibration Analysis", in *The Engineering Uses of Holography*, Proc. Symp. Glasgow 1968, ed. by E. R. Robertson, J. M. Harvey (University Press, Cambridge 1970) pp. 381–396

4.16 B. M. Watrasiewicz, P. Spicer: Vibration analysis by stroboscopic holography. Nature *217*, 1142–1143 (1968)

4.17 J. Cl. Binder: "Appareillage en interférométrie holographique: enregistreur séquentiel d'hologrammes en double exposition; dispositifs pour l'étude de déformations", Thesis, Université de Besançon (1971)

4.18 H. J. Caulfield: Multiplexing double exposure holographic interferograms. Appl. Opt. *11*, 2711–2712 (1972)

4.19 R. Dändliker, E. Marom, F. M. Mottier: Wavefront sampling in holographic interferometry. Opt. Commun, *6*, 368–371 (1972)

4.20 P. Hariharan, Z. S. Hegedus: Simple multiplexing technique for double-exposure hologram interferometry. Opt. Commun. *9*, 152–155 (1973)

4.21 T. R. Hsu, R. Lewak: Measurements of thermal distortion of composite plates by holographic interferometry. Exp. Mech. *16*, 182–187 (1976)

4.22 R. F. Wuerker, L. O. Heflinger: "Pulsed Laser Holography", in *The Engineering Uses of Holography*, Proc. Symp. Glasgow 1968, ed. by E. R. Robertson, J. M. Harvey (University Press, Cambridge 1970) pp. 99–114

4.23 J. Fujimoto, M. Yoneyama: Short exposure and biasing holography. Appl. Opt. *16*, 811–812 (1977)

4.24 J.-Ch. Viénot: Sur quelques essais d'interprétation quantitative des hologrammes dans l'étude des contraintes. Nouv. Rev. Opt. *1*, 91–106 (1970)

4.25 H. Rottenkolber: "Holographie 73", Teil 1. Laser Elektro-Opt. *5*, 27–29, n°1 (1973)

4.26 J. Ebbeni: Comparaison des différentes méthodes d'interférométrie holographique. Mécanique Matér. Elec. *57*, 30–36, n°291 (1974)

4.27 J. M. Burch: Holographic measurement of displacement and strain, an introduction. J. Strain Anal. *9*, 1–3 (1974)

4.28 J. M. Burch: "Outlines of Optical Metrology", in *The Engineering Uses of Coherent Optics*, Proc. Symp. Glasgow 1975, ed. by E. R. Robertson (University Press, Cambridge 1976) pp. 1–22

4.29 P. M. Boone: "Measurement of Displacement, Strain and Stress by Holography", in *The Engineering Uses of Coherent Optics*, Proc. Symp. Glasgow 1975, ed. by E. R. Robertson (University Press, Cambridge 1976) pp. 81–98

4.30 J. D. Briers: The interpretation of holographic interferograms. Opt. Quantum Electron. *8*, 469–501 (1976)

4.31 A. Felske: Anwendungen von Lasern in der Meßtechnik, eine Übersicht. Techn. Messen atm *43*, 293–304 (1976)

4.32 D. C. Holloway, W. F. Ranson, C. E. Taylor: A neoteric interferometer for use in holographic interferometry. Exp. Mech. *12*, 461–465 (1972)

4.33 J. E. Sollid: Holographic interferometry applied to measurements of small static displacements of diffusely reflecting surfaces. Appl. Opt. *8*, 1587–1595 (1969)

4.34 J. E. Sollid: "A Comparison of Two Methods of Measuring In-Plane Surface Displacements Using Holographic Interferometry", in *Applications de l'Holographie*, Proc. Symp. Besançon 1970, ed. by J.-Ch. Viénot, J. Bulabois, J. Pasteur (Univ. de Besançon 1970) paper 4.8

4.35 J. E. Sollid: Translational displacements versus deformation, displacements in double-exposure holographic interferometry. Opt. Commun. *2*, 282–288 (1970)

4.36 A. E. Ennos: Measurement of in-plane surface strain by hologram-interferometry. J. Sci. Instrum. *1*, 731–734 (1968)
4.37 C. M. Vest: Comment on: holographic interferometry applied to measurements of small static displacements of diffusely reflecting surfaces. Appl. Opt. *12*, 612–613 (1973)
4.38 D. Bijl, R. Jones: "On Tri-Hologram Interferometry and the Measurement of Elastic Anisotropies Using an Adapted Form of Cornu's Technique", in *Applications de l'Holographie,* Proc. Symp. Besançon 1970, ed. by J.-Ch. Viénot, J. Bulabois, J. Pasteur (Univ. de Besançon 1970) paper 5.3
4.39 K. Shibayama, H. Uchiyama: Measurement of three-dimensional displacements by hologram interferometry. Appl. Opt. *10*, 2150–2154 (1971)
4.40 Y. Y. Hung, C. P. Hu, D. R. Henley, C. E. Taylor: Two improved methods of surface-displacement measurements by holographic interferometry. Opt. Commun. *8*, 48–51 (1973)
4.41 M. R. Wall: "Zero Motion Fringe Identification", in *Applications de l'Holographie,* Proc. Symp. Besançon 1970, ed. by J.-Ch. Viénot, J. Bulabois, J. Pasteur (Univ. de Besançon 1970) paper 4.9
4.42 U. Köpf: Fringe order determination and zero motion fringe identification in holographic displacement measurements. Opt. Laser Techn. *5*, 111–113 (1973)
4.43 N. Abramson: The holo-diagram. V: a device for practical interpreting of hologram interference fringes. Appl. Opt. *11*, 1143–1147 (1972)
4.44 C. A. Sciammarella, J. A. Gilbert: Strain analysis of a disk subjected to diametral compression by means of holographic interferometry. Appl. Opt. *12*, 1951–1956 (1973)
4.45 W. M. Ewers, W. Fritzsch, K. Grünewald, H. Wachutka: Bestimmung dreidimensionaler Verformungsfelder mit Hilfe der holographischen Interferometrie. Optik *40*, 57–68 (1974)
4.46 T. Matsumoto, K. Iwata, R. Nagata: Measurement of deformation in a cylindrical shell by holographic interferometry. Appl. Opt. *13*, 1080–1084 (1974)
4.47 E. B. Aleksandrov, A. M. Bonch-Bruevich: Investigation of surface strains by the hologram technique. Sov. Phys.-Tech. Phys. *12*, 258–265 (1967)
4.48 P. W. King III: Holographic interferometry technique utilizing two plates and relative fringe orders for measuring microdisplacements. Appl. Opt. *13*, 231–233 (1974)
4.49 J. W. C. Gates: Holographic measurement of surface distortion in three dimensions. Opt. Technol. *1*, 247–250 (1969)
4.50 H. Kohler: Untersuchungen zur quantitativen Analyse der holographischen Interferometrie. Optik *39*, 229–235 (1974)
4.51 H. Kohler: Interferenzliniendynamik bei der quantitativen Auswertung holographischer Interferogramme. Optik *47*, 9–24 (1977)
4.52 H. Kohler: Ein neues Verfahren zur quantitativen Auswertung holographischer Interferogramme. Optik *47*, 135–152 (1977)
4.53 H. Kohler: Zur Vermessung holographisch-interferometrischer Verformungsfelder. Optik *47*, 271–282 (1977)
4.54 H. Kohler: General formulation of the holographic-interferometric evaluation methods. Optik *47*, 469–475 (1977)
4.55 H. E. Gascoigne: "Fringe Formation and Interpretation in Holographic Interferometry", Final Report, Grant GK-5423 (University of Utah 1972)
4.56 J. N. Butters: "Application of Holography to Instrument Diaphragm Deformations and Associated Topics", in *The Engineering Uses of Holography,* Proc. Symp. Glasgow 1968, ed. by E. R. Robertson, J. M. Harvey (University Press, Cambridge 1970) pp. 151–169
4.57 J. Der Hovanesian, J. Varner: "Methods for Determining the Bending Moments in Normally Loaded Thin Plates by Hologram Interferometry", in *The Engineering Uses of Holography,* Proc. Symp. Glasgow 1968, ed. by E. R. Robertson, J. M. Harvey (University Press, Cambridge 1970) pp. 173–184
4.58 I. K. Leadbetter, T. Allan: "Holographic Examination of the Prebuckling Behaviour of Axially Loaded Cylinders", in *The Engineering Uses of Holography,* Proc. Symp. Glasgow 1968, ed. by E. R. Robertson, J. M. Harvey (University Press, Cambridge 1970) pp. 185–195

4.59 T. D. Dudderar: Applications of holography to fracture mechanics. Exp. Mech. *9,* 281–285 (1969)
4.60 I. Yamaguchi, H. Saito: Application of holographic interferometry to the measurement of Poisson's ratio. Jpn. J. Appl. Phys. *8,* 768–771 (1969)
4.61 I. K. Leadbetter: "Holography in Engineering Practice", in *Applications de l'Holographie,* Proc. Symp. Besançon 1970, ed. by J.-Ch. Viénot, J. Bulabois, J. Pasteur (Univ. de Besançon 1970) paper 17.1
4.62 T. Tsuruta, Y. Itoh: "Holographic Interferometry for Rotating Subject", in *Applications de l'Holographie,* Proc. Symp. Besançon 1970, ed. by J.-Ch. Viénot, J. Bulabois, J. Pasteur (Univ. de Besançon 1970) paper 17.6
4.63 R. M. Grant, J. F. Delorme: "Essais non-destructifs par holographie", in *Applications de l'Holographie,* Proc. Symp. Besançon 1970, ed. by J.-Ch. Viénot, J. Bulabois, J. Pasteur (Univ. de Besançon 1970) paper 17.8
4.64 H. Saito, I. Yamaguchi, T. Nakajima: "Application of Holographic Interferometry to Mechanical Experiments", in *Applications of Holography,* ed. by E. S. Barrekette, T. J. Watson, W. E. Kock, T. Ose, J. Tsujiuchi, G. W. Stroke (Plenum Press, New York, London 1971) pp. 105–126
4.65 T. R. Hsu, R. G. Moyer: Application of holography in high-temperature displacement measurements. Exp. Mech. *12,* 431–432 (1972)
4.66 N. H. Abramson, H. Bjelkhagen: Industrial holographic measurements. Appl. Opt. *22,* 2792–2796 (1973)
4.67 F. Albe, P. Smigielski, H. Fagot: Application effective de l'interférométrie holographique par double exposition à l'étude des déformations de céramiques dues à l'impact d'un projectile. Opt. Commun. *8,* 369–371 (1973)
4.68 K. Grünewald, D. Kaletsch, V. Lehmann, H. Wachutka: Holographische Interferometrie und deren quantitative Auswertung, demonstriert am Beispiel zylindrischer GfK-Rohre. Optik *37,* 102–109 (1973)
4.69 N. L. Hecht, J. E. Minardi, D. Lewis, R. L. Fusek: Quantitative theory for predicting fringe pattern formation in holographic interferometry. Appl. Opt. *12,* 2665–2676 (1973)
4.70 H. Michael: Quantitative deformation measurements of diffusely reflecting objects with the aid of holography. Appl. Opt. *12,* 1111–1113 (1973)
4.71 H. Steinbichler: "Holographie 73", Teil 2: qualitative Beurteilung von Konstruktionsmerkmalen mit der holographischen Interferometrie. Laser Elektro-Optik *5,* 9–11, n°2 (1973)
4.72 A. Ajovalasit, S. Carollo, M. Tschinke: "Holographic Analysis of Thermal Deformations in a Bimetallic Cylindrical Joint", in *Proc. of the 5th Intern. Conf. on Exp. Stress Anal.,* Udine 1974 (CISM, Udine 1974) pp. 4.59–65
4.73 A. I. Tselikov, B. A. Morosov, I. M. Makeev, A. W. Sergeev, A. I. Surkov: "Study of Metal Plastic Working Processes and of Deformations of Metallurgical Machine Components with the Application of a Holographic Interferometry Technique", in *Proc. of the 5th Intern. Conf. on Exp. Stress Anal.,* Udine 1974 (CISM, Udine 1974) pp. 4.72–79
4.74 P. Hariharan, Z. S. Hegedus: Measurement of symmetrical and antisymmetrical deformations by hologram interferometry. Opt. Commun. *11,* 127–131 (1974)
4.75 H. J. Raterink, R. L. van Renesse: Investigation of holographic interferometry, applied for the detection of cracks in large metal objects. Optik *40,* 193–200 (1974)
4.76 H. Spetzler, C. H. Scholz, C. P. J. Lu: Strain and creep measurements on rocks by holographic interferometry. Pure Appl. Geophys. *112/3,* 571–582 (1974)
4.77 H. Spetzler, M. Meyer, G. Boucher: Thermal expansion measurements on molten levitated aluminum by holographic interferometry. High Temp. High Pressures *6,* 529–532 (1974)
4.78 R. K. Erf, R. M. Gagosz, J. P. Waters: "Holography in a Factory Environment", in *The Engineering Uses of Coherent Optics,* Proc. Symp. Glasgow 1975, ed. by E. R. Robertson (University Press, Cambridge 1976) pp. 23–46
4.79 E. R. Robertson, J. Der Hovanesian, W. King: "The Application of Holography to the Membrane Analogy for Torsion", in *The Engineering Uses of Coherent Optics,* Proc.

Symp. Glasgow 1975, ed. by E. R. Robertson (University Press, Cambridge 1976) pp. 47–58
4.80 E. Archbold: "The History of a Holographic Non-Destructive Test Procedure", in *The Engineering Use of Coherent Optics*, Proc. Symp. Glasgow 1975, ed. by E. R. Robertson (University Press, Cambridge 1976) pp. 59–72
4.81 E. Roth: "Holographic Interferometric Investigations of Plastic Parts", in *Laser 75 Optoelectronics*, Proc. Conf. Munich 1975, ed. by W. Waidelich (IPC Science and Technology Press, Guilford 1976) pp. 168–174
4.82 F. Pejša, M. Dittrich: "Holographische Konstruktionsoptimierung", in *Laser 75 Optoelectronics*, Proc. Conf. Munich 1975, ed. by W. Waidelich (IPC Science and Technology Press, Guilford 1976) pp. 195–197
4.83 M. C. Collins, C. E. Watterson: Surface-strain measurements on a hemispherical shell using holographic interferometry. Exp. Mech. *15*, 128–132 (1975)
4.84 K. Grünewald, W. Fritzsch, H. Wachutka: Quantitative holographische Verformungsmessungen an Kunststoff- und GFK-Bauteilen. Materialprüf. *17*, 69–72 (1975)
4.85 A. R. Hunter, T. M. Morton: Application of holographic interferometry to predict long-time torsional relaxation. Exp. Mech. *15*, 153–160 (1975)
4.86 F. Mayinger, W. May: Zerstörungsfreie Materialprüfung mit Hilfe der optischen Holographie. VGB Kraftwerkstechnik *56*, 334–340 (1976)
4.87 H. D. Meyer, H. A. Spetzler: Material properties using holographic interferometry. Exp. Mech. *16*, 434–438 (1976)
4.88 A. I. Tselikov, B. A. Morozov, O. G. Lisin, V. S. Aistov: Use of holographic interferometry for studying elastic and plastic deformation. Strength Mater. *8*, 728–733 (1976)
4.89 Y. Katzir, A. A. Friesem, Z. Rav-Noy: "Real and Non-Real Time Holographic Non-Destructive Testing", in *Applications of Holography and Optical Data Processing*, Proc. Conf. Jerusalem 1976, ed. by E. Marom, A. A. Friesem, E. Wiener-Avnear (Pergamon Press, Oxford, New York 1977) pp. 279–288
4.90 G. Von Bally: "Holographic Analysis of Tympanic Membrane Vibrations in Human Temporal Bone Preparations Using a Double Pulsed Ruby Laser System", in *Applications of Holography and Optical Data Processing*, Proc. Conf. Jerusalem 1976, ed. by E. Marom, A. A. Friesem, E. Wiener-Avnear (Pergamon Press, Oxford, New York 1977) pp. 593–602
4.91 D. C. Holloway, A. M. Patacca, W. L. Fourney: Application of holographic interferometry to a study of wave propagation in rock. Exp. Mech. *17*, 281–289 (1977)
4.92 H. K. Liu, R. L. Kurtz: A practical method for holographic interference fringe assessment. Opt. Eng. *16*, 176–186 (1977)
4.93 G. Ramachandra Reddy, V. Venkateswara Rao: Determination of amplitudes and phase of a vibration by double exposure holography. Atti Fond. Giorgio Ronchi *32*, 32–35 (1977)
4.94 G. Wernicke: Ermittlung von Verformungszuständen mit der holographischen Interferometrie. Feingerätetechnik *26*, 79–82 (1977)
4.95 W. Schmidt, A. Vogel, D. Preußler: Holographic contour mapping using a dye laser. Appl. Phys. *1*, 103–109 (1973)
4.96 M. Grosmann, P. Meyrueis: "Dimensional Metrology of Length Standards by Holographic Interferometry with Phase Heterodynage", in *Proc. 1st Europ. Conf. on Opt. Appl. to Metrol.*, Strasbourg 1977, ed. by M. Grosmann, P. Meyrueis, SPIE, Vol. 136 (SPIE, Washington 1978) pp. 92–100
4.97 N. Jolly, J. Poirier: "Double Exposure Holographic Interferometry: Application to Nondestructive Testing and to Breaking Point Mechanics", in *Proc. 1st Europ. Conf. on Opt. Appl. to Metrol.*, Strasbourg 1977, ed. by M. Grosmann, P. Meyrueis, SPIE, Vol. 136 (SPIE, Washington 1978) pp. 101–106
4.98 J. T. Atkinson, M. J. Lalor: "Holographic Interferometry Applied to Minimal Wear Measurement", in *Proc. 1st Europ. Conf. on Opt. Appl. to Metrol.*, Strasbourg 1977, ed. by M. Grosmann, P. Meyrueis, SPIE, Vol. 136 (SPIE, Washington 1978) pp. 107–113

4.99 G. Cadoret: "Application of Holography to the Study of Structures and Materials", in *Proc. 1st Europ. Conf. on Opt. Appl. to Metrol.*, Strasbourg 1977, ed. by M. Grosmann, P. Meyrueis, SPIE, Vol. 136 (SPIE, Washington 1978) pp. 114–126

4.100 M. Blandin, C. Durou, H. Soulet: "Study by Holographic Interferometry of Dimensional Variability in Precision-Moulding Materials Used in Odontology", in *Proc. 1st Europ. Conf. on Opt. Appl. to Metrol.*, Strasbourg 1977, ed. by M. Grosmann, P. Meyrueis, SPIE, Vol. 136 (SPIE, Washington 1978) pp. 130–134

4.101 J. M. Caussignac: "Application of Holographic Interferometry to the Study of Structural Deformations in Civil Engineering", in *Proc. 1st Europ. Conf. on Opt. Appl. to Metrol.*, Strasbourg 1977, ed. by M. Grosmann, P. Meyrueis, SPIE, Vol. 136 (SPIE, Washington 1978) pp. 136–142

4.102 A. Felske: "Holographic Analysis of Oscillations in Squealing Disk Brakes", in *Proc. 1st Europ. Conf. on Opt. Appl. to Metrol.*, Strasbourg 1977, ed. by M. Grosmann, P. Meyrueis, SPIE, Vol. 136 (SPIE, Washington 1978) pp. 148–155

4.103 P. Smigielski, D. Cesario, C. Patanchon: "Testing by Holographic Interferometry of Solid Propergol Engines", in *Proc. 1st Europ. Conf. on Opt. Appl. to Metrol.*, Strasbourg 1977, ed. by M. Grosmann, P. Meyrueis, SPIE, Vol. 136 (SPIE, Washington 1978) pp. 181–185

4.104 J. D. Dubourg: "Application of Holographic Interferometry to Testing of Spun Structures", in *Proc. 1st Europ. Conf. on Opt. Appl. to Metrol.*, Strasbourg 1977, ed. by M. Grosmann, P. Meyrueis, SPIE, Vol. 136 (SPIE, Washington 1978) pp. 186–191

4.105 P. Meyrueis, M. Pharok, J. Fontaine: "Holographic Interferometry in Osteosynthesis", in *Proc. 1st Europ. Conf. on Opt. Appl. to Metrol.*, Strasbourg 1977, ed. by M. Grosmann and P. Meyrueis, SPIE, Vol. 136 (SPIE, Washington 1978) pp. 202–205

4.106 P. Jacquot, L. Pflug: Application de l'holographie à l'étude d'un outil de coupe. Bull. Tech. Suisse Romande *104*, 13–17 (1978)

4.107 K. A. Stetson: The use of an image derotator in hologram interferometry and speckle photography of rotating objects. Exp. Mech. *18*, 67–73 (1978)

4.108 T. Matsumoto, K. Iwata, R. Nagata: Measuring accuracy of three-dimensional displacements in holographic interferometry. Appl. Opt. *12*, 961–967 (1973)

4.109 L. Ek, K. Biedermann: Analysis of a system for hologram interferometry with a continuously scanning reconstruction beam. Appl. Opt. *16*, 2535–2542 (1977)

4.110 L. Ek, K. Biedermann: "Hologram Interferometry with a Continuously Scanning Reconstruction Beam", in *Applications of Holography and Optical Data Processing*, Proc. Conf. Jerusalem 1976, ed. by E. Marom, A. A. Friesem, E. Wiener-Avnear (Pergamon Press, Oxford, New York 1977) pp. 233–239

4.111 S. K. Dhir, J. P. Sikora: An improved method for obtaining the general-displacement field from a holographic interferogram. Exp. Mech. *7*, 323–327 (1972)

4.112 S. K. Dhir, J. P. Sikora: "Holographic Determination of Surface Stresses in an Arbitrary Three-Dimensional Object", in *Digest of Papers*, Intern. Optical Computing Conf., Zurich 1974 (IEEE, New York 1974) pp. 79–83

4.113 R. J. Pryputniewicz: Determination of the sensitivity vectors directly from holograms. J. Opt. Soc. Am. *67*, 1351–1353 (1977)

4.114 B. D. Hansche, C. G. Murphy: Holographic interferogram analysis from a single view. Appl. Opt. *13*, 630–635 (1974)

4.115 W. G. Gottenberg: Some applications of holographic interferometry. Exp. Mech. *8*, 405–410 (1968)

4.116 P. M. Boone, R. Verbiest: Application of hologram interferometry to plate deformation and translation measurements. Opt. Acta *16*, 555–567 (1969)

4.117 R. C. Sampson: Holographic-interferometry applications in experimental mechanics. Exp. Mech. *10*, 313–320 (1970)

4.118 A. D. Wilson: Holographically observed torsion in a cylindrical shaft. Appl. Opt. *9*, 2093–2097 (1970)

4.119 A. D. Wilson: Inplane displacement of a stressed membrane with a hole measured by holographic interferometry. Appl. Opt. *10*, 908–912 (1971)

4.120 J. Tsujiuchi, N. Takeya, K. Matsuda: Mesure de la déformation d'un objet par interférométrie holographique. Opt. Acta *16*, 709–722 (1969)

4.121 J. Tsujiuchi, K. Matsuda: "Utilisation des franges d'égale inclinaison en interférométrie holographique", in *Applications de l'Holographie*, Proc. Symp. Besançon 1970, ed. by J.-Ch. Viénot, J. Bulabois, J. Pasteur (Univ. de Besançon 1970) paper 4.5

4.122 J. W. C. Gates: "Measurement of Displacements in Three Dimensions from a Single Hologram", in *Applications de l'Holographie*, Proc. Symp. Besançon 1970, ed. by J.-Ch. Viénot, J. Bulabois, J. Pasteur (Univ. de Besançon 1970) paper 4.4

4.123 P. M. Boone, L. C. De Backer: Determination of three orthogonal displacement components from one double-exposure hologram. Optik *37*, 61–81 (1973)

4.124 H. Steinbichler: "Beitrag zur quantitativen Auswertung von holographischen Interferogrammen", Thesis, Techn. Univ. Munich (1973)

4.125 H. Steinbichler, H. Rottenkolber, E. Mönch: "Holographie 73", Teil 4: quantitative Auswertung von holographischen Interferogrammen. Laser Elektro-Optik *5*, 9–15, n°5 (1973)

4.126 C. H. F. Velzel: Holographische Verformungsanalyse. Philips Tech. Rundsch. *35*, 29–41 (1975)

4.127 G. Tribillon, J. M. Fournier: Large-sized holographic interferometry in real image. Opt. Acta *24*, 893–896 (1977)

4.128 T. Matsumoto, K. Iwata, R. Nagata: Distortionless recording in double-exposure holographic interferometry. Appl. Opt. *12*, 1660–1662 (1973)

4.129 V. Fossati-Bellani, A. Sona: Measurement of three-dimensional displacements by scanning a double-exposure hologram. Appl. Opt. *13*, 1337–1341 (1974)

4.130 V. Fossati-Bellani: "Automatic Measurement of 3-d Displacements by Using the Scanning Technique in Double Exposure Holograms", in *Applications of Holography and Optical Data Processing*, Proc. Conf. Jerusalem 1976, ed. by E. Marom, A. A. Friesem, E. Wiener-Avnear (Pergamon Press, Oxford, New York 1977) pp. 225–231

4.131 N. Abramson: "The Holo-Diagram, a Practical Device for the Making and the Evaluation of Holograms", in *The Engineering Uses of Holography*, Proc. Symp. Glasgow 1968, ed. by E. R. Robertson, J. M. Harvey (University Press, Cambridge 1970) pp. 45–55

4.132 N. Abramson: The holo-diagram: a practical device for making and evaluating holograms. Appl. Opt. *8*, 1235–1240 (1969)

4.133 N. Abramson: The holo-diagram. II: a practical device for information retrieval in hologram interferometry. Appl. Opt. *9*, 97–101 (1970)

4.134 N. Abramson: The holo-diagram. III: a practical device for predicting fringe patterns in hologram interferometry. Appl. Opt. *9*, 2311–2320 (1970)

4.135 N. Abramson: "The Holo-Diagram: A Practical Device for Prediction of Fringe-Pattern in Hologram Interferometry", in *Applications de l'Holographie*, Proc. Symp. Besançon 1970, ed. by J.-Ch. Viénot, J. Bulabois, J. Pasteur (Univ. de Besançon 1970) paper 4.7

4.136 N. Abramson: The holo-diagram. IV: a practical device for simulating fringe patterns in hologram interferometry. Appl. Opt. *10*, 2155–2161 (1971)

4.137 N. Abramson: The holo-diagram. VI: practical device in coherent optics. Appl. Opt. *11*, 2562–2571 (1972)

4.138 N. Abramson: Fundamental resolution of optical systems. Optik *39*, 141–149 (1973)

4.139 N. Abramson: "Interferometric Information", in *Applications of Holography and Optical Data Processing*, Proc. Conf. Jerusalem 1976, ed. by E. Marom, A. A. Friesem, E. Wiener-Avnear (Pergamon Press, Oxford, New York 1977). pp. 269–278

4.140 W. Jüptner, K. Ringer, H. Welling: Auswertung von Interferenzstreifensystemen bei holografischer Translations- und Dehnungsmessung. Optik *38*, 437–448 (1973)

4.141 J. Janta, M. Miler: Model interferogram as an aid for holographic interferometry. J. Optics (Paris) *8*, 301–307 (1977)

4.142 M. J. Landry, C. M. Wise: Automatic data reduction of certain holographic interferograms. Appl. Opt. *12*, 2320–2327 (1973)

4.143 R. Dändliker, B. Ineichen, F. M. Mottier: High resolution hologram interferometry by electronic phase measurement. Opt. Commun. *9*, 412–416 (1973)

4.144 R. Dändliker, B. Ineichen, F. M. Mottier: "Electronic Processing of Holographic Interferograms", in *Digest of Papers*, Intern. Optical Computing Conf., Zurich 1974 (IEEE, New York 1974) pp. 69–72
4.145 R. Dändliker, B. Eliasson, B. Ineichen, F. M. Mottier: "Quantitative Determination of Bending and Torsion Through Holographic Interferometry", in *The Engineering Uses of Coherent Optics*, Proc. Symp. Glasgow 1975, ed. by E. R. Robertson (University Press, Cambridge 1976) pp. 99–117
4.146 B. Ineichen, R. Dändliker, J. Mastner: "Accuracy and Reproducibility of Heterodyne Holographic Interferometry", in *Applications of Holography and Optical Data Processing*, Proc. Conf. Jerusalem 1976, ed. by E. Marom, A. A. Friesem, E. Wiener-Avnear (Pergamon Press, Oxford, New York 1977) pp. 207–212
4.147 R. Dändliker: "Quantitative Strain Measurement Through Holographic Interferometry", in *Applications of Holography and Optical Data Processing*, Proc. Conf. Jerusalem 1976, ed. by E. Marom, A. A. Friesem, E. Wiener-Avnear (Pergamon Press, Oxford, New York 1977) pp. 169–181
4.148 R. Dändliker: Strain and stress analysis through heterodyne holographic interferometry. Brown Boveri Res. Rep. KLR 77–107C (1977)
4.149 R. Dändliker: "Heterodyne Holographic Interferometry: A Review", in *Proc. 1st Europ. Conf. on Opt. Appl. to Metrol.*, Strasbourg 1977, ed. by M. Grosmann, P. Meyrueis, SPIE, Vol. 136 (SPIE, Washington 1978) p. 215
4.150 F. Lanzl, M. Schlüter: "Video-Electronic Analysis of Holographic Interferograms", in *Proc. 1st Europ. Conf. on Opt. Appl. to Metrol.*, Strasbourg 1977, ed. by M. Grosmann, P. Meyrueis, SPIE, Vol. 136 (SPIE, Washington 1978) pp. 166–171
4.151 F. Gori, G. Guattari: Time-ordered double-exposure holography. Opt. Commun. *5*, 359–361 (1972)
4.152 N. G. Vlasov, A. E. Shtan'ko: Determination of order number and sign of interference fringes. Sov. Phys.-Tech. Phys. *21*, 111–112 (1976)
4.153 P. C. Gupta, A. K. Aggarwal: Simultaneous detection of direction of motion and fringe order determination in holographic displacement measurement. Appl. Opt. *15*, 2961–2962 (1976)
4.154 S. Toyooka: Holographic interferometry with increased sensibility for diffusely reflecting objects. Appl. Opt. *16*, 1054–1057 (1977)
4.155 P. M. Boone: Holographic determination of in-plane deformation. Opt. Technol. *2*, 94–98 (1970)
4.156 A. R. Luxmoore, C. House: "In-Plane Strain Measurements by a Three Beam Holographic Method", in *Applications de l'Holographie*, Proc. Symp. Besançon 1970, ed. by J.-Ch. Viénot, J. Bulabois, J. Pasteur (Univ. de Besançon 1970) paper 5.2
4.157 J. Ebbeni: "Combinaison d'une méthode de moiré et d'une méthode holographique pour déterminer l'état de déformation d'un objet diffusant", in *Proc. of the 5th Intern. Conf. on Exp. Stress Anal.*, Udine 1974 (CISM, Udine 1974) pp. 4.20–25
4.158 C. A. Sciammarella, J. A. Gilbert: A holographic-moiré technique to obtain separate patterns for components of displacement. Exp. Mech. *16*, 215–220 (1976)
4.159 J. A. Gilbert, C. A. Sciammarella: "Extension to Three Dimensions of a Holographic-Moiré Technique to Separate Patterns Corresponding to Components of Displacement", in *SESA Spring Meeting*, Silver Spring, May 1976 (SESA, Westport 1976)
4.160 C. P. Hu, J. L. Turner, C. E. Taylor: Holographic interferometry: compensation for rigid-body motion. Appl. Opt. *15*, 1558–1564 (1976)
4.161 C. H. F. Velzel: Contours of equal in-plane displacement in holographic interferometry. Opt. Commun. *7*, 302–304 (1973)
4.162 G. Tribillon, J. F. Miles: Approche du déplacement moyen d'un front d'onde à partir d'un interférogramme holographique. Opt. Commun. *11*, 123–126 (1974)
4.163 R. E. Rowlands, J. A. Jensen, K. D. Winters: Differentiation along arbitrary orientations. Exp. Mech. *18*, 81–86 (1978)
4.164 K. A. Stetson: The argument of the fringe function in hologram interferometry of general deformations. Optik *31*, 576–591 (1970)

4.165 M. Dubas, W. Schumann: Sur la détermination holographique de l'état de déformation à la surface d'un corps non-transparent. Opt. Acta *21*, 547–562 (1974)
4.166 D. Bijl, R. Jones: A new theory for the practical interpretation of holographic interference patterns resulting from static surface displacements. Opt. Acta *21*, 105–118 (1974)
4.167 K. A. Stetson: Fringe interpretation for hologram interferometry of rigid-body motions and homogeneous deformations. J. Opt. Soc. Am. *64*, 1–10 (1974)
4.168 R. J. Pryputniewicz, K. A. Stetson: Holographic strain analysis: extension of fringe-vector method to include perspective. Appl. Opt. *15*, 725–728 (1976)
4.169 K. A. Stetson: "Vibration Measurement by Holography", in *The Engineering Uses of Holography*, Proc. Symp. Glasgow 1968, ed. by E. R. Robertson, J. M. Harvey (University Press, Cambridge 1970) pp. 307–331
4.170 K. A. Stetson: A rigorous treatment of the fringes of hologram interferometry. Optik *29*, 386–400 (1969)
4.171 N. E. Molin, K. A. Stetson: Measurement of fringe loci and localization in hologram interferometry for pivot motion, in-plane rotation and in-plane translation. Optik *31*, 157–177, 281–291 (1970)
4.172 N. E. Molin, K. A. Stetson: Fringe localization in hologram interferometry of mutually independent and dependent rotations around orthogonal, nonintersecting axes. Optik *33*, 399–422 (1971)
4.173 K. A. Stetson: Fringe vectors and observed-fringe vectors in hologram interferometry. Appl. Opt. *14*, 272–273 (1975)
4.174 K. A. Stetson: Homogeneous deformations: determination by fringe vectors in hologram interferometry. Appl. Opt. *14*, 2256–2259 (1975)
4.175 K. A. Stetson: Use of fringe vectors in hologram interferometry to determine fringe localization. J. Opt. Soc. Am. *66*, 626–627 (1976)
4.176 K. A. Stetson: Holographic strain analysis by fringe-localization planes. J. Opt. Soc. Am. *66*, 627 (1976)
4.177 R. J. Pryputniewicz: "Holographic Analysis of Body Deformation", Thesis, University of Connecticut, Storrs (1976)
4.178 T. Tsuruta, N. Shiotake, Y. Itoh: Hologram interferometry using two reference beams. Jpn. J. Appl. Phys. *7*, 1092–1100 (1968)
4.179 T. Tsuruta, N. Shiotake, Y. Itoh: Formation and localization of holographically produced interference fringes. Opt. Acta *16*, 723–733 (1969)
4.180 J.-Ch. Viénot, Cl. Froehly, J. Monneret, J. Pasteur: "Hologram Interferometry: Surface Displacement Fringe Analysis as an Approach to the Study of Mechanical Strains and Other Applications to the Determination of Anisotropy in Transparent Objects", in *The Engineering Uses of Holography*, Proc. Symp. Glasgow 1968, ed. by E. R. Robertson, J. M. Harvey (University Press, Cambridge 1970) pp. 133–150
4.181 C. Froehly, J. Monneret, J. Pasteur, J.-Ch. Viénot: Etude des faibles déplacements d'objets opaques et de la distorsion optique dans les lasers à solide par interférométrie holographique. Opt. Acta *16*, 343–362 (1969)
4.182 J. Monneret: Exploitation des systèmes d'interférence observables en interférométrie holographique par mesure des déplacements angulaires de l'onde diffractée par l'objet. Opt. Commun. *2*, 159–162 (1970)
4.183 J. Monneret: "Exploitation des systèmes d'interférence observables en interférométrie holographique d'objets opaques diffusants", in *Applications de l'Holographie*, Proc. Symp. Besançon 1970, ed. by J.-Ch. Viénot, J. Bulabois, J. Pasteur (Univ. de Besançon 1970) paper 4.3
4.184 J. Monneret: "Etude théorique et expérimentale des phénomènes observables en interférométrie holographique, interprétation des interférogrammes et applications à la métrologie des microdéplacements", Thesis, Université de Besançon (1973)
4.185 P. Jacquot: "Analyse de l'information contenue dans un interférogramme en double exposition: étude de quelques procédés sur des exemples concrets", Thesis, Université de Besançon (1973)

4.186 W. T. Welford: Fringe visibility and localization in hologram interferometry. Opt. Commun. *1*, 123–125 (1969)
4.187 W. T. Welford: Fringe visibility and localization in hologram interferometry with parallel displacement. Opt. Commun. *1*, 311–314 (1970)
4.188 W. T. Welford: "Geometrical Optics Treatment of Fringe Localization in Two-Beam Hologram Interferometry", in *Applications de l'Holographie,* Proc. Symp. Besançon 1970, ed. by J.-Ch. Viénot, J. Bulabois, J. Pasteur (Univ. de Besançon 1970) paper 4.2
4.189 C. H. F. Velzel: Fringe contrast and fringe localization in holographic interferometry. J. Opt. Soc. Am. *60*, 419–420 (1970)
4.190 A. F. Fercher, R. Torge: Apertureinflüsse in der Hologramm-Interferometrie. Optik *30*, 521–526 (1970)
4.191 W. H. Steel: Fringe localization and visibility in classical hologram interferometers. Opt. Acta *17*, 873–881 (1970)
4.192 S. Walles: Visibility and localization of fringes in holographic interferometry of diffusely reflecting surfaces. Ark. Fys. *40*, 299–403 (1970)
4.193 S. Walles: On the concept of homologous rays in holographic interferometry of diffusely reflecting surfaces. Opt. Acta *17*, 899–913 (1970)
4.194 M. A. Machado Gama: Fringe localization and visibility in hologram and classical broad source interferometry. Opt. Commun. *8*, 362–365 (1973)
4.195 W. Schumann: Some aspects of the optical techniques for strain determination. Exp. Mech. *13*, 225–231 (1973)
4.196 M. Dubas, W. Schumann: On direct measurements of strain and rotation in holographic interferometry using the line of complete localization. Opt. Acta *22*, 807–819 (1975)
4.197 M. Dubas: "Sur l'analyse expérimentale de l'état de déformation à la surface d'un corps opaque par interférométrie holographique, en particulier à l'aide de la localisation des franges", Thesis, EPF Zurich (1976)
4.198 I. Přikryl: Localization of interference fringes in holographic interferometry. Opt. Acta *21*, 675–681 (1974)
4.199 J. Leroy: Localisation des franges en interférométrie holographique. Nouv. Rev. Opt. *6*, 329–337 (1975)
4.200 H. Kreitlow: "Untersuchung quantitativer Zusammenhänge in der holographischen Interferometrie insbesondere im Hinblick auf eine Auswertung holographischer Interferenzmuster", Thesis, Techn. Univ. Hannover (1976)
4.201 J. C. Charmet: "Holographie appliquée à l'analyse des structures sous contrainte: détermination de l'état de déformation, en particulier par l'étude du contraste de l'interférogramme", Thesis, Université P. et M. Curie, Paris (1977)
4.202 J. C. Charmet, F. Montel: Interférométrie holographique sur objets diffusants: application de la mesure du contraste à la détermination des gradients de déplacement. Rev. Phys. Appl. *12*, 603–610 (1977)
4.203 J. Ebbeni, J. C. Charmet: Strain components obtained from contrast measurement of holographic fringe patterns. Appl. Opt. *16*, 2543–2545 (1977)
4.204 I. Yamaguchi: Fringe loci and visibility in holographic interferometry with diffuse objects. I. Fringes of equal inclination. Opt. Acta *24*, 1011–1025 (1977)
4.205 P. Beckmann, A. Spizzichino: *The Scattering of Electromagnetic Waves from Rough Surfaces* (Pergamon Press, Oxford 1963)
4.206 L. I. Goldfischer: Autocorrelation function and power spectral density of laser-produced speckle patterns. J. Opt. Soc. Am. *55*, 247–253 (1965)
4.207 H. H. Hopkins, H. Tiziani: "Speckling in Diffraction Patterns and Optical Images Formed with the Laser", in *Applications de l'Holographie,* Proc. Symp. Besançon 1970, ed. by J.-Ch. Viénot, J. Bulabois, J. Pasteur (Univ. de Besançon 1970) opening conf.
4.208 S. Lowenthal, H. Arsenault, D. Joyeux: "Image Formation with Diffuse Coherent Light: Speckling Determination for Static or Moving Diffusers", in *Applications de l'Holographie,* Proc. Symp. Besançon 1970, ed. by J.-Ch. Viénot, J. Bulabois, J. Pasteur (Univ. de Besançon 1970) paper 2.2

4.209 S. Lowenthal, H. Arsenault: Image formation for coherent diffuse objects: statistical properties. J. Opt. Soc. Am. *60,* 1478–1483 (1970)
4.210 A. A. Michelson: Interference phenomena in a new form of refractometer. Phil. Mag. & J. Sc. (London) *13* (5th series), 236–242 (1882)
4.211 J. Macé de Lépinay, Ch. Fabry: Théorie générale de la visibilité des franges d'interférence. J. de Phys. *10* (2e série), 5–20 (1891)
4.212 G. Bruhat: *Optique,* 6th ed. (Masson, Paris 1965)
4.213 R. Jones: An experimental verification of a new theory for the interpretation of holographic interference patterns resulting from static surface displacements. Opt. Acta *21,* 257–266 (1974)
4.214 K. Piwernetz: A posteriori compensation for rigid-body motion in holographic interferometry by means of a Moiré-technique. Opt. Acta *24,* 201–210 (1977)
4.215 T. R. Hsu: Large-deformation measurements by real-time holographic interferometry. Exp. Mech. *14,* 408–411 (1974)
4.216 E. Champagne, L. Kersch: Control of holographic interferometric fringe patterns. J. Opt. Soc. Am. *59,* 1535 (1969)
4.217 L. A. Kersch: Advanced concepts of holographic nondestructive testing. Mater. Eval. *29,* 125–129, 140 (1971)
4.218 E. B. Champagne: "Holographic Interferometry Extended", in *Digest of Papers,* Intern. Optical Computing Conf., Zurich 1974 (IEEE, New York 1974) pp. 73–74
4.219 P. M. de Larminat, R. P. Wei: A fringe-compensation technique for stress analysis by reflection holographic interferometry. Exp. Mech. *16,* 241–248 (1976)
4.220 P. M. de Larminat, R. P. Wei: Normal surface displacement around a circular hole by reflection holographic interferometry. Exp. Mech. *18,* 74–80 (1978)
4.221 V. J. Corcoran, R. W. Herron, J. G. Jamarillo: Generation of a hologram from a moving target. Appl. Opt. *5,* 668–669 (1966)
4.222 G. S. Ballard: Double-exposure holographic interferometry with separate reference beams. J. Appl. Phys. *39,* 4846–4848 (1968)
4.223 J. Politch, J. Shamir, J. Ben Uri: Some characteristics of multibeam holography. Opt. Laser Techn. *3,* 226–228 (1971)
4.224 J. Surget: Schéma d'holographie à deux sources de référence. Nouv. Rev. Opt. *5,* 201–217 (1974)
4.225 J. Surget: "Two Reference Beam Holographic Interferometry for Aerodynamic Flow Studies", in *Applications of Holography and Optical Data Processing,* Proc. Conf. Jerusalem 1976, ed. by E. Marom, A. A. Friesem, E. Wiener-Avnear (Pergamon Press, Oxford, New York 1977) pp. 183–192
4.226 R. Dändliker, E. Marom, F. M. Mottier: Two-reference-beam holographic interferometry. J. Opt. Soc. Am. *66,* 23–30 (1976)
4.227 W. Schumann, M. Dubas: On the holographic interferometry used for deformation analysis, with one fixed and one movable reconstruction source. Optik *47,* 391–404 (1977)
4.228 G. E. Sommargren: Double-exposure holographic interferometry using common-path reference waves. Appl. Opt. *16,* 1736–1741 (1977)
4.229 R. F. C. Kriens: "Some Considerations on the Quantitative Interpretation of Holographic Interferograms", in *Proc. 1st Europ. Conf. on Opt. Appl. to Metrol.,* Strasbourg 1977, ed. by M. Grosmann, P. Meyrueis, SPIE, Vol. 136 (SPIE, Washington 1978) pp. 156–165
4.230 G. L. Rogers: Wavelength tolerances in frozen-fringe holography. J. Opt. Soc. Am. *61,* 784–788 (1971)
4.231 B. Ineichen, U. Kogelschatz, R. Dändliker: Schlieren diagnostics and interferometry of an arc discharge using pulsed holography. Appl. Opt. *12,* 2554–2556 (1973)
4.232 G. W. Stroke, A. E. Labeyrie: Interferometric reconstruction of phase objects using diffuse "coding" and two holograms. Phys. Lett. *20,* 157–158 (1966)
4.233 M. De, L. Sevigny: Three-beam holographic interferometry. Appl. Opt. *6,* 1665–1671 (1967)

4.234 J. W. C. Gates: Holographic phase recording by interference between reconstructed wavefronts from separate holograms. Nature *220,* 473–474 (1968)
4.235 O. Bryngdahl: Shearing interferometry by wavefront reconstruction. J. Opt. Soc. Am. *58,* 865–871 (1968)
4.236 P. M. Boone: "Surface Deformation Measurements Using Deformation Following Holograms", in *Applications de l'Holographie,* Proc. Symp. Besançon 1970, ed. by J.-Ch. Viénot, J. Bulabois, J. Pasteur (Univ. de Besançon 1970) paper 5.1
4.237 P. M. Boone: "The Use of Reflection Holograms in Holographic Interferometry", in *Digest of Papers,* Intern. Optical Computing Conf., Zurich 1974 (IEEE, New York 1974) pp. 75–78
4.238 P. M. Boone: Use of reflection holograms in holographic interferometry and speckle correlation for measurement of surface-displacement. Opt. Acta *22,* 579–589 (1975)
4.239 D. B. Neumann, R. C. Penn: Off-table holography. Exp. Mech. *15,* 241–244 (1975)
4.240 N. Abramson: Sandwich hologram interferometry. 2: some practical calculations. Appl. Opt. *14,* 981–984 (1975)
4.241 N. Abramson: Sandwich hologram interferometry. 3: contouring. Appl. Opt. *15,* 200–205 (1976)
4.242 N. Abramson: "Sandwich Holography: An Analogue Method for the Evaluation of Holographic Fringes", in *The Engineering Uses of Coherent Optics,* Proc. Symp. Glasgow 1975, ed. by E. R. Robertson (University Press, Cambridge 1976) pp. 631–645
4.243 N. Abramson: Sandwich hologram interferometry. 4: holographic studies of two milling machines. Appl. Opt. *16,* 2521–2531 (1977)
4.244 H. Bjelkhagen: Pulsed sandwich holography. Appl. Opt. *16,* 1727–1731 (1977)
4.245 N. Abramson, H. Bjelkhagen: Pulsed sandwich holography. 2: practical application. Appl. Opt. *17,* 187–191 (1978)
4.246 P. Hariharan, Z. S. Hegedus: Two-hologram interferometry: a simplified sandwich technique. Appl. Opt. *15,* 848–849 (1976)
4.247 P. Hariharan: Hologram interferometry: identification of the sign of surface displacements. Opt. Acta *24,* 989–990 (1977)

Chapter 5. Second Derivatives of the Displacement and of the Optical-Path Difference

5.1 W. Blaschke: *Vorlesungen über Differentialgeometrie I* (Springer, Berlin 1927)
5.2 W. A. Nash: *Bibliography on Shells and Shell-like Structures* (Dep. of the Navy, David Taylor Model Basin, Dep. of Eng. Mech. 1954–1956)
5.3 W. Z. Wlassov: *Allgemeine Schalentheorie und ihre Anwendung in der Technik,* translated from Russian (Akademie Verlag, Berlin 1958)
5.4 K. Girkmann: *Flächentragwerke,* 5th ed. (Springer, Vienna 1959)
5.5 V. V. Novozhilov: *The Theory of Thin Shells* (Noordhoff, Groningen 1959)
5.6 S. Timoshenko, S. Woinowsky-Krieger: *Theory of Plates and Shells* (Mc Graw-Hill, New York 1959)
5.7 W. Flügge: *Stresses in Shells* (Springer, Berlin, Göttingen, Heidelberg 1960)
5.8 A. L. Gol'denveizer: *Theory of Elastic Thin Shells* (Pergamon Press, Oxford 1961)
5.9 P. M. Naghdi: "The Theory of Shells and Plates", in *Encyclopedia of Physics,* ed. by S. Flügge, Vol. 6a (Springer, Berlin, Heidelberg, New York 1972) pp. 425–640
5.10 H. S. Rutten: *Theory and Design of Shells on the Basis of Asymptotic Analysis* (Rutten & Kruisman, Voorburg 1973)
5.11 P. Seide: *Small Elastic Deformations of Thin Shells* (Noordhoff, Leyden 1975)
5.12 W. Schumann: "Ausgewählte Kapitel der Platten- und Schalentheorie. Vorlesung an der ETH Zürich", Lecture notes (1976)
5.13 J. L. Sanders: "An Improved First Approximation Theory of Shells", NASA Rpt. *24* (1959)
5.14 W. T. Koiter: "A Consistent First Approximation in the General Theory of Thin Elastic Shells", in *The Theory of Thin Elastic Shells,* Proc. Symp. Delft 1959, ed. by W. T. Koiter (North-Holland, Amsterdam 1960) pp. 12–33

5.15 W. T. Koiter: "Foundations and Basic Equations of Shell Theory, a Survey of Recent Progress", in *Theory of Thin Shells*, Proc. 2nd Symp. Copenhagen 1967, ed. by F. I. Niordson (Springer, Berlin, Heidelberg, New York 1969) pp. 93–105

5.16 C. R. Steele: A geometric optics solution for the thin shell equation. Int. J. Eng. Sci. *9*, 681–704 (1971)

5.17 W. Wüthrich: "Sur l'aplanissement élastique d'ouvrages à courbure initiale dans le cadre d'un problème aux déplacements finis", Thesis, EPF Zurich (1972)

5.18 G. Teichmann, W. Schumann: Some remarks concerning the linear theory of heterogeneous orthotropic shells. Acta Mech. *25*, 291–299 (1977)

5.19 G. Teichmann: "Quelques aspects généraux de la théorie linéaire des coques orthotropes et inhomogènes, en particulier en vue d'utiliser un principe de variation mixte", Thesis, EPF Zurich (1979)

5.20 C. A. Sciammarella, J. A. Gilbert: Holographic interferometry applied to the measurement of displacements of the interior points of transparent bodies. Appl. Opt. *15*, 2176–2182 (1976)

5.21 O. Vilnay: Measuring the displacement field inside a stressed sample. Appl. Opt. *16*, 813–814 (1977)

Author Index

Reference numbers appear in square brackets, followed by page numbers indicating the specific location of each citation.

Abraham, R. [2.1] 3
Abramowitz, I.A. [3.74, 75] 60
Abramson, N. [3.83; 4.43, 66, 131–139, 240–243, 245] 69, 83, 84, 87, 89, 143, 150, 166
Adkins, J.E. [2.10] 27
Aggarwal, A.K. [4.153] 88
Aistov, V.S. [4.88] 84
Ajovalasit, A. [4.72] 84
Albe, F. [4.67] 84
Aleksandrov, E.B. [4.47] 83
Allan, T. [4.58] 84
Archbold, E. [4.15, 80] 75, 84
Armstrong, J.A. [3.46] 48
Arsenault, H.H. [3.78; 4.208, 209] 60, 99, 108
Atkinson, J.T. [4.98] 84

Ballantyne, J.M. [3.74] 60
Ballard, G.S. [4.222] 133
Beckmann, P. [4.205] 99
Belvaux, Y. [3.40] 44
Ben Uri, J. [4.223] 133
Biedermann, K. [3.22; 4.109, 110] 39, 84, 86
Bijl, D. [4.38, 166] 82, 96, 98, 129
Binder, J.-Cl. [4.17] 76
Bjelkhagen, H. [4.66, 244, 245] 84, 143
Blandin, M. [4.100] 84
Blaschke, W. [5.1] 154
Bonch-Bruevich, A.M. [4.47] 83
Boone, P.M. [4.29, 116, 123, 155, 236–238] 76, 84, 85, 89, 143
Born, M. [3.5] 39, 48, 59, 63, 64, 98, 99, 108, 112, 116
Boucher, G. [4.77] 84
Briers, J.D. [4.30] 76
Brillouin, L. [2.6] 3, 18
Brooks, R.E. [4.1, 10] 75
Brown, G. M. [3.45] 45
Bruhat, G. [4.212] 116
Brumm, D. [3.44; 4.4] 45, 75
Bryngdahl, O. [4.235] 141
Buinov, G.N. [3.61, 62] 48
Burch, J.M. [4.2, 8, 27, 28] 75, 76, 82
Burckhardt, C.B. [3.14] 39, 98
Butters, J.N. [3.13; 4.56] 39, 84, 89

Cadoret, G. [4.99] 84
Carollo, S. [4.72] 84
Caulfield, H.J. [3.12; 4.18] 39, 76
Caussignac, J.M. [4.101] 84
Cesario, D. [4.103] 84
Champagne, E.B. [3.52, 53; 4.216, 218] 48, 53, 132, 133
Charmet, J.C. [4.201–203] 98, 100, 108, 112, 114, 130, 131
Collier, R.J. [3.14; 4.3] 39, 75, 98
Collins, M.C. [4.83] 84
Corcoran, V.J. [4.221] 133

Dameron, D. [3.35] 44
Dändliker, R. [3.24, 34, 38, 43; 4.19, 143–149, 226, 231] 39, 44, 45, 76, 87, 90, 129, 131, 133, 139, 141, 169
De, M. [4.233] 140
De Backer, L.C. [4.123] 85
Delorme, J.F. [4.63] 84
Der Hovanesian, J. [4.57, 79] 84
Dhir, S. K. [4.111, 112] 84, 90
Dittrich, M. [4.82] 84
Doherty, E.T. [4.3] 75
Dubourg, J.D. [4.104] 84
Dudderar, T.D. [4.59] 84
Durou, C. [3.41; 4.100] 44, 84

Ebbeni, J. [4.26, 157, 203] 76, 89, 98, 100, 103, 108, 114, 126, 130
Ek, L. [4.109, 110] 84, 86
Eliasson, B. [4.145] 87, 129, 169
Ennos, A.E. [4.8, 15, 36] 75, 81, 82, 83
Erf, R.K. [3.21; 4.78] 39, 84, 88
Eringen, A.C. [2.12, 13] 27
Ewers, W.M. [4.45] 83, 84

Fabry, Ch. [4.211] 109
Fagot, H. [4.67] 84

Felske, A. [4.31, 102] 76, 84
Fercher, A.F. [4.90] 98, 110, 112, 118
Flügge, W. [5.7] 154
Fontaine, J. [4.105] 84
Fossati-Bellani, V. [4.129, 130] 86, 87
Fourney, W.L. [4.91] 84
Fournier, J.M. [4.127] 86
Françon, M. [3.9, 80] 39, 48, 65
Friesem, A.A. [4.89] 84
Fritzsch, W. [4.45, 84] 83, 84
Froehly, C. [4.180, 181] 98, 105, 116, 118, 126
Fujimoto, J. [4.23] 76
Funkhouser, A. [3.44; 4.4] 45, 75
Fusek, R.L. [4.69] 84

Gabor, D. [3.1–4, 44; 4.4] 39, 41, 45, 75
Gagosz, R.M. [4.78] 84
Gascoigne, H.E. [4.55] 84
Gates, J.W.C. [4.49, 122, 234] 83, 85, 141
Gilbert, J.A. [4.44, 158, 159; 5.20] 83, 84, 89, 169
Girkmann, K. [5.4] 154
Gizatullin, R.K. [3.61] 48
Gol'denveizer, A.L. [5.8] 154
Goldfischer, L.I. [4.206] 99
Goodman, J.W. [3.8] 39
Gori, F. [4.151] 88
Gottenberg, W.G. [4.115] 84
Grant, R.M. [3.45; 4.63] 45, 84
Green, A.E. [2.5, 10] 3, 8, 9, 11, 18, 20, 27
Groh, G. [3.16] 39
Grosmann, M. [4.96] 84
Grünewald, K. [4.45, 68, 84] 83, 84
Guattari, G. [4.151] 88
Guggenheimer, H. [2.8] 3, 18, 119, 154
Gupta, P.C. [4.153] 88

Haines, K.A. [3.47; 4.9, 11] 48, 75, 97, 98, 100, 103, 105, 109, 115, 126, 127, 130
Hansche, B.D. [4.114] 84
Hariharan, P. [3.63; 4.20, 74, 246, 247] 48, 76, 84, 143, 151, 169
Hecht, N.L. [4.69] 84
Heflinger, L.O. [4.1, 10, 22] 75, 76
Hegedus, Z.S. [4.20, 74, 246] 76, 84, 143, 151, 169
Helstrom, C.W. [3.71] 50, 60, 62
Henley, D.R. [4.40] 82, 83
Herron, R.W. [4.221] 133
Hildebrand, B.P. [4.9, 11] 75, 97, 98, 100, 103, 105, 109, 115, 126, 127, 130
Holloway, D.C. [4.32, 91] 78, 84
Hopkins, H.H. [3.67; 4.207] 48, 59, 99
Horman, M.H. [4.5] 75
Hot, J.-P. [3.41] 44
House, C. [4.156] 89
Hsu, T.R. [4.21, 65, 215] 76, 84, 132
Hu, C.P. [4.40, 160] 82, 83, 89, 132
Hung, Y.Y. [4.40] 82, 83
Hunter, A.R. [4.85] 84

Ineichen, B. [3.34, 38, 43; 4.143–146, 231] 44, 45, 87, 90, 129, 139, 169
Itoh, Y. [4.62, 178, 179] 84, 98, 109, 112, 115, 118, 126, 130, 133, 137
Iwata, K. [4.46, 108, 128] 83, 84, 86

Jacquot, P. [4.106, 185] 84, 98, 105
Jamarillo, J.G. [4.221] 133
Janta, J. [4.141] 87
Jensen, J.A. [4.163] 90
Jolly, N. [4.97] 84
Jones, R. [4.38, 166, 213] 82, 96, 98, 126, 129
Joyeux, D. [4.208] 99
Jüptner, W. [4.140] 87

Kaletsch, D. [4.68] 84
Katz, J. [3.36–38] 44
Katzir, Y. [4.89] 84
Kersch, L.A. [4.216, 217] 132, 133
Kiemle, H. [3.10] 39, 48, 53
King, P.W. [4.48] 83
King, W. [4.79] 84
Kock, W.E. [3.19] 39
Kogelschatz, U. [4.231] 139
Kohler, H. [4.50–54] 83, 84, 86
Koiter, W.T. [5.14, 15] 154, 156
Köpf, U. [4.42] 83
Kozma, A. [3.31] 44
Kreitlow, H. [4.200] 98, 100
Kriens, R.F.C. [4.229] 133
Kurtz, R.L. [4.92] 84

Labeyrie, A.E. [4.4, 14, 232] 75, 140
Lalor, M.J. [4.98] 84
Landry, M.J. [4.142] 87
Lanzl, F. [4.150] 87
Larminat, P.M. de [4.219, 220] 132, 141
Latta, J.N. [3.57, 76] 48, 60
Laugwitz, D. [2.9] 3, 154
Leadbetter, I.K. [4.58, 61] 84
Lefèvre, R. [3.41] 44
Lehmann, V. [4.68] 84
Leigh, D.C. [2.2] 3, 5, 8, 9, 27
Leith, E.N. [3.28–30, 47] 43, 48
Lenk, H. [3.11] 39
Leroy, J. [4.199] 98, 109, 115, 120, 124
Lewak, R. [4.21] 76

Lewis, D. [4.69] 84
Lin, L.H. [3.14] 39, 98
Lisin, O.G. [4.88] 84
Liu, H.K. [4.92] 84
Lowenthal, S. [4.208, 209] 99, 108
Lu, C.P.J. [4.76] 84
Lu, S. [3.12] 39
Lukosz, W. [3.26, 27, 69, 70] 42, 50, 53
Lurie, M. [3.42] 45
Luxmoore, A.R. [4.156] 89

Macé de Lépinay, J. [4.211] 109
Machado Gama, M.A. [4.194] 98, 109
Makeev, I.M. [4.73] 84
Maréchal, A. [3.68, 79] 48, 59, 63, 65
Marom, E. [3.37, 38; 4.19, 226] 44, 76, 133, 141
Marquet, M. [3.48] 48
Massey, N.G. [3.53] 48
Mastner, J. [4.146] 87
Matsuda, K. [4.120, 121] 85, 97, 98, 109, 126, 127
Matsumoto, T. [4.46, 108, 128] 83, 84, 86
May, W. [4.86] 84
Mayinger, F. [4.86] 84
Meier, R.W. [3.49, 50] 48
Menzel, E. [3.17] 39
Meyer, H.D. [4.87] 84
Meyer, M. [4.77] 84
Meyrueis, P. [4.96, 105] 84
Michael, H. [4.70] 84
Michelson, A.A. [4.210] 109
Miler, M. [3.58; 4.141] 48, 87
Miles, J.F. [3.59, 60, 162] 48, 60, 61, 89
Minardi, J.E. [4.69] 84
Mirandé, W. [3.17] 39
Molin, N.E. [4.171, 172] 98, 120, 124
Mönch, E. [4.125] 85, 87
Monneret, J. [4.180–184] 98, 99, 100, 103, 105, 108, 110, 116, 118, 126, 130, 131, 141, 150, 166
Montel, F. [4.202] 98, 130, 131
Morozov, B.A. [4.73, 88] 84
Morton, T. M. [4.85] 84
Mottier, F.M. [3.43; 4.19, 143–145, 226] 45, 76, 87, 90, 129, 133, 141, 169
Moyer, R.G. [4.65] 84
Murphy, C.G. [4.114] 84
Mustafin, K.S. [3.61, 62] 48

Nagata, R. [4.46, 108, 128] 83, 84, 86
Naghdi, P.M. [5.9] 154
Nakajima, T. [4.64] 84, 166
Nash, W.A. [5.2] 154
Nassenstein, H. [4.12] 75
Neumann, D.B. [3.51; 4.239] 48, 143
Noll, W. [2.14] 27
Novozhilov, V.V. [5.5] 154

Offner, A. [3.73] 60

Pasteur, J. [4.180, 181] 98, 105, 116, 118, 126
Patacca, A.M. [4.91] 84
Patanchon, C. [4.103] 84
Pejša, F. [4.82] 84
Penn, R.C. [4.239] 143
Pennington, K.S. [4.3] 75
Pflug, L. [4.106] 84
Pharok, M. [4.105] 84
Piwernetz, K. [4.214] 132
Poirier, J. [4.97] 84
Politch, J. [4.223] 133
Powell, R.L. [4.6, 7, 13] 75
Prager, W. [2.11] 23, 27, 52, 159, 161
Preussler, D. [4.95] 84
Přikryl, I. [3.39, 77; 4.198] 44, 60, 98 105, 109, 115, 118, 126
Pryputniewicz, R.J. [4.113, 168, 177] 84, 96, 98, 129

Ramachandra Reddy, G. [4.93] 84
Ranson, W.F. [4.32] 78
Raterink, H.J. [4.75] 84
Rav-Noy, Z. [4.89] 84
Renesse, R.L. van [4.75] 84
Restrick, R. [3.44] 45
Reynolds, G.O. [3.7] 39
Richter, W. [3.55] 48
Ringer, K. [4.140] 87
Robertson, E.R. [4.79] 84
Rogers, G.L. [4.230] 139
Röss, D. [3.10] 39, 48, 53
Roth, E. [4.81] 84
Rottenkolber, H. [4.25, 125] 76, 85, 87
Rowlands, R.E. [4.163] 90
Royer, H. [3.15, 48] 39, 48
Rutten, H.S. [5.10] 154, 156

Saito, H. [4.60, 64] 84, 166
Sampson, R.C. [4.117] 84
Sanders, J.L. [5.13] 154, 156
Schlüter, M. [4.150] 87
Schmidt, W. [4.95] 84
Scholz, C.H. [4.76] 84
Sciammarella, C.A. [4.44, 158, 159; 5.20] 83, 84, 89, 169
Seide, P. [5.11] 154, 156
Sergeev, A.W. [4.73] 84
Sevigny, L. [4.233] 140

Shamir, J. [4.223] 133
Shibayama, K. [4.39] 82
Shiotake, N. [4.178, 179] 98, 109, 112, 115, 118, 126, 130, 133, 137
Shtan'ko, A.E. [4.152] 88
Sikora, J.P. [4.111, 112] 84, 90
Smigielski, P. [3.15; 4.67, 103] 39, 84
Smith, H.M. [3.20, 23] 39, 53
Smith, R.W. [3.65, 66] 48
Sokolnikoff, I.S. [2.7] 3
Sollid, J.E. [4.33–35] 81
Sommargren, G.E. [4.228] 133
Sona, A. [4.129] 86, 87
Soulet, H. [4.100] 84
Spetzler, H.A. [4.76, 77, 87] 84
Spicer, P. [4.16] 75
Spizzichino, A. [4.205] 99
Steel, W.H. [4.191] 98, 109
Steele, C.R. [5.16] 154
Steinbichler, H. [4.71, 124, 125] 84, 85, 87
Stetson, K.A. [3.54; 4.6, 7, 13, 107, 164, 167–176] 48, 75, 84, 93, 96, 98, 100, 103, 105, 109, 114, 115, 116, 120, 124, 126, 129, 130, 131
Storck, E. [3.56] 48
Stroke, G.W. [3.6, 44, 45; 4.4, 14, 232] 39, 45, 75, 140
Surjet, J. [4.224, 225] 133
Surkov, A.I. [4.73] 84
Sweatt, W.C. [3.72] 53

Takeya, N. [4.120] 85, 97, 98, 109, 126, 127
Taylor, C.E. [4.32, 40, 160] 78, 82, 83, 89, 132
Teichmann, G. [5.18, 19] 154, 155, 160
Timoshenko, S. [5.6] 154, 156
Tiziani, H. [4.207] 99
Torge, R. [4.190] 98, 110, 112, 118
Toyooka, S. [4.154] 88
Tribillon, G. [4.127, 162] 86, 89
Truesdell, C. [2.3, 14, 15] 3, 21, 27
Tschinke, M. [4.72] 84
Tselikov, A.I. [4.73, 88] 84
Tsujiuchi, J. [4.120, 121] 85, 97, 98, 109, 126, 127
Tsuruta, T. [4.62, 178, 179] 84, 98, 109, 112, 115, 118, 126, 130, 133, 137
Turner, J.L. [4.160] 89, 132

Uchiyama, H. [4.39] 82
Upatnieks, J. [3.28–30, 47] 43, 48

VanLigten, R.F. [3.32, 33] 44
Varner, J. [4.57] 84
Velis, J.B. de [3.7] 39
Velzel, C.H.F. [4.126, 161, 189] 85, 89, 98, 108, 109, 118, 127
Venkateswara Rao, V. [4.93] 84
Verbiest, R. [4.116] 84
Vest, C.M. [3.35; 4.37] 44, 81
Viénot, J.-Ch. [3.15; 4.24, 180, 181] 39, 76, 98, 105, 116, 118, 126
Villagio, P. [2.4] 3, 27
Vilnay, O. [5.21] 169
Vlasov, N.G. [4.152] 88
Vogel, A. [4.95] 84
Von Bally, G. [4.90] 84

Wachutka, H. [4.45, 68, 84] 83, 84
Wall, M.R. [4.41] 83
Walles, S. [4.192, 193] 96, 98, 100, 103, 105, 107, 108, 109, 111, 112, 113, 114, 116, 126, 164
Wang, C.C. [2.3] 3, 21, 27
Waters, J.P. [4.78] 84
Watrasiewicz, B.M. [4.16] 75
Watterson, C.E. [4.83] 84
Wei, R.P. [4.219, 220] 132, 141
Weingärtner, I. [3.17] 39
Welford, W.T. [4.186–188] 98, 115, 118
Welling, H. [4.140] 87
Wernicke, G. [4.94] 84
Wilson, A.D. [4.118, 119] 84
Wilton, R.J. [4.8] 75
Winters, K.D. [4.163] 90
Wise, C.M. [4.142] 87
Wlassov, W.Z. [5.3] 154
Woinowsky-Krieger, S. [5.6] 154, 156
Wolf, E. [3.5] 39, 48, 59, 63, 64, 98, 99, 108, 112, 116
Wuerker, R.F. [4.1, 10, 22] 75, 76
Wüthrich, A. [3.69, 70] 50
Wüthrich, W. [5.17] 154

Yamaguchi, I. [4.60, 64, 204] 84, 98, 107, 108, 110, 112, 166
Yoneyama, M. [4.23] 76
Young, M. [3.25] 39
Yu, F.T.S. [3.18] 39

Zerna, W. [2.5] 3, 8, 9, 11, 18, 20, 27

Subject Index

Aberration
  astigmatism 52, 53, 59, 63, 68
  coma 54
  geometrical theory of 48, 53, 59
  spherical 54
  transverse ray 59
  wave 53
Amplitude
  complex 40
  total 44, 107
  transmittance 41
Aperture
  circular 63, 111, 130
  function 102
  rectangular 113
  slit 114, 130
Astigmatism 52, 53, 59, 63, 68

Base vectors
  contravariant 4
  covariant 4, 15
Boundary conditions, static 161

Cauchy-Green tensor 28
Central ray 103
Center of collineation 21
Christoffel symbol 11
Codazzi-Mainardi, equations of 159
Collineation center 21
Coma 54
Compatibility 161
Components
  contravariant 5, 7
  covariant 5, 7
  physical 5, 8
Condition
  compatibility 161
  integrability 158
  of interference identity 49, 60, 66, 133, 138, 141, 144, 146
  of stationary behavior 22, 50, 58, 60, 109, 117
  static boundary 161
Configuration, reference 27, 32, 57, 76, 134
Conjugate image wave field 42, 46, 54, 56
Conjugate wave 42
Connections between surfaces 21
Constitutive equations 163
Contrast of the fringes 108
Coordinates
  convected 21
  curvilinear in space 3
  curvilinear on a surface 15
Correlation
  auto- 99, 107
  cross- 107
Curl 11
Curvature
  change 155
  Gaussian 19
  geodesic 21
  mean 19
  normal 18
  of a fringe 165
  of a surface 18
  of the line of complete localization 119
  "tensor" 18

Decomposition
  additive 31, 35
  multiplicative 29
  of a tensor 15
  of a vector 15
  polar 29, 34
Deformation 28
  gradient 28, 34
Derivative
  covariant 11
  of a tensor 11, 20
  of a vector 10, 19
  of combinations 12
  operator of partial 26
  three-dimensional operator 10
  two-dimensional operator 18, 25
  with a fixed center 13
Difference
  finite 90, 129, 169
  optical-path 72, 78, 134, 138, 139, 142, 144, 146
  phase 49, 66

Dilatation
  angular 29
  linear 28, 29
Displacement vector
  first derivative of 28
  measurement of 80, 127, 150
  mechanical 28, 77, 133, 156
  "optical" 57, 68, 69, 133, 141, 143, 147
  second derivative of 153
  sign of 89
Divergence 11
Double exposure 45, 75
Dyadic product 6
Dynamic method 83

Electro-optical apparatus 87
Ensemble average 99
Equilibrium, equations of 162
Evanescent 50
Expected value 99
Exposure 41
  double 45, 75
Exterior part 15, 20

Filtering effect 102
Finite differences 90, 129, 169
Focal line 63
Fourier transform 89, 101
Frequency
  of light 39
  spatial 101
Fringe
  contrast 108
  control term 134
  curvature of 165
  direction 96, 97, 134
  modification 132
  order, see Fringe-order
  spacing 96, 129, 134, 140, 142, 145, 149
  vector 96
  visibility 108
Fringe-order 78
  absolute 81
  derivative of 92
  difference 83
  relative 83
  zero 81, 82
Frozen-fringe 75
Fundamental form of a surface
  first 16
  second 18

Gauss'
  curvature 19
  equation 159
  image 53
  theorem 158
Gradient 10

Heterodyne holographic interferometry 87
Holo-diagram 87
Hologram 41
  movement of 65, 141
  sandwich 69, 143
Homologous rays 105, 109, 110, 113
Hooke's law 164
Huygens-Fresnel principle 99

Identity
  interference 49, 60, 66, 133, 138, 141, 144, 146
  tensor 9
Image
  aberrations of 52, 53, 59, 63, 68
  conjugate 42, 46, 54, 56
  Gaussian 53
  primary 42, 46, 48, 56
  real 43, 52, 85
  virtual 42, 52
Inclination 34, 35
Integrability 158
Integral transformation 101
Intensity 40, 107
Interference identity 49, 60, 66, 133, 138, 141, 144, 146
Interior
  part 15, 20
  strain in the 169
Involution 160

Kinematic relation
  in space 29, 31
  in the neighborhood of a surface 163
  on a surface 35, 37
Kirchhoff's stress tensor 161
Kronecker symbol 4

Laplace operator 13
Lateral part 58, 68, 70
Line of complete localization 110, 117
Linear transformations see Tensors
Live-fringe 75
Localization
  absolute 117
  complete 109, 115, 130, 137, 150
  non-localization 126
  partial 110, 130, 136
  standard case of partial 113, 125, 130, 136, 149

Longitudinal part 59, 69

Matrix 81
Modification of the fringes
  by one hologram 141
  by one reconstruction source 133
  by sandwich-holography 143
Moiré 89
Motion, equations of 162
Multiplexing 76

Normality
  condition 156
  theorem 116, 131, 137, 150

Optical-path difference 72, 78, 134, 138, 139, 142, 144, 146
  first derivative of 92
  second derivative of 164

Permutation tensor
  three-dimensional 9
  two-dimensional 17
Phase difference 49, 66
Pivot rotation 34, 35
Plane wave 100
Poisson's ratio 164, 166
Primary image wave field 42, 46, 48, 56
Product
  dyadic 6
  tensor 6
  triadic 8
  vector 10
Projection
  double 25
  normal 14, 16
  oblique 23
  super- 14
Pupil
  circular 63, 111, 130
  rectangular 113
  slit 114, 130

Ray
  central 103
  homologous 105, 109, 110, 113
  tracing equation 60
Real-time 75, 132
Reconstruction 41
Recording 39
Reference configuration 27, 32, 57, 76, 134
Reference source 40
  modification of 47, 56, 59
Reflection function 98

Rotation
  determination of 127, 150
  tensor 29, 31, 35
  vector 31
Rough surface 98

Sandwich holography 69, 143
Sensitivity vector 79, 82, 128
  of the fringe control 142, 144
Shadow 105
Slit aperture 114, 130
Spatial frequency 101
Spherical
  aberration 54
  wave 39, 99
Standard
  case of partial localization 113, 125, 130, 136, 149
  holographic interferometry 76
Static method 81
Stationary behavior 22, 50, 58, 60, 109, 117
Stokes'theorem 157
Strain
  determination of 90, 127, 150, 169
  shearing 29
  tensor 29, 31, 34
Stress
  tensor 161
  vector 161
Stretch tensor 29
Stroboscopic illumination 75
Surface
  curved 15
  of partial localization 110, 120, 125

Tangent plane 15
Tensor 6
  curvature 18
  curvature change 155
  decomposition of 15
  derivative of 11, 20
  metric 16
  product 6
  second-order 6
  third-order 8
Time-average 76
Trace 7
Transformation
  integral 101
  linear, see Tensors
Transpose 6
  partial 8
Triadic product 8

Vector, see also Displacement
decomposition of 15
derivative of 10, 19
product 10
sensitivity 79, 82, 128
Visibility of the fringes 108
with a circular pupil 111
with a rectangular pupil 113

Wave
aberration 53
conjugage image 42, 46, 54, 56
object 40
plane 100
primary image 42, 46, 48, 56
reference 40
spherical 39, 99
Wavelength 39
change of 47, 56, 65, 137

Young's modulus 164

Zero-order fringe 82

Holographic Recording Materials

Editor: H. M. Smith
1977. 96 figures, 17 tables. XIII, 252 pages
(Topics in Applied Physics, Volume 20)
ISBN 3-540-08293-X

Contents

H. M. Smith: Basic Holographic Principles. – *K. Biedermann:* Silver Halide Photographic Materials. – *D. Meyerhofer:* Dichromated Gelatin. – *D. L. Staebler:* Ferroelectric Crystals. – *R. C. Duncan Jr., D. L. Staebler:* Inorganic Photochromic Materials. – *J. C. Urbach:* Thermoplastic Hologram Recording. – *R. A. Bartolini:* Photoresists. – *J. Bordogna, S. A. Keneman:* Other Materials and Devices.

Laser Speckle and Related Phenomena

Editor: J. C. Dainty
1975. 133 figures. XIII, 286 pages
(Topics in Applied Physics, Volume 9)
ISBN 3-540-07498-8

Contents:

J. C. Dainty: Introduction. – *J. W. Goodman:* Statistical Properties of Laser Speckle Patterns. – *G. Parry:* Speckle Patterns in Partially Coherent Light. – *T. S. McKechnie:* Speckle Reduction. – *M. Françon:* Information Processing Using Speckle Patterns. – *A. E. Ennos:* Speckle Interferometry. – *J. C. Dainty:* Stellar Speckle Interferometry. – Additional References with Titles. – Subject Index.

Optical and Infrared Detectors

Editor: R. J. Keyes
1977. 115 figures, 13 tables. XI, 305 pages
(Topics in Applied Physics, Volume 19)
ISBN 3-540-08209-3

Contents:

R. J. Keyes: Introduction. – *P. W. Kruse:* The Photon Detection Process. – *E. H. Putley:* Thermal Detectors. – *D. Long:* Photovoltaic and Photoconductive Infrared Detectors. – *H. R. Zwicker:* Photoemissive Detectors. – *A. F. Milton:* Charge Transfer Devices for Infrared Imaging. – *M. C. Teich:* Nonlinear Heterodyne Detection.

Optical Data Processing

Applications
Editor: D. Casasent
1978. 170 figures, 2 tables. XIII, 286 pages
(Topics in Applied Physics, Volume 23)
ISBN 3-540-08453-3

Contents:

D. Casasent, H. J. Caulfield: Basic Concepts. – *B. J. Thompson:* Optical Transforms and Coherent Processing Systems With Insights From Cristallography. – *P. S. Considine, R. A. Gonsalves:* Optical Image Enhancement and Image Restoration. – *E. N. Leith:* Synthetic Aperture Radar. – *N. Balasubramanian:* Optical Processing in Photogrammetry. – *N. Abramson:* Nondestructive Testing and Metrology. – *H. J. Caulfield:* Biomedical Applications of Coherent Optics. – *D. Casasent:* Optical Signal Processing.

Springer-Verlag
Berlin
Heidelberg
New York